THE ROLE OF PROBIOTICS IN GUT HEALTH

THE ROLE OF PROBIOTICS IN GUT HEALTH

Edited by

Dr. Vaibhao Kisanrao Lule

Delve
PUBLISHING

www.delvepublishing.com

The Role of Probiotics in Gut Health

Dr. Vaibhao Kisanrao Lule

Delve Publishing
4164 Lakeshore Road
Burlington, ON L7L 1A4
Canada
www.delvepublishing.com

Email: orders@arclereducation.com

ISBN: 978-1-77956-438-2 (Hardcover)

ABOUT THE EDITOR

Dr. Vaibhao Kisanrao Lule is an Assistant Professor in the Department of Dairy Microbiology of College of Dairy Technology, Udgir (Latur), which is a constituent college of Maharashtra Animal and Fishery Sciences University, Nagpur (India). He has eight year of experience as an academician. He did his B. Tech. (Dairy Technology) from MAFSU, Nagpur followed by M. Tech. and Ph. D. (Dairy Microbiology) from the ICAR-National Dairy Research Institute (NDRI), Karnal (Haryana-132001), India. He has several peer reviewed national and international publications with good impact factors to his credit. He is actively involved in research on therapeutic benefits of probiotics and fermented milks. He attended several National and International conferences/seminars on probiotics, beneficial microbes, food safety and bio-preservation of milk and dairy foods at national and international platforms. He bestowed with several awards like, ICAR-Junior Research Fellow, ICAR-Senior Research Fellow, Recipient of Travel Fellowship award from Centre for International Co-operation in Science (CICS), Chennai and Department of Biotechnology (DBT), Government of India.

TABLE OF CONTENTS

LIST OF FIGURES

LIST OF TABLES

PREFACE

Humans are exposed to numerous pathogens daily, with bacteria being a significant component. For decades, probiotic microorganisms have been incorporated into various diets because of their beneficial impacts on health. Probiotics refer to foods and supplements that include beneficial microbes, such as specific strains of bacteria and yeast, which inhabit the gut and offer a range of potential health advantages. Research into the use of probiotic bacteria for treating intestinal disorders is ongoing, with clinical studies and laboratory experiments providing deeper insights into their effects on gut microbiome conditions. Evidence from these studies indicates that probiotics can help alleviate various gastrointestinal issues and enhance overall well-being. This book focuses on the role of probiotics and commensal microbes, exploring their potential in managing and improving gut microbiome-related health conditions.

The textbook *The Role of Probiotics in Gut Health* offers a comprehensive exploration of how probiotics impact various aspects of gastrointestinal health and overall well-being. This book is structured into nine distinct chapters, each focusing on a unique facet of probiotics and their effects on gut health.

The first chapter provides a foundational overview of probiotics and prebiotics, explaining their definitions, differences, and significance in maintaining gut health. It sets the stage for understanding how these beneficial microorganisms contribute to a balanced gut microbiota and supports the overall health of the digestive system.

Following this, the book examines the mechanisms by which probiotic yeasts function. This chapter highlights the specific ways in which these yeasts interact with the gut environment, including their influence on microbial balance and their potential therapeutic benefits.

The third chapter explores the broader role of probiotics in digestive health. It discusses how probiotics can help manage common digestive disorders, enhance nutrient absorption, and contribute to overall digestive well-being. This chapter emphasizes the practical applications of probiotics in treating and preventing gastrointestinal issues.

Expanding beyond digestive health, the fourth chapter examines the relationship between the gut biome and mental health. It presents evidence on how the gut microbiota can affect mood, cognitive function, and mental well-being, illustrating the gut-brain connection and the potential of probiotics in supporting mental health.

The book also addresses the maternal-fetal gut microbiota axis in the fifth chapter. This section explores how the gut microbiota of a mother can influence the development and health of her fetus, shedding light on the significance of maternal gut health during pregnancy and its long-term effects on the child. Immune health is another critical area covered in the sixth chapter. It discusses how probiotics can modulate the immune system, enhance immune responses, and provide protection against infections and inflammatory conditions. This chapter underscores the role of probiotics in supporting a robust and balanced immune system.

The seventh chapter focuses on gut microbiome composition, detailing the diverse microbial communities present in the gut and their roles in maintaining health. It explores factors that influence microbial diversity and how changes in microbiome composition can impact health and disease.

In the eighth chapter, the book investigates how probiotics affect nutrient absorption. It highlights the mechanisms through which probiotics enhance the absorption of essential nutrients and the implications for overall nutritional health and dietary practices.

Finally, the ninth chapter examines the relationship between the gut microbiome, aging, and longevity. It discusses how the gut microbiota changes with age, the impact of these changes on health, and how probiotics might contribute to healthier aging and extended longevity.

The Role of Probiotics in Gut Health provides an in-depth look at the multifaceted roles of probiotics in maintaining and improving gut health. It combines scientific research with practical insights, offering valuable information for both professionals and individuals interested in the benefits of probiotics for digestive and overall health.

CHAPTER 1

INTRODUCTION TO PROBIOTICS, PREBIOTICS AND GUT HEALTH

Contents

INTRODUCTION

The advantages that probiotics, prebiotics, and synbiotics have for health have been the focus of much research over the last several years. It has been shown that these dietary supplements, referred to as functional foods, can change, modify, and restore the natural intestinal flora. When taken in sufficient quantities, probiotics—also known as good bacteria—are live microorganisms that provide a number of health advantages. The primary function of these advantageous microorganisms is to support a balanced gut flora, which is essential for healthy digestion and general wellbeing. The intricate community of bacteria, fungi, and other microorganisms that live in the digestive tract is known as the gut microbiota, and it is important for immune system performance, nutrient absorption, and the defense against pathogens. People can improve the health of their gut by adding probiotics to their diet through supplements or fermented foods like yogurt, kefir, and sauerkraut. This may help with conditions like constipation, diarrhea, and irritable bowel syndrome (IBS). Additionally, new research emphasizes the connection—the gut-brain axis—and links better mental health to a balanced gut microbiota.

1.1. WHAT ARE PROBIOTICS?

Microbes are essential to our existence. They are crucial in preventing the growth of pathogenic organisms in and on the body, which could result in infections, in addition to being used in industry to produce a variety of materials like antibiotics and water purification agents. Amount 10 times as many bacteria are found in the colon than in all of the host's mammalian cells. As soon as the baby is born, the mother's vaginal and intestinal flora begin to colonize the gastrointestinal tract (GIT). Non-pathogenic *E. coli* strains. The intestinal resident bacteria, *Bifidobacterium*, *Lactobacillus*, and *E. coli*, inhibit the growth of putrefactive (proteolytic) bacteria like *Clostridium* and *Shigella*, which cause diarrhea. Eli Metchnikoff was the first to report this phenomenon. It is predicated on his theory that "it is possible to adopt measures to modify the flora in our bodies and to replace the harmful microbes by useful microbes because intestinal microbes are dependent on food." Kollath defined the term "probiotic" for the first time in 1953. As of right now, the World Health Organization (WHO) defines probiotics as live microorganisms when given in sufficient amounts (Hill et al., 2014; Hogan et al., 2018).

1.1.1. Need for Probiotics

Microorganisms make up the normal flora of the intestine, but a large number of these organisms are necessary to keep pathogenic organisms like *Clostridium* in check. There are numerous reasons why the body's natural flora could be upset. Bad eating practices, drinking water that has been chlorinated, illness, stress, and the use of antibiotics can all cause havoc in the GIT by killing off beneficial bacteria and fostering the growth of harmful ones. Issues arise when the proportion of beneficial bacteria to harmful bacteria is reduced.

The use of antibiotics, particularly broad-spectrum ones, is the most significant factor. Antibiotics that pass through the enterohepatic circulation or are not fully absorbed enter the large intestine and destroy probiotic species that are vulnerable. This causes the pathogenic strains to multiply and causes pathological conditions like bacterial dysentery or, in more severe cases, pseudomembranous colitis. The bacterium *Clostridium difficile* is the source of this colon infection. Fever, diarrhea with an unpleasant odor, and stomach pain are the symptoms of the illness. It can be lethal and serious, leading to toxic megacolon. The widespread use of antibiotics with a wide spectrum of applications is the primary cause of this illness. The normal bacterial flora of the colon is altered by the use of broad-spectrum antibiotics like cephalosporins and clindamycin. More specifically, any bacteria that survive in the intestine will have less competition for resources and space when the antibiotic eradicates other, rival bacteria. Overall, this allows certain bacteria to grow far more widely than they would normally. One such kind is *Clostridium difficile*.

1.1.2. Postulated Mechanism of Action of Probiotics

- Probiotics maintain the balance of gut microbiota, thereby averting the production of inflammatory mediators—the body's continual reaction to potentially dangerous antigens that compromise intestinal integrity.
- Stimulates IgA production, which plays an important role in mucosal immunity.
- Increased production of Interferon-γ can activate macrophages, neutrophils, and Natural Killer (NK) cells, enhancing their capacity to eliminate microorganisms and tumor cells. Additionally, it can induce the expression of Major Histocompatibility Complex (MHC) class I and II proteins, thereby improving antigen presentation.
- By competing with other bacteria for adhesion sites and increasing mucus production, gut microflora keeps other bacteria from sticking around.
- Probiotics also fight with pathogens for nutrients that they would otherwise consume. For instance, a probiotic that consumes monosaccharides may inhibit the growth of *Clostridium difficile*, which is necessary for the organism to flourish.

- The bacteriocin and defensins that are produced by the normal microflora are linked to the antimicrobial activity that the normal intestinal microflora develops against enter virulent bacteria, which helps to contribute to the barrier effect.
- Bifidobacteria and *Lactobacilli* generate potent acids, i.e., lactic and acetic, which lower the pH of the intestines (5.8). Lowering the pH of the GIT may have the effect of converting toxic ammonia (NH_3) into the non-diffusible ammonium ion (NH^+), which may lower blood ammonia levels. Bifidobacteria growth rate was largely unaffected by low pH, but *E. coli* and *C. perfringens* growth was pH 5.0 and pH 4.5 completely inhibited. Propionic and butyric acid, which probiotics produce, encourage the growth and differentiation of intestinal epithelial cells (IECs).
- Specific goblet cells secrete glycoproteins that condense into viscous gels that ensnare microorganisms and prevent them from diffusing too far into the intestinal epithelium.
- Additionally, these bacteria produce digestive enzymes that may help dairy products digest and tolerate lactose, such as lysozyme, lactase, and casein phosphatase (Hill et al., 2014; Hogan et al., 2018).

1.1.3. Functions of Probiotics

Numerous benefits of probiotics contribute to preserving and improving general health. They primarily support the development of beneficial bacteria and impede the spread of pathogens, aiding in the maintenance of a balanced gut microbiota. For efficient digestion and nutrient absorption, this balance is essential. Additionally, probiotics contribute to the strengthening of the gut barrier, which lessens inflammation and boosts immunity by preventing toxins and pathogens from leaking into the bloodstream.

They generate compounds that further enhance gut health and support the immune system, such as short-chain fatty acids (SCFAs) and antimicrobial peptides (AMPs). Additionally, by reestablishing a healthy gut flora and enhancing bowel regularity, probiotics can help manage digestive disorders like diarrhea, constipation, and irritable bowel syndrome (IBS). Certain strains have been demonstrated to improve the stability of the digestive environment and lessen gastrointestinal distress symptoms. Because probiotics can affect the gut-brain axis, a communication network that connects the gut and the brain, they may also have positive effects on mood and cognition.

1.1.4. Properties of Probiotics

A potential probiotic strain is anticipated to possess specific desirable characteristics in order for it to be able to exert its beneficial effects. The ones currently determined by *in vitro* tests are:

- acid and bile tolerance which seems to be crucial for oral administration;
- adhesion to mucosal and epithelial surfaces, an important property for successful immune modulation, competitive exclusion of pathogens, as well as prevention of pathogen adhesion and colonization;
- antimicrobial activity against pathogenic bacteria; and
- bile salt hydrolase activity.

However, there is ongoing discussion regarding the usefulness of these parameters due to relevant issues, differences between *in vivo* and *in vitro* experiments, and a lack of standardization in operating procedures. Target population and target physiologic function specific studies are the best ways to determine a strain's properties, as there are no particular parameters that are necessary for all probiotic applications. In terms of the finished product, the colony forming units per gram of product is an important parameter, and the probiotic dose levels should be based on those that have been found to be effective in human studies. It is widely acknowledged that probiotic products work, even though there is currently insufficient information regarding the minimum effective concentrations.

1.1.5. Types of Probiotics

Probiotics most commonly found in the market belong to the genera *Lactobacilli* and Bifidobacteria. Numerous well-known probiotics have similar health benefits, such as enhancing immunity and warding off infections. Certain probiotic species or strains may provide unique advantages. When taken in sufficient amounts, live microorganisms known as probiotics have been shown to have positive effects on health. With so many probiotics on the market these days, it can be difficult to choose the best one for your unique requirements. Gaining knowledge about the most popular probiotics and their physiological effects may be beneficial (Hill et al., 2014; Hogan et al., 2018).

1.1.5.1. Understanding Probiotic Species and Strain Names

There are several categories for probiotics, such as genus, species, and strain. While species are groups that share characteristics with one another within a genus, genus is the more general category used for classification. Groups of dog breeds with similar characteristics, such as Spaniels or Retrievers, would be considered species within the dog genus if we were to imagine a genus containing all domestic cats and another containing all domestic dogs. Within the Retriever group, strains would be akin to distinct dog breeds, such as Labrador and Golden Retrievers, each with unique traits. The many probiotic species within each genus can have different effects on the body, just as retrievers are recognized for specific qualities. However, they are typically grouped by characteristics that have a major impact on health, such as the capacity to fight specific pathogens or generate vitamins. All living things, including probiotics, have scientific names that consist of their genus (the first part of the name, written in capital letters), species (the second part of the name, written in lowercase letters), and subspecies (abbreviated subsp.), if appropriate. Probiotics and other microorganisms can be further categorized into strains, which are expressed alphanumerically and may have unique properties.

Example: *Lactobacillus* acidophilus W55 belongs to the genus *Lactobacillus*, the species acidophilus, and the strain W55.

1.1.5.2. Common Categories of Probiotics

Probiotics are commonly found in the following genera: *Lactobacillus*, *Bifidobacterium*, *Saccharomyces*, *Streptococcus*, *Enterococcus*, *Escherichia*, and *Bacillus*; the most frequently used genera are *Lactobacillus* and *Bifidobacterium*. Benefits to the immune system, dental cavities, vaginal and urinary tract infections, and eczema in children are among the common advantages of these beneficial bacteria. Additional research has demonstrated the extensive benefits of probiotics, such as decreased incidence and severity of upper respiratory tract infections and favorable effects on lipid, blood sugar, and body fat measurements.

The effect that probiotics have on gut health may be their most well-known benefit. They have the following effects on the gastrointestinal (GI) tract: improved digestion, increased absorption of certain micronutrients, and decreased incidence and severity of diarrhea due to different causes. Depending on the genera and species of probiotic, different health benefits may result. Different strains within a species may also have different outcomes. Probiotics that target the microbiota can improve immunity, decrease inflammation, strengthen the intestinal barrier, and inhibit pathogens and harmful bacteria. Furthermore, certain probiotic strains have been connected to advantages for the brain, such as enhanced mood, focus, and memory (Hill et al., 2014; Hogan et al., 2018).

1.1.5.3. Bifidobacteria

Bifidobacteria are widespread in the environment and in the intestines of many animals, including humans. They are the first colonizers of the infant gut microbiota.

Some common species include:

- B. bifidum
- B. animalis
- B. breve
- B. longum
- B. adolescentis

Bifidobacteria have been linked to beneficial effects on gut health, such as better lactose digestion. Additionally, this probiotic may lessen the symptoms of IBS and prevent or lessen diarrhea (IBS). They have been connected to higher lipid levels as well.

B. breve, *B. longum* subsp. *infantis*, *B. bifidum* and *B. longum* subsp. *longum* are the most common *Bifidobacteria* in human infants. Newborns receiving breast milk have higher levels of these helpful bacteria in their gut. These probiotics may have an impact on metabolism and have been linked to a stronger intestinal barrier and defense against infection.

Certain Bifidobacteria species, such as Bifidum, *B. infantis*, *B. breve*, and *B. longum*, can synthesize B, C and K vitamins. They have long been employed in the food processing industry, especially in the culture of fermented dairy products. They might also help with mineral absorption by reducing stomach acidity.

1.1.5.4. Lactobacilli

Since ancient times, lactic acid-producing *Lactobacillus* bacteria have been utilized in food processing, especially for the culture of dairy products. The beneficial effects of *Lactobacilli* on health have been associated with the production of lactic acid and other metabolites.

Some common species include:

- L. acidophilus
- L. plantarum
- L. rhamnosus
- L. reuteri
- L. casei
- *L. delbrueckii* subsp. *Bulgaricus*

- *L. gasseri*
- *L. fermentum*
- *L. johnsonii*
- *L. paracasei*
- *L. salivarius*

Lactobacilli are significant oral, vaginal, and digestive tract colonizers. Certain species have been associated with enhanced lactose and micronutrient digestion. They have also been linked to a lower incidence of vaginal infections and diarrhea brought on by antibiotics. Lower cholesterol, better weight loss in obese people, and the prevention or treatment of eczema in children are among the other possible health benefits of *Lactobacillus* species.

1.1.5.5. Saccharomyces

There are various yeast species that belong to the *Saccharomyces* genus. Certain species can be found in plant material or soil, while others have undergone hybridization for commercial purposes like brewing. The GI system of humans contains them as well.

Saccharomyces boulardii is a yeast probiotic that has been extensively researched, primarily for its application in gastrointestinal disorders. It is a resilient species that can withstand higher temperatures and the low pH found in the GI tract more so than other species. Though it has long been used in baking and brewing, the closely related yeast *Saccharomyces cerevisiae* does not seem to have the same health advantages and can even be harmful to patients with weakened immune systems. Therefore, in patients who are critically ill or have compromised immune systems, yeast-based probiotics should be avoided.

1.1.5.6. Soil-Based Probiotics

Bacillus species are the predominant microbiota in the environment and are primarily found in soil, though they can also be found in water, air, some fermented foods, and the GITs of humans and animals. Exposure to these bacteria has decreased in developed areas due to changes in lifestyle, such as increased hygiene and consumption of processed foods. Spore-forming probiotics, or bacteria, are renowned for their resilience in hostile environments.

Bacillus species can be found in the fermented foods kimchi, natto, and some fish sauces, as well as juices and raw fruits and vegetables.

Common species of Bacilli include *B. subtilis*, *B. clausii*, and *B. coagulans*. Commercially available strains are mainly from the well-researched subtilis species. There are also household chemicals available with Bacilli probiotics due to their antimicrobial effects.

Bacilli are beneficial because they help restore the gut microbiota, have anti-inflammatory and anti-pathogenic properties, produce vitamins such as B vitamins, carotenoids, and vitamin K, and secrete digestive enzymes. Additionally, they have been demonstrated to combat urinary tract and GI infections. According to one study, an older adult's immune system was strengthened and a different strain of beneficial bacteria significantly increased when exposed to a *B. coagulans* strain (Hill et al., 2014; Hogan et al., 2018).

1.1.5.7. Common Probiotic Species

Among the common genera of probiotics, some species have been studied for their potential health benefits more thoroughly than others. The following are among the most well-studied:

- ***Bifidobacterium animalis***: This species is often found in the environment and in mammals' GITs, including those of humans. Studies involving *B. animalis subspecies B. lactis*, or mixtures containing it, have reported significant reductions in total and low-density lipoprotein (LDL) cholesterol levels. Moreover, *B. lactis* may help with constipation.

  Originally isolated from yogurt for use as a supplement, the common strain *B. animalis subsp. lactis* HN019 has mostly been researched for its potential benefits to the digestive system. It has been discovered that this strain lowers GI pathogens, strengthens the intestinal barrier, and lowers the prevalence of infectious diarrhea. Patients who are constipated may also experience an increase in bowel movements.
- ***Bifidobacterium bifidum***: This probiotic, which is mostly found in the GI tract and is beneficial there, is the species most frequently used to culture fermented milk products. Adding *B. bifidum* and *Streptococcus thermophilus* to baby formula has been associated with a decreased incidence of infectious diarrhea. Further research is necessary, but some studies have suggested that *B. bifidum* may play a role in preventing colon cancer. In a clinical trial, the strain *B. bifidum* MIMBb75 was found to significantly reduce IBS symptoms.
- ***Bifidobacterium longum**:* This species of Bifidobacteria, which is prevalent in fermented milk products, is derived from the human gut. In brief, it generates vitamins, such as C and B1, B6, and B12. It might also aid in fortifying the intestinal barrier and warding off infections.

B. longum subsp. *infantis*, also known as *B. infantis*, has been found in some studies to improve symptoms of IBS. It is also an important colonizer of the newborn gut microbiome. In one study, the *B. infantis* strain EVC001 was found to reduce pathogenic bacteria and intestinal inflammation in breastfed infants.

- ***Lactobacillus acidophilus***: This well-known probiotic is frequently used in fermented milk products and has been studied for many years for potential health benefits. The human mouth, vagina, and GI tract contain it. It has been demonstrated that *L. acidophilus* strains are effective in treating vaginal bacterial infections. In lab tests, *L. acidophilus* has also been shown to inhibit the growth of colon cancer, but further investigation is required. According to certain research, *L. acidophilus* or multi-strain products containing it lowered the levels of both LDL and total cholesterol. A typical strain is *L. acidophilus* NCFM. It has been associated with a decrease in bloating in IBS patients.
- ***Lactobacillus casei:*** This probiotic, which is present in the human mouth, vagina, and GI tract, is also employed in food fermentation. According to certain studies, this species lowers the risk of acute infectious diarrhea and diarrhea linked to antibiotic use. In certain studies, it has also been shown to inhibit the growth of colon cancer, but more investigation is required. The strain *L. casei* DN-11401 was found to significantly reduce the incidence of antibiotic-associated diarrhea (AAD).
- ***Lactobacillus paracasei***: The closely related *L. paracasei* species has also been used in fermented food production. Along with *L. rhamnosus*, this species was linked with a lower incidence of eczema in children. The strain *L. paracasei* subspecies paracasei F19 lessened bloating in patients being treated for diverticular disease.
- ***Lactobacillus plantarum:*** This probiotic colonizes the vagina, GI tract, and mouth of humans. Additionally, a large range of fermented foods contain it. Certain studies have reported that *L. plantarum* lowers LDL and total cholesterol levels. The well-researched strain *L. plantarum* 299v has been shown to decrease IBS symptoms. It has also been found to increase iron absorption and may prevent iron deficiency in those at risk.
- ***Lactobacillus reuteri:*** *L. reuteri* appears to have a number of health advantages and has been detected in the human GI tract, urinary system, skin, and breast milk. Some strains of it have been found to protect against vaginal candida infections, and its antibacterial properties are well established. *L. reuteri* has also been shown to help with a number of digestive health issues, such as acute diarrhea,

relief from infant colic and IBS symptoms, and possible cholesterol lowering. It has been discovered that the candida species responsible for vaginal yeast infections are blocked by the strain *L. reuteri* RC-14. It was discovered that the *L. reuteri* DSM 17938 strain shortens the duration of infectious diarrhea.

The strain *L. reuteri* NCIMB 30242 was found in one study to reduce levels of total and LDL cholesterol in adults with elevated levels. This strain was also shown to increase vitamin D levels. Researchers have suggested that gut microbiome changes may alter vitamin D status.

- ***Lactobacillus rhamnosus**:* This probiotic species is used to ferment dairy products and is closely related to *L. casei* and *L. paracasei.* This well-researched probiotic is present in the vagina and GI system. It appears to be capable of a great range of actions. Certain strains of it have been found in studies to lower the risk of acute infectious diarrhea and antibiotic-associated diarrhea, as well as to exhibit both antifungal and antibacterial activity in the vagina. There is also evidence connecting this species to a decreased occurrence of eczema in children. One study found that giving an *L. rhamnosus* strain supplement to obese women (but not men) caused them to lose a significant amount of weight. One of the most popular probiotic strains, *L. rhamnosus* GG, has been demonstrated to guard against diarrhea due to a variety of causes and is capable of fighting candida and GI infections. It has been discovered that the strain *L. rhamnosus* GR-1 inhibits the species of Candida responsible for vaginal infections.
- ***Saccharomyces boulardii**: Saccharomyces boulardii* has anti-pathogenic effects and has been found to reduce the risk of acute diarrhea, including antibiotic-associated diarrhea, in children and adults and to fight H. Pylori infections. It may also alleviate symptoms of IBS. It is resistant to most antibiotics.

Specifically, the strain *Saccharomyces boulardii* CNCM I-745 has been shown to prevent gastroenteritis and AAD.

1.1.5.8. The Power of Multi-Strain Formulations

Various bacterial strains carry out distinct bodily functions. Even more intriguingly, research has revealed that probiotic strains cooperate to improve one another's abilities. These groups can occasionally even carry out tasks that none of the strains could complete on their own. Consider a sports team: You must carefully balance your offense and defense and assign players to various positions and roles in order to have a winning team. You would not think to select only players for your offense or defense because the group's output would be negatively impacted. The same holds true for probiotics. In many cases, it is more advantageous to provide your gut flora with a variety of strains and species of probiotics rather than a probiotic with a very high CFU count of one probiotic strain in order to help restore and maintain optimal balance. For this reason, every Omni-Biotic formulation is designed to be multi-species, multi-strain. The careful blending of the probiotic strains takes particular health goals into consideration. Therefore, nine probiotic strains are included in Omni-Biotic Stress Release to support the gut flora and the gut-brain axis during stressful situations. A few strains in this mixture aid with inflammation management.

1.1.6. Health Benefits of Probiotics

The benefits of probiotics, such as improved intestinal health, improved immunological response, lower serum cholesterol, and cancer prevention, are becoming more and more supported by research. These health characteristics vary depending on the strain and are influenced by the different mechanisms outlined previously. While some of the health benefits are well-established, others need more research to be confirmed. Probiotics have been shown to be effective in treating acute diarrheal illnesses, preventing AAD, and enhancing lactose metabolism; however, there is not enough data to support their use in treating other clinical conditions.

1.1.6.1. Antibiotic-Associated Diarrhea

Antibiotic therapy frequently causes mild or severe episodes of diarrhea because it suppresses the normal microflora, promoting the overgrowth of pathogenic or opportunistic strains. The spectrum of associated conditions includes pseudomembranous colitis and diarrhea without mucosal abnormalities. The latter, more severe form of diarrhea, is linked to antibiotic use and is caused by *Clostridium difficile*, which can develop cytotoxic strains under such conditions. The condition is named for the fibrinopurulent material's plaque-like adhesion to the damaged mucosal layer. If left untreated, symptoms may include leukocytosis, vomiting, fever, diarrhea, and abdominal distention. Complications can include toxic megacolon and perforation. Probiotic therapy, involving the administration of *Lactobacillus rhamnosus* and *Saccharomyces boulardii*, has been utilized in clinical settings. Probiotic use is associated with a lower risk of antibiotic-associated diarrhea (AAD), according to several studies. A recent meta-analysis of evidence on probiotics for preventing and treating AAD concluded that probiotic administration, particularly with *L. rhamnosus*, *L. casei*, and the yeast *S. boulardii* (which are the probiotics predominantly included in most trials), is associated with a reduced risk of the condition.

1.1.6.2. Infectious Diarrhea

The most well-known health benefits of probiotic microorganisms are likely the treatment and prevention of infectious diarrhea. The most frequent cause of acute infantile diarrhea worldwide and a major contributor to infant mortality is rotavirus. Since it has been demonstrated that the absorption of antigens is more enhanced in germ-free mice than in normal mice, the normal microflora appears to play a significant role in the host response to the infection. The virus replicates in the highly differentiated absorptive columnar cells of the small intestinal epithelium. The addition of probiotics to baby formulas has been used to treat existing illnesses as well as prevent rotaviral infections. Probiotics like *L. rhamnosus* GG, *L. reuteri*, *L. casei* Shirota, and *B. animalis* Bb12 have been demonstrated in well-controlled clinical trials to be effective in reducing the length of acute rotavirus diarrhea. Of these, *L. rhamnosus* GG and *B. animalis* Bb12 have been shown to be particularly beneficial. Among the suggested mechanisms are the generation of compounds that directly inhibit, the enhancement of the immune system, and the competitive blockage of receptor site signals controlling secretory and motility defenses.

1.1.6.3. Lactose Intolerance

A genetically determined beta-galactosidase deficiency that prevents lactose from being hydrolyzed into the monosaccharides glucose and galactose is known as lactose intolerance. Osmotic diarrhea results from the breakdown of undigested lactose by bacterial enzymes in the large intestine. The causes of acquired beta-galactosidase deficiency are usually reversible and include pelvic radiation therapy, which damages the mucosa, rotavirus infection, which infects lactase-producing cells, and short bowel syndrome. Those who are lactose intolerant experience nausea, diarrhea, and upset stomach after consuming milk or milk products. While following traditional yogurt recipes, *S. thermophilus* and *L. delbrueckii* ssp. Increased beta-galactosidase activity is one reason why Bulgaricus species are even more effective in this regard. Probiotics are said to improve lactose metabolism, though this effect appears to depend more on which strain and at what concentration. Therefore, doctors should think about probiotic supplementation as a therapeutic option because some patients have responded well to it.

1.1.6.4. Probiotics and Allergy

Recent data indicates that early-life exposure to bacteria may have a protective effect against allergies; in this regard, probiotics may offer a safe substitute for the microbial stimulation required for an infant's developing immune system. They also enhance the function of the mucosal barrier, which is thought to help regulate the allergic response. The quantitative and qualitative differences in intestinal microbiota between children and infants with allergies and those who are healthy, the former showing colonization by a more adult-like type of microflora, support the role of intestinal microbiota in allergy. It appears that atopic dermatitis and food allergies are specifically affected by these probiotic effects. Hereditary predisposition and exposure to environmental allergens are important factors in the pathogenesis of the latter, a common chronic relapsing skin disorder that affects children and infants. A limited number of strains have been tested for their efficacy in the treatment and prevention of allergy in infants. In a recent study of breast-fed infants suffering from atopic eczema B. lactis and *L. rhamnosus* GG were found to be effective in decreasing the eczema severity. Furthermore, *L. rhamnosus* GG has been found successful in preventing the occurrence of atopic eczema in high-risk infants, when supplied prenatally to selected mothers who had at least one first-degree relative with atopic eczema, allergic rhinitis, or asthma.

1.1.6.5. Probiotics and Cardiovascular Diseases

The first people to raise the possibility of probiotic consumption having an impact on lipid metabolism were Mann and Spoerry. They discussed how lactic acid bacteria (LAB) and fermented milks have less of an impact on lowering cholesterol. Among the probiotic strains with documented hypocholerolemic effects are *L. bulgaricus*, *L. reuteri*, and *B. coagulans*. Research on humans fed milk containing *L. acidophilus* L1 showed a noteworthy decrease in serum cholesterol levels. Two possible causes for the hypocholesterolemic effect of probiotics are: (i) a significant conversion of cholesterol into bile acids; (ii) a decrease in hydroxyl-methyl-glutaryl-Coenzyme-A reductase in the liver. Furthermore, probiotic enzymes can also facilitate the enzymatic deconjugation of bile acids; and (iii) Probiotics can eliminate cholesterol by integrating themselves into the membranes of growing cells. Such *in vitro*-based claims require verification through *in vivo* studies.

1.1.6.6. Cancer and Probiotics

There is proof that eating foods high in probiotic bacteria may help lower the incidence of cancer. Although the precise processes are still being studied, research has shown that some *Lactobacillus* and *Bifidobacterium* spp. reduce the amounts of carcinogenetic enzymes that the colonic flora produces by improving the host's immune system, producing antimutagenic organic acids, and restoring intestinal permeability and microflora balance. It is well known that *L. acidophilus* delays the initiation of colon cancers.

1.2. SOURCES OF PROBIOTICS

Dietary supplements and fermented foods are the two ways to increase the amount of good bacteria in your stomach. Foods that are fermented are the most organic source. Probiotic supplements, which are usually available over-the-counter, should only be used as directed by your doctor to treat conditions. Regular use is not advised. Furthermore, the FDA does not oversee supplements in the same way that it does drugs.

Probiotics have no recommended daily intake, so it is impossible to determine which fermented foods to eat in what quantity. Thus, the general recommendation is to incorporate as many fermented foods as you can into your diet on a daily basis.

One of the earliest methods of food preservation is fermentation. For centuries, people have been fermenting food and beverages, such as wine and beer. Lacto-fermentation is the process by which naturally occurring bacteria break down the sugar and starch in food to produce lactic acid in fermented foods. This procedure produces an environment that not only helps to preserve the food but also encourages the growth of different kinds of good bacteria, B vitamins, and beneficial enzymes along with omega-3 fatty acids. An alternative perspective on fermentation is that it changes one kind of food into another. For example, milk can be used to make yogurt, cheeses, and sour cream; cabbage can be turned into sauerkraut; cucumbers can be turned into pickles; and soybeans can be turned into miso. Not every fermented food has probiotics in it. Certain food processing techniques, such as canning and baking, destroy or render probiotics inactive, as is the case with wine and beer. But the majority of fermented foods also contain probiotics. Fermented foods have a strong taste and smell, which some people may find unpleasant. This is the only possible drawback. The various species contribute to the distinct flavors and textures of fermented foods.

Because yogurt has a flavor and texture that appeals to most Western palates, it is regarded as the most valuable player among probiotic foods. Depending on the brand of yogurt, there may be differences in the quantity and kind of bacteria. Yogurt products can have anywhere between 90 billion and 500 billion CFU of probiotics per serving. The term "colony-forming units" (CFU) refers to the number of bacteria that can split and form colonies. Check the label for the words "live and active cultures."

Yogurt is easy to add to your diet. Besides having it for breakfast or a midday snack, you can substitute yogurt whenever you use mayonnaise in egg salad or potato salad, or in almost any baking recipe. Yogurt also can be the basis for sauces, salad dressings, or marinades (Hill et al., 2014; Hogan et al., 2018).

1.2.1. Food Sources of Probiotics

Foods that contain natural probiotics include some types of:

- **Yogurt**: It is a well-known source of probiotics, which are beneficial bacteria that maintain a balanced environment in your digestive system. According to studies, probiotics may lessen the symptoms of lactose intolerance. They might also aid in controlling diarrhea, gas, and other digestive issues. Brands with specific probiotics may cost more, but any with live and active cultures could be beneficial.
- **Sauerkraut**: Select the unpasteurized variety. Most supermarket brands undergo pasteurization, which destroys beneficial bacteria. Vitamins that strengthen the immune system and help prevent infection are abundant in both sauerkraut and kimchi, a spicy Korean dish that is similar to it.
- **Miso Soup:** This fermented soybean paste is a popular breakfast food in Japan that can help stimulate your digestive system. Miso that has been infused with probiotics is frequently used to make a sour soup that is rich in B vitamins and antioxidants and low in calories.
- **Soft Cheeses**: While probiotics aid in digestion, not all of them can make it through your stomach and intestines. Gouda and other fermented soft cheeses have strains that are resilient enough to survive, according to research.
- **Kefir**: Shepherds in the Caucasus Mountains, which separate Asia and Southeast Europe, are said to have discovered that the milk they carried had a tendency to ferment into a bubbly drink. Similar to yogurt, kefir is thick, creamy, and tart. It also contains some beneficial yeast strains and its own strains of probiotic bacteria.
- **Sourdough Bread**: Next time you are assembling a sandwich, take note of what is containing your cheese and cold cuts. The well-known sourdough bread from San Francisco contains a probiotic that might aid with digestion.
- **Acidophilus Milk:** One of the easiest ways to get probiotics is to use this kind of milk, which has been fermented with bacteria. You may see it labeled as sweet acidophilus milk. Buttermilk, too, is rich in probiotics.
- **Sour Pickles**: If you are looking for pickles that will give you probiotics, go for naturally fermented varieties that were not pickled with vinegar. Sour pickles may aid in digestion by fostering the growth of beneficial bacteria in a solution of sea salt and water.

- **Tempeh***:* This Indonesian burger, which is made from fermented soybeans as a base, produces a natural antibiotic that combats specific bacteria. Tempeh has a lot of protein as well. Its flavor is frequently described as nutty, smoky, and mushroom-like. It can be marinated and substituted for meat in recipes.

1.2.2. Supplements

Probiotics can be found as supplements, but there is not enough data to say that a particular brand or kind will treat a certain illness just yet. Probiotics are governed by the Food and Drug Administration (FDA) for therapeutic uses. They do not, however, keep an eye on claims regarding the probiotic contents of probiotic supplements because they do not view them as pharmaceuticals. One can never be certain of a product's exact contents. Furthermore, the duration of a probiotic's bodily activity is not always known. Because of this, one cannot be certain of a product's expected efficacy or the condition it may treat.

1.3. PROBIOTIC PRODUCTS AND FORMULATIONS

Microorganisms known as probiotics have the potential to improve the health of human populations, especially those using antibiotics or those whose nutritional needs are not always met. Given that probiotic formulations consist of bacteria or yeasts, it is critical to follow appropriate validation protocols.

1.3.1. Probiotic Formulations and their Contents

Probiotic theory has been around for about a century, and Nobel Laureate Elie Metchnikoff was the first to formally discuss it. Probiotic science has advanced in the last few decades, resulting in an improved understanding of the relationship between probiotics, gut microbiota, and the host, as well as increased efficiency and mode of action. In spite of this, there are still some safety claims, and research on their health benefits is ongoing.

Although there are many different kinds of probiotic strains, *Lactobacilli*, *Saccharomyces boulardii*, *Bacillus coagulans*, and *Bifidobacterium* are the most widely used strains. Other microorganisms that are employed are the enterococci strains, such as *Enterococcus faecium* SF68, and the *Escherichia coli* strain Nissle 1917.

Among the *Lactobacilli,* the species *L. rhamnosis, L. reuteri, L. acidophilus* groups and certain strains of *L. casei* are commonly used. These are also among the most studied probiotics, despite *Bacillus* species being perhaps more popular in use where they are added to milk products. Probiotic microorganisms can be used on their own or in combinations, which may confer different effects.

1.3.2. Validation Guidelines

Probiotics are still largely untrusted in large part because they are viewed more like dietary supplements than like medications. This implies that they are not required to demonstrate efficacy or safety, and they are subject to less regulatory scrutiny. Although there are no laws that must be followed, the WHO has provided guidelines for assessing the effectiveness and possible advantages of probiotics. These regulations, which are upheld by specific businesses and geographical areas, guarantee that validation calls for multiple stages akin to those of medical validation. These procedures involve identifying the strain, functionally characterizing the strain (or strains) for safety and probiotic qualities, validating the health benefits of the strain in human studies, and truthfully labeling the strain's content and efficacy claims when it is on the shelf.

Scientists evaluate probiotics in research according to similar standards. Scientists recommended in the 1990s that strains be identified first, ideally with high accuracy using molecular taxonomy. Genetic stability and details about the microorganism's nutrition and physiology should be discovered. It is important to evaluate the environment of the GIT to determine the probiotic's possible impact. Next, a clinical trial or animal testing should be conducted to determine the efficacy. The results' publication is crucial, especially in the field of research. Only once each of these phases has been completed should a probiotic be taken into consideration for use and commercial production.

1.3.3. Limitations and Dangers

There is not much proof that probiotics are healthy, even in spite of WHO guidelines. There are times when information is lacking, such as the right dosages for positive effects, and it is unclear how probiotics work to produce these potentially advantageous effects. This is particularly true for the industry, as there has not been much research done on longer shelf lives.

Although adverse effects from probiotics themselves are uncommon, they have happened. The most typical probiotic adverse effect is gastrointestinal distress, such as bloating. Some patient groups, such as those with compromised immune systems, have experienced an acceleration of complications due to the presence of *Lactobacillus* GG and *S. boulardii*. Pregnant women, newborns, and the elderly are among those who are more susceptible to probiotic infection because of their compromised immune systems. The possibility of antibiotic resistance spreading to pathogenic bacteria is a major worry. Natural resistance to the widely used antibiotic vancomycin exists in a number of *Lactobacillus* strains. It is feared that their widespread use will make it easier for other, potentially harmful microorganisms in the gut environment to acquire the resistance.

1.4. DEFINITION OF PREBIOTICS

In 1995, Glenn Gibson and Marcel Roberfroid presented the prebiotics concept for the first time. A non-digestible food ingredient known as a prebiotic has been defined as one that improves host health by selectively promoting the growth and/or activity of one or a small number of bacteria in the colon. For more than 15 years, this definition remained largely unaltered. Prebiotics, as defined by this definition, are limited to a small number of carbohydrate group compounds, including lactulose, GOS, short and long-chain fructans (FOS and inulin), and others. Dietary prebiotics were defined in 2008 at the 6th Meeting of the International Scientific Association of Probiotics and Prebiotics (ISAPP) as an ingredient that has been selectively fermented and causes specific changes in the gastrointestinal microbiota's composition and/or activity, thereby improving host health.

1.4.1. Characteristics and Classification of Prebiotics

A compound is considered prebiotic if it meets the following criteria: (i) it must be resistant to the stomach's acidic pH, unable to be hydrolyzed by mammalian enzymes, and unable to be absorbed in the GIT; (ii) it must be fermentable by the intestinal microbiota; and (iii) it must be able to selectively stimulate the growth and/or activity of intestinal bacteria, improving the host's health in the process. The following two characteristics can be used to separate fiber from prebiotics derived from carbohydrates, even though not all prebiotics are carbohydrates: (i) fibers are carbohydrates with a degree of polymerization (DP) equal to or greater than 3; and (ii) they cannot be hydrolyzed by the small intestine's natural enzymes. Remember that the fermentability or solubility of the fiber is not essential. Prebiotics come in a wide variety of forms. A subset of the carbohydrate groups, primarily oligosaccharide carbohydrates (OSCs), make up the majority of them. While most of the pertinent literature focuses on OSCs, there are some data points that indicate prebiotics are more diverse than just carbohydrates.

1.4.1.1. Fructans

A fructan is a polymer made up of many fructose molecules that is used to store carbohydrates. They are referred to as fructoligosaccharides (FOS) and inulin, respectively, and can have either a short or long chain length. The 'O' in the acronym FODMAP is accounted for by fructans and galactans. a variety of foods, such as wheat, ripe bananas, garlic, onions, leek, and artichokes. As a "functional ingredient," inulin, a long-chain fructan, can be included in processed foods. Inulin serves as a prebiotic in addition to raising the food product's fiber content. It is possible to alter the amount of fructans in food by cooking, drying, processing, sour-doughing, pickling, activating, and sprouting foods that already contain fructans.

Some examples:

- Dried fruit (increased levels)
- Rice cakes, corn cakes, cornflakes (increased)
- Sourdough bread (decreased)
- Pickled onions (decreased)
- Activated cashews (decreased)

1.4.1.2. Galacto-Oligosaccharides

Galacto-oligosaccharides (GOS), the product of lactose extension, are classified into two subgroups: (i) the GOS with excess galactose at C3, C4, or C6; and (ii) the GOS manufactured from lactose through enzymatic trans-glycosylation. The end product of this reaction is mainly a mixture of tri- to pentasaccharides with galactose in (1→6), (1→3), and (1→4) linkages. This type of GOS is also termed as trans-galacto-oligosaccharides or TOS. GOSs have a strong stimulating effect on *lactobacilli* and Bifidobacteria. Infant bifidobacteria have a high incorporation rate with GOS.

Though not as much as Bifidobacteria, GOS also stimulates Enterobacteria, *Bacteroidetes*, and Firmicutes. Some GOSs are made from the lactose isomer lactulose. These GOSs derived from lactulose are regarded as prebiotics as well. Other forms of GOS, known as raffinose family oligosaccharides (RFO), are based on sucrose extension in addition to these types. The impact of RFO on the gut microbiota remains unclear at this time.

1.4.1.2. Starch and Glucose-Derived Oligosaccharides

Resistant starch (RS) is a type of starch that cannot be broken down in the upper GIT. Because RS produces a high amount of butyrate, which is known to promote health, it has been proposed that RS belong in the prebiotic category. Different Firmicute groups exhibit the highest incorporation with the highest amount of RS.

An *in vitro* study showed that *Bacteroides thetaiotaomicron*, *Eubacterium rectale*, and *Ruminococcus bromii* could all degrade RS to a lesser extent. However, in the absence of *R. bromii*, RS degradation is not possible in the mixed bacterial and fecal incubations. One oligosaccharide derived from glucose is polydextrose. It is made up of glycosidic linkages and glucan with numerous branches. Though it has not been verified yet, there is some evidence that it can stimulate Bifidobacteria.

1.4.1.3. Other Oligosaccharides

A polysaccharide called pectin is the source of some oligosaccharides. Pectic oligosaccharides (POS) are a particular kind of oligosaccharide. Based on the extension of either rhamnose (rhamnogalacturonan I) or galacturonic acid (homogalacturonan), they function. Methyl esterification can be used to replace the carboxyl groups, and either C2 or C3 of the structure can be acetylated. Various types of sugars (e.g., ferulic acid) or the side chains of arabinose, galactose, and xylose. Depending on where POSs are coming from, their structures differ greatly.

1.4.1.4. Non-Carbohydrate Oligosaccharides

While carbohydrates are more likely to fit the definition of prebiotics, some compounds, like cocoa-derived flavanols, are recommended to be classified as prebiotics but are not considered carbohydrates. Experiments conducted both *in vitro* and *in vivo* show that flavanols can stimulate LAB.

1.4.2. Production of Prebiotics

Prebiotics are essential to human health. Asparagus, sugar beet, garlic, chicory, onion, Jerusalem artichokes, wheat, honey, bananas, barley, tomatoes, rye, soybeans, human and cow's milk, peas, beans, and so on are among the dietary food products that naturally contain them and more recently, microalgae and seaweeds. Owing to their low food concentration, they are produced on an industrial scale. Lactose, sucrose, and starch are used as raw materials to make some prebiotics. Given that prebiotics are generally categorized as GOS or FOS on an industrial scale.

1.4.3. Mechanism of Action of Prebiotics

The community of gut microbiota has a vast genetic potential that is involved in numerous metabolic processes; regulating this potential could enhance the host's overall health. Prebiotics, which are short-chain carbohydrates that are indigestible by the digestive enzymes of humans and other animals and have a DP ranging from two to roughly 60, can be used to achieve this modulation. The ability to be selectively utilized sets prebiotics apart from other undigested dietary ingredients and compounds, like antibiotics, minerals, and vitamins, even though prebiotics are not the only substances capable of changing the intestinal environment. Prebiotics, which are primarily found in fruits and vegetables, can have a number of positive effects on health. Prebiotics have several benefits, some of which are as follows: they lower blood density lipoprotein levels, boost immunity, increase calcium absorption, maintain the proper pH balance in the intestines, and have a low-calorie content.

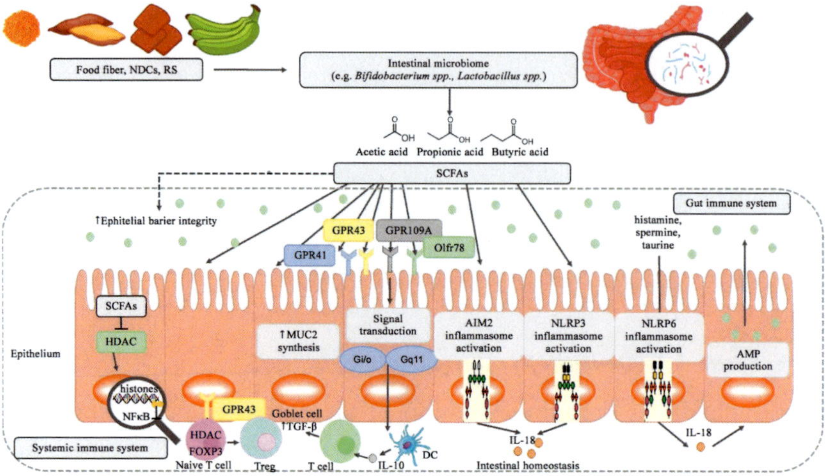

Figure 1.1. *Mechanism of prebiotic action. These mechanisms include the production of microbial metabolic products, noting short-chain fatty acids (SCFAs), the promotion of ion and trace element absorption (such as that of calcium, iron, and magnesium), a decrease in luminal pH, and the regulation of the immune system (increasing IgA production and modulating cytokine production).*

Source: Markowiak-Kopeć P, Śliżewska K. (2020). The effect of probiotics on the production of short-chain fatty acids by human intestinal microbiome. Nutrients; 12(4):1107. https://doi.org/10.3390/nu12041107.

According to recent research, the mechanisms by which prebiotics benefit the host (Figure 1.2) are mediated by microbial metabolic products, specifically SCFAs, the enhancement of ion and trace element absorption, including that of calcium, iron, and magnesium, and the modulation of cytokine production and increased production of IgA in the immune system. Prebiotics, which are primarily found in fruits and vegetables, can have a number of positive effects on health. The lowering of blood low density lipoprotein levels, immune system stimulation, improved calcium absorbability, preservation of the proper intestinal pH level, and low-calorie value are just a few benefits of those prebiotics that are worth mentioning. According to recent studies, the mechanisms by which prebiotics benefit the host (Figure 1.1) are mediated by microbial metabolic products, which include SCFAs, the enhancement of ion and trace element absorption, including that of calcium, iron, and magnesium, and the modulation of cytokine production and increased production of IgA in the immune system.

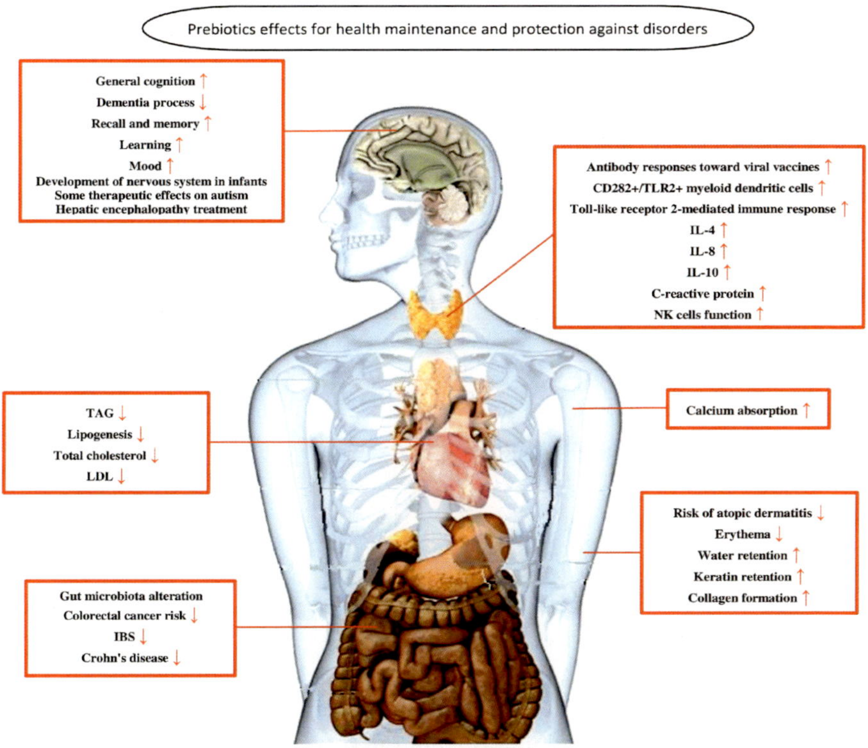

Figure 1.2. *A schematic diagram about potential mechanisms whereby probiotic bacteria might perform within the intestine. These mechanisms include antagonistic effects on various microorganisms, competitive adherence to the mucosa and epithelium (antimicrobial activity), increased mucus production and enhanced barrier integrity (enhancement of barrier function), and modulation of the human immune system (immunomodulation).*

Source: Davani-Davari D, Negahdaripour M, Karimzadeh I, Seifan M, Mohkam M, Masoumi SJ, Berenjian A, Ghasemi Y. (2019). Prebiotics: Definition, types, sources, mechanisms, and clinical applications. Foods; 8(3):92. https://doi.org/10.3390/foods8030092.

1.4.4. Health Benefits of Prebiotics

Prebiotics have a significant impact on human health, making them appealing and desirable agents to enhance human well-being against conditions like cancer, vascular disease, obesity, and mental disorders. Numerous studies have demonstrated the beneficial effects of prebiotics on human health; nonetheless, to validate the health claims, carefully planned, long-term clinical trials and genomics research are required. Through the identification of prebiotics' basic mechanisms, scientists hope to develop improved food supplements that will improve human health. One promising approach to managing and treating many major illnesses is the use of prebiotic dietary supplements to restore the natural balance of the gut microbiota.

1.4.4.1. Irritable Bowel Syndrome and Crohn's Disease

Prebiotics and Crohn's disease and IBS have been the subject of a few studies. In the absence of an organic cause, IBS, or IBS, is characterized by persistent abdominal pain and changed bowel habits. One kind of recurrent, chronic inflammatory bowel disease (IBD) that can affect any part of the gastrointestinal system, from the mouth to the anus, is Crohn's disease. There have been reports of decreased populations of Bifidobacteria and *Faecalibacterium prausnitzii*, as well as a lower ratio of *Bacteroides* to Firmicutes, in both Crohn's disease and IBS.

1.4.4.2. Colorectal Cancer (CRC)

The third most common cancer in the world, CRC develops through a series of stages, starting with a genetic mutation, progressing to adenomatous polyps, and ultimately resulting in invasive and metastatic cancer. It has been shown that by inducing apoptosis, prebiotic fermentation products like butyrate may have protective effects against both the risk and progression of CRC. Furthermore, symbiotic therapy (*Lactobacillus rhamnosus* and *Bifidobacterium* Lactis plus inulin) was shown in a clinical trial to lower the risk of CRC by reducing the rate of colorectal proliferation, which causes necrosis of the colonic cells and improves the integrity and function of the epithelial barrier.

1.4.4.3. Necrotizing Enterocolitis

Necrotizing enterocolitis (NEC) is an emergency gastrointestinal disorder that mainly affects preterm newborns and causes necrosis in parts of the colon. High rates of morbidity and mortality may result from it. Prebiotics like FOS and GOS have the ability to decrease the amount of harmful bacteria in preterm infants and promote the growth of the gut microbiota, which includes Bifidobacteria. Moreover, by increasing bowel motility and gastric emptying, SCFAs can increase feeding tolerance.

1.4.4.4. Prebiotics and the Immune System

Prebiotic consumption can strengthen immunity by boosting the number of beneficial microorganisms in the body. Studies on humans and animals have demonstrated that prebiotics can reduce the number of dangerous bacteria, specifically *Lactobacilli* and Bifidobacteria. Mannose, for instance, can lessen pathogen colonization by encouraging mannose adherence to *Salmonella*. mannose binds to *Salmonella* aq1 via type 1 fimbriae, or projections that resemble fingers. Furthermore, in the presence of OSCs, pathogens are unable to bind to the epithelium. Additionally, prebiotics can stimulate the production of immune-related molecules, particularly cytokines. It is interesting to note that metabolites of maternal prebiotics can cross the placenta and influence how the fetal immune system develops. Fugiwara et al. (2010) revealed that feeding FOS to a pregnant mouse model altered the microbiota of the offspring, which reduced the inflammation of their skin. By comparison, Shadid et al. in a placebo-controlled, randomized, double-blind study showed that prebiotic supplementation did not have any bifidogenic effects that could be passed down to offspring in humans. The details of well-known prebiotic effects on the immune systems are discussed below:

- **Oligofructose and Inulin Mixture**: The combination of inulin and oligofructans can enhance the body's immune system's response to viral vaccines, including those for measles and influenza.
- **FOS**: Studies have demonstrated that consuming FOS improves the body's ability to produce antibodies against the influenza vaccine. Additionally, there are fewer adverse effects from the influenza vaccination. This particular class of prebiotics also lessens fever associated with diarrhea in babies. Apart from these, fructans can increase serum levels of interleukin 4 (IL-4), CD282+/TLR2+ myeloid dendritic cells, and a toll-like receptor 2-mediated immune response in healthy volunteers. It can also reduce the need for antibiotics, lengthen the course of the illness, and the frequency of febrile seizures in infants. However, a different study showed that after consumption, there were no changes in serum immune cells, salivary immunoglobulin A (IgA), or the activation and proliferation of T and NK cells. It has been observed that FOS lowers an infant's risk of developing certain immune disorders, like atopic dermatitis. This kind of prebiotic inhibits phagocytosis and IL-6 expression in granulocytes and monocytes.
- **GOS**: Studies reported that GOS increased the blood level of interleukin 8 (IL-8), interleukin 10 (IL-10), and C-reactive protein in adults, but decreased IL-1. Further, NK cells functions improve by consuming GOS. In infants, GOS reduces the risk of atopic dermatitis and eczema.
- **Acidic Oligosaccharides (AOS)**: The possibility of atopic dermatitis is reduced by AOS in low-risk infants.

1.4.4.5. Prebiotics and the Nervous System

The "gut-brain axis" connects the GIT to the central nervous system (CNS). To enhance neural pruning, for example, prebiotics are given to piglets, which reduces gray matter. As outlined below, the gut microbiota influences the brain via the neural, endocrine, and immune pathways.

- **Neural Pathway**: Prebiotic fermentation products have the ability to influence the brain through the vagus nerve. Prebiotics like FOS and GOS can regulate neurotransmitters like d-serine, synaptic proteins, and brain-derived neurotrophic factors (BDNFs). NMDA receptor subunits, synaptophysin, and N-methyl-D-aspartate.
- **Endocrine Pathway**: The neuroendocrine hypothalamic-pituitary-adrenal (HPA) axis is one such pathway. Adrenocorticotropic hormone (ACTH) and corticosterone can be appropriately induced in mice by the growth of their microbiome. Furthermore, prebiotics regulate other hormones like plasma peptide YY.
- **Immune Pathway**: Prebiotics have been shown to impact various immune system components. By altering the activity and/or composition of gut microbiota, prebiotics can also affect mood, memory, learning, and certain psychiatric disorders in addition to neurological functions.
- **Mood**: Stress hormones have the power to influence behaviors linked to anxiety. ACTH and corticosterone levels were shown to rise in germ-free mice after they were subjected to regulated stress. Following administration of *Bifidobacterium infantis*, levels of ACTH and corticosterone returned to normal.

1.4.4.6. Prebiotics and Skin

Prebiotic supplementation has been demonstrated to lessen the incidence and severity of allergic skin conditions like atopic dermatitis. Compounds like phenols may be produced when gut microbes metabolize aromatic amino acids. These substances are absorbed through the skin. Patients with underlying kidney diseases may be toxic when exposed to phenols like p-cresol. GOS consumption, with or without probiotics like *Bifidobacterium breve*, can reverse the phenol-induced decrease in keratin and water in females.

1.4.4.7. Prebiotics and Cardiovascular System

According to the data, cardiovascular diseases (CVD) accounted for 30% of all deaths in the US in 2013. The primary cause of this expanding trend is the way that people's eating and living habits have changed. Consequently, a large number of researchers have looked into how consuming fiber and prebiotics affects CVD. Prebiotics may be beneficial in this way, but their specific benefits have not yet been proven. We outlined a few of the prebiotics' indirect effects on CVD in this section. Prebiotics have the ability to reduce inflammation, which lowers the risk of CVD. Prebiotic consumption has been shown in several studies to improve the lipid profile. Russo et al. showed that, rather than lowering cholesterol levels, eating inulin-enriched pasta (made up of 86% semolina, 1% inulin, and 3% durum wheat vital gluten) reduced TAG and lipogenesis in healthy people. On the other hand, Frochen and Beylot found that after six months, consuming 10 g/day of inulin-type fructans did not significantly alter the process of lipogenesis in the liver of healthy individuals (Hill et al., 2014; Hogan et al., 2018).

1.5. GUT HEALTH

Gut health has garnered significant attention in recent years as emerging research continues to uncover its profound impact on overall well-being and nutrition. The GI tract, also known as the gut, is a vital conduit between our bodies and the food we eat. The gut microbiome, a complex ecosystem made up of trillions of microorganisms, is essential to immune system function, metabolism, digestion, nutrition absorption, and even mental health. We will examine the complex relationship between gut health and nutrition in this thorough review, looking at how food preferences, the gut microbiome, and their interactions affect our overall health and well-being. The gut microbiome, a complex community of microorganisms living in the GI tract, is fundamental to gut health. This microbiome is made up of various microbes, including fungi, viruses, bacteria, and others that coexist in a delicate balance. For the body to remain healthy, the makeup and diversity of this microbial community are crucial. These microbes have a wide range of interactions with our bodies and have a big impact on our nutritional status.

Two of the gut's main jobs are digestion and nutrition absorption. The job of dissolving complex molecules into simpler forms that the body can absorb and use is carried out by the gut when we eat. Enzymes and other digestive components that cooperate with the gut microbiota during this process guarantee effective nutrient extraction. Through the intestinal lining, nutrients including vitamins, minerals, amino acids, and fatty acids are absorbed and distributed throughout the body, serving as vital building blocks for cell growth, repair, and energy production.

In addition, a significant fraction of the immune system in the body is housed in the gut; this group of cells is known as gut-associated lymphoid tissue (GALT). The immune response is crucially regulated by the gut microbiome, which also aids in distinguishing between potentially harmful bacteria and helpful microorganisms. An immune system that is well-balanced and healthy lowers the risk of infections and autoimmune diseases. Preventing toxic substances from entering the bloodstream also depends on the integrity of the gut barrier, a selective membrane that divides the GI tract from the rest of the body.

It is interesting to note that the effects of the gut microbiota go beyond the digestive tract. Recent studies indicate that it has a significant impact on metabolism and controlling weight. Dysbiosis, or an imbalance in gut flora, has been linked to diseases like metabolic syndrome and obesity. The way our bodies store and use the energy we get from food is influenced by certain gut microbes, which is why finding effective weight-management strategies revolves around them. Moreover, the concept of the gut-brain axis has become clear due to the complex relationship between the gut and the brain. Through this two-way communication channel, the gut microbiota and the brain can affect mood, stress levels, and even mental health. Interest in the potential connections between gut health and disorders like anxiety, depression, and neurodegenerative diseases has grown as a result of this interaction.

1.5.1. What Is Gut Health, Exactly?

In a nutshell, gut health is good health. Since your GIT is part of your gut, "gut health" describes the general state of that system, including the microbes that live there. After food is transported from your mouth to your stomach, your GI absorbs the nutrients and energy it contains and excretes the remaining material. That in and of itself is a fairly significant task, but research has revealed that our stomachs are responsible for much more than just breaking down that avocado toast you had for breakfast. In fact, we now know that our guts influence a wide range of other aspects of our health, including immunity, chronic illness, brain function, and even stress management. Although it may not sound very glamorous, 70% of our immune system is located in our stomach. It's crucial to take care of your digestive system to feel well because it's where we metabolize hormones, detoxify enzymes, neutralize pathogens, and produce neurotransmitters.

Furthermore, for your gut to function properly, it needs an army of bacteria, fungi, and other microorganisms. Fun fact: approximately 39 million microbial cells compared to 30 million human cells means that the microbes in and on your body may outnumber your human cells (though this is debatable). Despite making up only about 3% of our body weight, these tiny cells have a far greater impact on our general health than their physical location within our bodies. We will go into more detail about the microbiome—a system of bacteria and other microorganisms. Additionally, there are good and bad bugs in your stomach, just like in any community.

Not only do your gut bacteria aid in food digestion, but they also support your body's overall homeostasis. Research is just starting to uncover a connection between disease and the bacteria in your stomach. Although the exact relationship is still unknown, it is most likely symbiotic—disease is caused by bad bacteria in the gut, and vice versa. The aim of a healthy gut is to maintain the balance of "good" and "bad" bacteria that are present in all of us. For instance, there are bacteria in the gut that both cause and combat inflammation. As we can see from the swelling following an injury vs. prolonged joint inflammation). However, if this equilibrium is upset, it may create the conditions for illness to spread. Reduced vitamin C, an increased risk of allergies, diabetes, multiple sclerosis, and some cancers have all been related to specific gut bacteria. Studies have indicated a connection to neurological conditions like dementia, depression, and anxiety. People with different body sizes tend to have different gut health, which suggests that there may be a connection between gut health and an increased risk of obesity.

1.5.2. What Constitutes a Healthy Gut?

The diversity and richness of the bacteria present are essential for a healthy gut microbiome. An environment that is healthier and more resilient is produced by a greater variety of microbes in the gut, whereas symptoms of an unhealthy gut typically appear when there is insufficient diversity and balance. Even though genetics play a role, research indicates that diet has a far greater influence on the internal operations of the gut, meaning that what you eat actually does matter.

1.5.2.1. Tell Me More

The small intestine is where most nutrients, including proteins, fats, and carbs, are absorbed. But partially digested or undigested nutrients—like fiber, resistant starches, and some protein—can end up in the colon. Since the colon contains the majority of the microbiome, nutrients that enter the colon serve as food for the bacteria. In the colon, undigested protein and non-digestible carbohydrates are fermented to produce SCFAs and gases like carbon dioxide and methane.

1.5.2.2. What's So Special About Short-Chain Fatty Acids?

During the breakdown of non-digestible carbohydrates, gut bacteria primarily produce SCFAs, such as acetate, propionate, and butyrate. Taken as a whole, they have been demonstrated to affect hunger, energy levels, and fat storage, potentially regulating an individual's metabolism. Additionally, they are crucial in preserving the integrity of the intestinal lining, which lessens leaky gut and may even avert systemic inflammation. Additionally, they have been linked to a lower incidence of a few autoimmune conditions, such as multiple sclerosis, inflammatory arthritis, and IBD. By consuming more fiber, we can enhance the production of these beneficial compounds for our health.

1.5.2.3. Harmony is Key

To reap the benefits of gut health, the bacteria in the gut must be diverse and in balance. Unhealthy bacteria can take control of the symbiotic relationship when it is out of balance, resulting in dysbiosis. This produces metabolites that enter the bloodstream through the gut lining and spread inflammation to other areas of the body.

1.5.2.3.1. So, How Do You Maintain That Balance?

Foods rich in probiotics and prebiotics are essential for keeping the digestive system in good working order. Prebiotic fiber can be found in a variety of shelf-stable foods, including cereals, as well as high-fiber foods like whole grains, legumes, fruits, and vegetables. Probiotics are live bacteria and yeasts that help the body create and maintain a healthy colony of bacteria and other microorganisms in the gut. Fermented foods like yogurt, kefir, and kombucha contain probiotics.

1.5.2.4. Gut Health Innovation on the Rise

Research is ongoing to learn more about the effects of digestive health products on overall health and disease prevention as well as the microbiome. Concurrently, research and development endeavors persist in exploring novel and enhanced approaches to provide these wellness remedies to customers. Food scientists are rising to the challenge as new research becomes available and creating gut-friendly innovations for every product category.

1.5.2.5. Paving the Way to Personalized Nutrition

Since the initial sequencing of the human genome approximately two decades ago, the hopes of scientists regarding personalized nutrition have not materialized quickly. We now know that success depends on our ability to comprehend the microbiome, which is thought to contain 2 million genes as opposed to 23,000 in humans. An individual's gut microbiome contains an estimated 500–1,000 distinct species of bacteria, and no two microbiomes are alike. Research has indicated that individuals can be classified into two or three groups according to the presence of specific bacteria, even though each person's composition is unique. Scientists will be trying to build a spectrum of microbial states that humans fall into, fill in the gaps between these classifications, and develop more specialized classifications based on functional profiles as more research becomes available. Future developments in food product development, distribution, and packaging, as well as dietary planning and human health through personalization, are all made possible by the ongoing investigation of the connections among microbiota, diet, and health.

1.5.3. Why Is Gut Health Important?

Your stomach digests food and absorbs nutrients to keep your body functioning. In the medical field, there is a growing body of research on the significance of the gut to our general health. Every organ in our body is impacted by our gut microbiome, according to research.

It is understood that there are links between gut health and:

- the immune system
- mental health
- autoimmune diseases
- endocrine disorders – such as type 2 diabetes
- gastrointestinal disorders – such as irritable bowel syndrome and inflammatory bowel disease
- cardiovascular disease
- cancer
- sleep
- Digestion.

An important measure of the health of your microbiome is the diversity of your gut flora. Although more research is needed, it seems that gut health is closely related to overall health. Although each person's microbiome is different, there are some general guidelines regarding what is and is not healthy. "There is a diverse array of organisms in healthy people." Although fungi, viruses, and other microbes are also present, bacteria make up the majority of those organisms. "A sick person has significantly less diversity in their body and appears to have more of the bacteria linked to illness. Hetcht emphasizes the word "associate" because it is unclear to scientists whether gut bacteria affect gut health or disease risk influences bacteria first. She states that both are probably true. Although specific evidence of this connection's operation is still lacking, we are aware of it.

While some bacteria encourage inflammation, others work to combat it. These two types balance each other out when the digestive system functions properly. However, inflammatory bacteria can take control when that delicate balance is upset. They can also produce metabolites that enter the bloodstream through the gut lining and spread inflammation to other parts of the body. Certain bacteria in the stomach can also cause other illnesses. Research on humans and animals has connected certain bacteria to increased susceptibility to allergies and asthma, to chronic conditions like diabetes, heart disease, multiple sclerosis, and some types of cancer, and yet others to decreased immune function. Even anxiety, depression, and neurological diseases like dementia and schizophrenia have been connected to gut health. Additionally, the composition of gut microbiota differs in lean versus overweight individuals, indicating a potential role for gut microbiota in the development of obesity.

1.5.4. What Affects Gut Health?

Your birth story is just one of many variables that affect the bacterial composition of your gut in addition to the food you eat. Studies reveal that babies born vaginally have more varied microbiomes than babies born via cesarean section because of the exposure to various bacteria during the birth canal. It has also been demonstrated that breastfeeding promotes healthy gut flora. Your upbringing's surroundings are important as well. Hecht says, "Our society is way too clean." Within reason, increased exposure to bacteria and germs can improve our microbiomes. It is all good to go outside and play with animals and dig in the dirt. These are the kinds of things that promote gut health. Gut microbes can also be impacted by emotional stress. Researchers refer to this pathway as the "gut-brain axis," which allows signals from the gut to influence neurotransmitters in the brain and vice versa. Although research is still in its early stages, it appears that a person's mental health and microbiome can have some influence on one another.

Additionally, there is the influence of pharmaceuticals, which include over-the-counter pain relievers and medications for diabetes, acid reflux, and mental health issues; all of these have been connected to alterations in the microbiome. Antibiotics are the most well-known medications that alter the gut because, despite being prescribed to eradicate dangerous bacteria, they also eradicate all other types of bacteria.

1.5.4.1. Can You Tell if You Are Having Health Problems in Your Gut?

It is usually simple to determine when the microbiome is out of balance for whatever reason. Indices such as bloating, gas, diarrhea, stomach pain, or nausea are clear indicators that something is wrong with the gut. The imbalances usually correct themselves in a short period of time, but if they persist, a medical diagnosis and course of treatment may be necessary. Gastroenterologists are able to perform tests for particular microbiome-related disorders, such as an overabundance of a particular type of bacteria. However, an increasing number of medical professionals are identifying dysbiosis in the gut flora that does not result in any symptoms right away, at least not digestive ones. "You can have bacteria in your stomach that are linked to a higher risk of colon cancer, for example, but they are not overproducing gas or changing your motility or anything you'd notice." Commercial testing kits are available for those who are interested in learning more about their microbiome. They analyze a stool sample and reveal the strains of bacteria that are present. Nonetheless, medical professionals advise against purchasing the kits if you are looking for advice regarding your health. "We do not currently know enough to interpret those readouts as meaningful." "We are generally aware of what constitutes inflammation and non-inflammation."

1.5.4.2. How Can I Maintain My Gut Health?

It is not necessary to always be aware of what is happening in your stomach. Additionally, you are probably helping your microbiome as long as you are adhering to your doctor's recommendations for general health. We advise people to maintain a healthy gut by eating a balanced diet, drinking plenty of water, exercising frequently, and getting enough sleep each night. In a similar vein, behaviors detrimental to the heart, lungs, and brain, such as heavy alcohol consumption and cigarette smoking, can also negatively impact the microbiome. There is some evidence, though, that drinking red wine in moderation may be advantageous. Hecht advises against taking superfluous drugs and discussing with your doctor how your present medication schedule may be affecting your digestive system. Reducing consumption of dairy, red, processed, and refined sugars can also help with gut health. Getting the recommended 20 to 40 g of fiber per day, depending on your age and gender, can also help. Although the majority of Americans don't reach these recommendations, you can up your intake by including more fruits, vegetables, whole grains, nuts, and seeds in your diet.

Menon recommends keeping a food diary to track fiber intake, as well as GI symptoms related to food. A sudden switch to high-fiber foods can cause bloating, so introduce them gradually and keep track of how your body reacts to anything new. "Sometimes you can identify specific trigger foods that make you feel bad every time you eat them," and you can find alternatives that work better for you.

1.5.5. Signs of an Unhealthy Gut

Your gut microbiome can be affected by:

- stress
- too little sleep
- lack of physical activity
- eating too many ultra-processed foods
- smoking and drinking alcohol
- taking antibiotics.

The gut microbiome is also influenced by factors outside of our control, like our upbringing, age, mode of birth, and whether we were bottle- or breast-fed as infants.

While we cannot use one specific measure for our gut health, some signs that you may have poor gut health include:

- digestive symptoms – such as gas, bloating, constipation, diarrhea and heartburn
- sleep disturbances or fatigue
- mood/emotional state – such as high stress, low mood or anxiety
- high frequency of infectious illnesses – such as the common cold.

1.5.6. How to Improve Your Gut Health

Changes in diet and lifestyle may help you achieve better gut health. Foods high in dietary fiber can help maintain regularity, lower the risk of colon cancer, and nourish the good bacteria in our stomachs, all of which contribute to improved gut health. Wholefoods including fruits, vegetables, whole grains, legumes, and nuts may stop some bacteria that cause inflammation and illness from growing. Our way of life, which includes exercise, restful sleep, and lowering stress levels, is also beneficial to gut health.

1.5.6.1. Gut Health and Diet

What you eat affects the bacteria in your gut. Providing them with the proper nutrition is essential for maintaining a healthy gut microbiome. Eating a variety of fresh, whole foods, mostly from plant sources like fruits, vegetables, legumes, beans, nuts, and whole grains, is the best way to keep your microbiome healthy.

1.5.6.2. Eat a High Fiber Diet

There are numerous reasons why fiber is crucial for gut health. Fiber can have an impact on how well our gut works, including how nutrients are absorbed and digested, how quickly or slowly things pass through, and the consistency of our stools. Our gut bacteria's breakdown of fiber can also result in the production of significant compounds that may have an impact on the emergence of gastrointestinal disorders like colon cancer. Beyond just helping the gut, fiber also lowers the risk of heart disease and type 2 diabetes, among other health benefits. Plant-based foods are the only ones that contain fiber. Adult Australian women should strive to consume at least 25 g of fiber daily, while men should consume 30 g.

Foods that are high in fiber include:

- vegetables
- beans and legumes
- fruit
- bread and cereals
- nuts and seeds.

Prebiotic fibers, which are not found in all high-fiber foods, may be especially helpful for our gut microbiome, as they can act as a fertilizer for the healthy bacteria in our gut.

They are found in some types of:

- vegetables – for example, leek, onion and garlic
- legumes – for example, chickpeas, beans and lentils
- whole grains – for example, rye bread, barley and oats
- nuts – for example, pistachios, cashews and almonds.

1.5.6.3. Eat a Diverse Range of Food

Eating a wide range of fruits and vegetables ensures you are including a whole range of vitamins, minerals and nutrients in your diet.

1.5.6.4. Limit Ultra-Processed Foods

To support the health of your gut, eat foods that are as close to their natural state as possible. Although practically every food has undergone some processing, it is preferable to consume minimally processed foods. These foods maintain their nutritional value and are typically free of added sugar, salt, unhealthy fats, and additives like artificial sweeteners and emulsifiers that can negatively affect your gut health. Fruits, vegetables, whole grains, eggs, seafood, poultry, and lean red meat are examples of unprocessed foods. Deli meats like ham and salami, a lot of breakfast cereals, prepared meals, sweet desserts, and a lot of packaged snacks like chips are examples of ultra-processed foods.

1.5.6.5. Drink Water

The best liquid to consume is water, which also improves gut health. In order for your body to absorb nutrients from food, water helps with the breakdown process. Additionally, water helps to soften stools, which helps to ward off constipation. Increased diversity of bacteria in the gut may also be associated with a high-water intake.

1.5.7. Eat Foods Rich in Polyphenols

Polyphenols are plant compounds that may beneficially impact our gut microbiome.

Foods rich in polyphenols include:

- herbs and spices
- colorful fruits and vegetables
- nuts and seeds
- green and black tea
- coffee
- cocoa and dark chocolate.

1.5.7.1. Eat Slowly

Chewing your food thoroughly and eating slowly may reduce digestive discomfort such as gas, pain and bloating.

1.5.7.2. Eat Fermented Foods

Foods that have undergone fermentation have had their sugars broken down by bacteria and yeast.

Fermented foods include:

- yoghurt
- kimchi
- sauerkraut
- kefir
- kombucha
- tempeh.

While research into fermented foods is limited, the bacteria found in some fermented foods have been linked with digestive health and other benefits.

1.5.8. Gut Health and Breastfeeding

A healthy gut microbiome that an infant develops through breastfeeding may help shield them from certain diseases later in life.

1.5.8.1. Gut Health and Exercise

Walking and cycling are examples of regular cardiovascular exercise that can activate the gut's muscles to move digestive contents throughout the body. Exercise has a beneficial impact on the gut microbiota as well.

1.5.8.2. Gut Health and Stress

Your gut health can be impacted by stress. To keep your gut healthy, it is critical to take care of your mental and physical health. Your gut health, mental health, and what you eat are all related. Reduce your stress by doing regular exercise, getting enough sleep, interacting with people, practicing relaxation techniques, and maintaining a healthy diet.

1.5.8.3. Gut Health and Sleep

Insufficient sleep, whether in quantity or quality, can affect your gut microbiota and cause discomfort in your digestive system.

1.5.8.4. Gut Health and Probiotic Supplements

Rather than using supplements, it is preferable to improve your gut health through diet and other lifestyle choices. Wholefoods contain a multitude of nutrients that are too numerous to fit into a single supplement. Additionally, nutrients in food have beneficial interactions with one another that are not possible for supplements to mimic. Probiotic supplements are of interest to a large number of people. It is usually not necessary to take a probiotic for gut health if you are in good health. Research has occasionally been done to support the use of probiotics; however, much like with medications, the type of probiotic you take will depend on the particular health condition you are trying to treat. It is a good idea to consult your general practitioner and an accredited practicing dietitian before taking probiotics or any other supplement to find out which one might be safe and effective.

1.5.8.5. Gut Health and Antibiotics

Antibiotics have a negative effect on your gut microbiome even though they can be very important and useful. When you have an infection or disease, antibiotics work to destroy the bad bacteria, but in the process, they can also eliminate some of the good bacteria in your stomach.

1.5.8.6. Myths About Gut Health

Gut health cannot be achieved by miracle treatments. Scientific research has not shown that any one food or product can quickly restore the balance of the gut microbiota. Furthermore, there is no proof from science that colon cleansing is helpful or enhances health in any way. Although the field of study on gut health is still relatively young, our knowledge of this intricate subject is growing. Avoid believing anything about gut health that is not supported by research. The greatest information we currently have suggests concentrating on eating healthfully and following the advice on this page.

1.6. IMPORTANCE OF GUT HEALTH FOR OVERALL WELL-BEING

A healthy GIT is widely recognized as one of the best ways to improve general health and wellness, and anyone who has ever experienced digestive problems understands the damaging effects they can have on your body. There is an easy explanation for this. The primary route via which your body can absorb the nutrients it requires is through your digestive system. It serves as a disease fighter, a communication hub between your immune and nervous systems, and the recipient of over 80% of your body's serotonin and immune system components. It is obvious why gut health is crucial for preserving general health, but you might not know exactly what gut health entails. Your overall health is based on your gut health, which is defined as the equilibrium between beneficial and harmful bacteria in your digestive system. The microbiome, also referred to as the gut flora, is the collective term for the trillions of bacteria, fungi, and viruses that inhabit your body. While certain bacteria have been connected to health issues, others are critical for maintaining a healthy weight, heart, and immune system. You can quickly start to when something goes wrong in your digestive system.

Prebiotics are substances that the good bacteria in your stomach eat, and they can be found in foods like oats, bananas, berries, and asparagus. Probiotics, on the other hand, are live bacteria that are naturally present in some foods and dietary supplements. Probiotic-rich foods include fermented foods like kefir, sauerkraut, kimchi, and pickles, as well as plain yogurt. Drinking lots of water is also crucial because it facilitates the digestion and processing of food. This enables your body to take in the nutrients it requires to function at its peak. Additionally, water is absorbed by your small and large intestines and subsequently enters your bloodstream. Your stool then becomes solid in your large intestine, which facilitates its passage through your digestive tract. Limiting the quantity of "bad" foods you eat is another crucial step in keeping your digestive system healthy. Diabetes and leptin resistance may result from diets heavy in sugar and fat because they negatively impact gut flora. These may result in fatty liver disease and are contributing factors to diabetes and obesity. You might additionally experience stomach discomfort even if you follow a healthy diet (American Gastroenterological Association, 2020).

REFERENCES

1. Allen, S. J., Martinez, E. G., Gregorio, G. V., & Dans, L. F. (2010). Probiotics for treating acute infectious diarrhoea. *Cochrane Database of Systematic Reviews*, *2010*, CD003048.
2. American Gastroenterological Association. (2020). Consideration for use of probiotics in gastrointestinal diseases: Clinical decision support tool. *Gastroenterology, 159*(3), 706.
3. Avena-Woods, C. (2017). Overview of atopic dermatitis. *American Journal of Managed Care, 23*(S5), S115–S123.
4. Blaabjerg, S., Artzi, D. M., & Aabenhus, R. (2017). Probiotics for the prevention of antibiotic-associated diarrhea in outpatients: A systematic review and meta-analysis. *Antibiotics, 6*(4), 21.
5. Caffarelli, C., Cardinale, F., Povesi-Dascola, C., Dodi, I., Mastrorilli, V., & Ricci, G. (2015). Use of probiotics in pediatric infectious diseases. *Expert Review of Anti-Infective Therapy, 13*(12), 1517–1535.
6. Casas, I. A., & Dobrogosz, W. J. (2000). Validation of the probiotic concept: *Lactobacillus reuteri* confers broad-spectrum protection against disease in humans and animals. *Microbial Ecology in Health and Disease, 12(4), 247–285.*
7. Derikx, L. A., Dieleman, L. A., & Hoentjen, F. (2016). Probiotics and prebiotics in ulcerative colitis. *Best Practice & Research Clinical Gastroenterology, 30*(1), 55–71.
8. Didari, T., Mozaffari, S., Nikfar, S., & Abdollahi, M. (2015). Effectiveness of probiotics in irritable bowel syndrome: Updated systematic review with meta-analysis. *World Journal of Gastroenterology, 21*(10), 3072–3084.
9. FAO/WHO. (2001). Report on Joint FAO/WHO Expert Consultation on Evaluation of Health and Nutritional Properties of Probiotics in Food Including Powder Milk with Live Lactic Acid Bacteria.
10. Feizizadeh, S., Salehi-Abargouei, A., & Akbari, V. (2014). Efficacy and safety of *Saccharomyces boulardii* for acute diarrhea. *Pediatrics, 134*(1), e176–e191.
11. Food and Drug Administration. (2018). Policy regarding quantitative labeling of dietary supplements containing live microbials: Guidance for industry.
12. Ford, A. C., Quigley, E. M., Lacy, B. E., Lembo, A. J., Saito, Y. A., Schiller, L. R., et al. (2014). Efficacy of prebiotics, probiotics, and synbiotics in irritable bowel syndrome and chronic idiopathic constipation: Systematic review and meta-analysis. *American Journal of Gastroenterology, 109*(10), 1547–1561.

13. Freedman, S. B., Williamson-Urquhart, S., Farion, K. J., Gouin, S., Willan, A. R., Poonai, N., et al. (2018). Multicenter trial of a combination probiotic for children with gastroenteritis. *New England Journal of Medicine, 379*(20), 2015–2026.
14. Ghouri, Y. A., Richards, D. M., Rahimi, E. F., Krill, J. T., Jelinek, K. A., & DuPont, A. W. (2014). Systematic review of randomized controlled trials of probiotics, prebiotics, and synbiotics in inflammatory bowel disease. *Clinical and Experimental Gastroenterology, 7*, 473–487.
15. Gibson, G. R., Hutkins, R., Sanders, M. E., Prescott, S. L., Reimer, R. A., Salminen, S. J., et al. (2017). Expert consensus document: The International Scientific Association for Probiotics and Prebiotics (ISAPP) consensus statement on the definition and scope of prebiotics. *Nature Reviews Gastroenterology & Hepatology, 14*(8), 491–502.
16. Grover, S., et al. (2012). Probiotics for human health: New innovations and emerging trends. *Gut Pathogens, 4(1), 15.*
17. Guarino, A., Ashkenazi, S., Gendrel, D., Lo Vecchio, A., Shamir, R., Szajewska, H., et al. (2014). European Society for Pediatric Gastroenterology, Hepatology, and Nutrition/European Society for Pediatric Infectious Diseases evidence-based guidelines for the management of acute gastroenteritis in children in Europe: Update 2014. *Journal of Pediatric Gastroenterology and Nutrition, 59*(1), 132–152.
18. Hendry, G. (1987). The ecological significance of fructan in a contemporary flora. *New Phytologist, 106*(S1), 201–216.
19. Hill, C., Guarner, F., Reid, G., et al. (2014). The International Scientific Association for Probiotics and Prebiotics consensus statement on the scope and appropriate use of the term probiotic. *Nature Reviews Gastroenterology & Hepatology, 11*(8), 506–514.
20. Hogan, D. E., Ivanina, E. A., & Robbins, D. H. (2018). Probiotics: A review for clinical use. *Gastroenterology & Endoscopy News*, 1–7.
21. Huang, R., Ning, H., Shen, M., Li, J., Zhang, J., & Chen, X. (2017). Probiotics for the treatment of atopic dermatitis in children: A systematic review and meta-analysis of randomized controlled trials. *Frontiers in Cellular and Infection Microbiology, 7*, Article 392.
22. Iheozor-Ejiofor, Z., Kaur, L., Gordon, M., Baines, P. A., Sinopoulou, V., & Akobeng, A. K. (2020). Probiotics for maintenance of remission in ulcerative colitis. *Cochrane Database of Systematic Reviews, 2020*(3), CD007443.
23. Jafarnejad, S., Shab-Bidar, S., Speakman, J. R., Parastui, K., Daneshi-Maskooni, M., & Djafarian, K. (2016). Probiotics reduce the risk of

antibiotic-associated diarrhea in adults (18–64 years) but not the elderly (>65 years): A meta-analysis. *Nutrition in Clinical Practice, 31*(4), 502–513.

24. Kailasapathy, K., & Chin, J. (2000). Survival and therapeutic potential of probiotic organisms with reference to *Lactobacillus acidophilus* and *Bifidobacterium* spp. *Immunology and Cell Biology, 78*(1), 80–88.

25. Kaur, L., Gordon, M., Baines, P. A., Iheozor-Ejiofor, Z., Sinopoulou, V., & Akobeng, A. K. (2020). Probiotics for induction of remission in ulcerative colitis. *Cochrane Database of Systematic Reviews, 2020*(3), Article CD005573.

26. Kerry, R. G., Patra, J. K., Gouda, S., Park, Y., Shin, H. S., & Das, G. (2018). Benefaction of probiotics for human health: A review. *Journal of Food and Drug Analysis, 26*(3), 927–939.

27. Ki Cha, B., Mun Jung, S., Hwan Choi, C., Song, I. D., Woong Lee, H., Joon Kim, H., et al. (2012). The effect of a multispecies probiotic mixture on the symptoms and fecal microbiota in diarrhea-dominant irritable bowel syndrome: A randomized, double-blind, placebo-controlled trial. *Journal of Clinical Gastroenterology, 46*(3), 220–227.

28. Lamb, C. A., Kennedy, N. A., Raine, T., Hendy, P. A., Smith, P. J., et al. (2019). British Society of Gastroenterology consensus guidelines on the management of inflammatory bowel disease in adults. *Gut, 68*(Suppl. 3), s1–s106.

29. Li, L., Han, Z., Niu, X., Zhang, G., Jia, Y., & Zhang, S. (2018). Probiotic supplementation for prevention of atopic dermatitis in infants and children: A systematic review and meta-analysis. *American Journal of Clinical Dermatology.*

30. Lorenzo-Zuniga, V., Llop, E., Suarez, C., Alvarez, B., Abreu, L., Espadaler, J., et al. (2014). I.31, a new combination of probiotics, improves irritable bowel syndrome-related quality of life. *World Journal of Gastroenterology, 20*(27), 8709–8716.

31. Makrgeorgou, A., Leonardi-Bee, J., Bath-Hextall, F. J., Murrell, D. F., Tang, M. L., Roberts, A., et al. (2018). Probiotics for treating eczema. *Cochrane Database of Systematic Reviews, 2018*(11), CD006135.

32. McKenzie, Y. A., Bowyer, R. K., Leach, H., Gulia, P., Horobin, J., O'Sullivan, N. A., et al. (2016). British Dietetic Association systematic review and evidence-based practice guidelines for the dietary management of irritable bowel syndrome in adults (2016 update). *Journal of Human Nutrition and Dietetics, 29*(5), 549–575.

33. National Institutes of Health. (2018). Dietary Supplement Label Database.

34. Pandey, K. R., et al. (2015). Probiotics, prebiotics, and synbiotics: A review. *Journal of Food Science and Technology, 52(12), 7577–7587.*

35. Preidis, G. A., Weizman, A. V., Kashyap, P. C., & Morgan, R. L. (2020). AGA technical review on the role of probiotics in the management of gastrointestinal disorders. *Gastroenterology, 159*(3), 708–738.

36. Sairenji, T., Collins, K. L., & Evans, D. V. (2017). An update on inflammatory bowel disease. *Primary Care, 44*(4), 673–692.

37. Sanders, M. E. (2008). Clinical use of probiotics: What physicians need to know. *American Family Physician, 78*(9), 1026.

38. Sanders, M. E. (2015). Probiotics in 2015: Their scope and use. *Journal of Clinical Gastroenterology, 49*(Suppl. 1), S2–S6.

39. Schnadower, D., Tarr, P. I., Casper, T. C., Gorelick, M. H., Dean, J. M., O'Connell, K. J., et al. (2018). *Lactobacillus rhamnosus* GG versus placebo for acute gastroenteritis in children. *New England Journal of Medicine, 379*(20), 2002–2014.

40. Silverman, M. A., Konnikova, L., & Gerber, J. S. (2017). Impact of antibiotics on necrotizing enterocolitis and antibiotic-associated diarrhea. *Gastroenterology Clinics of North America, 46*(1), 61–76.

41. Staudacher, H. M., & Whelan, K. (2016). Altered gastrointestinal microbiota in irritable bowel syndrome and its modification by diet: Probiotics, prebiotics, and the low FODMAP diet. *Proceedings of the Nutrition Society, 75*(3), 306–318.

42. Su, G. L., Ko, C. W., Bercik, P., Falck-Ytter, Y., Sultan, S., et al. (2020). AGA clinical practice guidelines on the role of probiotics in the management of gastrointestinal disorders. *Gastroenterology, 159*(2), 697–705.

43. Szajewska, H., & Horvath, A. (2018). *Lactobacillus rhamnosus* GG in the primary prevention of eczema in children: A systematic review and meta-analysis. *Nutrients, 10*(9).

44. Szajewska, H., & Kolodziej, M. (2015). Systematic review with meta-analysis: *Saccharomyces boulardii* in the prevention of antibiotic-associated diarrhea. *Alimentary Pharmacology & Therapeutics, 42*(7), 793–801.

45. Szajewska, H., Guarino, A., Hojsak, I., Indrio, F., Kolacek, S., Shamir, R., et al. (2014). Use of probiotics for management of acute gastroenteritis: A position paper by the ESPGHAN Working Group for Probiotics and Prebiotics. *Journal of Pediatric Gastroenterology and Nutrition, 58*(4), 531–539.

46. Szajewska, H., Skorka, A., Ruszczynski, M., & Gieruszczak-Bialek, D. (2013). Meta-analysis: *Lactobacillus GG* for treating acute gastroenteritis

in children—Updated analysis of randomized controlled trials. *Alimentary Pharmacology & Therapeutics, 38*(5), 467–476.

47. Van den Ende, W. (2013). Multifunctional fructans and raffinose family oligosaccharides. *Frontiers in Plant Science, 4*, 247.
48. Vasudha, S., & Mishra, H. N. (2013). Non-dairy probiotic beverages. *International Food Research Journal, 20*(1), 7–15.
49. Vyas, U., & Ranganathan, N. (2012). Probiotics, prebiotics, and synbiotics: Gut and beyond. *Gastroenterology Research and Practice, 2012*, 1–16.
50. Watson, R. R., & Preedy, V. R. (2010). *Bioactive Foods in Promoting Health: Probiotics and Prebiotics.* Academic Press.
51. Webb, G. P. (2011). *Dietary Supplements and Functional Foods.* John Wiley & Sons.
52. Wilkins, T., & Sequoia, J. (2017). Probiotics for gastrointestinal conditions: A summary of the evidence. *American Family Physician, 96*(3), 170–178.
53. Williams, E. A., Stimpson, J., Wang, D., Plummer, S., Garaiova, I., Barker, M. E., et al. (2009). Clinical trial: A multistrain probiotic preparation significantly reduces symptoms of irritable bowel syndrome in a double-blind placebo-controlled study. *Alimentary Pharmacology & Therapeutics, 29*(1), 97–103.
54. Zhao, M., Shen, C., & Ma, L. (2018). Treatment efficacy of probiotics on atopic dermatitis, zooming in on infants: A systematic review and meta-analysis. *International Journal of Dermatology, 57*(6), 635–641.
55. Zhuang, X., Xiong, L., Li, L., Li, M., & Chen, M. (2017). Alterations of gut microbiota in patients with irritable bowel syndrome: A systematic review and meta-analysis. *Journal of Gastroenterology and Hepatology, 32*(1), 28–38.
56. Zmora, N., Zilberman-Schapira, G., Suez, J., Mor, U., Dori-Bachash, M., Bashiardes, S., et al. (2018). Personalized gut mucosal colonization resistance to empiric probiotics is associated with unique host and microbiome features. *Cell, 174*(6), 1388–1405.e21.

CHAPTER 2

MECHANISMS OF ACTION OF PROBIOTIC YEASTS

Contents

INTRODUCTION

Probiotic yeasts, mainly *Saccharomyces boulardii*, influence the gut microbiome and provide health benefits through a variety of methods. Enhancing the function of the gut barrier is one of their primary actions. These yeasts create metabolites and proteins that strengthen the connections between intestinal epithelial cells (IECs), decreasing intestinal permeability and obstructing the translocation of toxic substances and pathogens. Furthermore, through their interactions with the gut-associated lymphoid tissue (GALT), probiotic yeasts have an immunomodulatory effect. They bolster the immune response against pathogenic organisms by inducing the production of protective cytokines and boosting the activity of mucosal immune cells. The competitive exclusion of pathogens is another significant mechanism. Probiotic yeasts attach themselves to the intestinal mucosa and reduce pathogen colonization and overgrowth by outcompeting harmful bacteria for resources and binding sites. In addition, they generate antimicrobial agents like hydrogen peroxide and organic acids, which stop the growth of harmful bacteria. Probiotic yeasts also contribute to the preservation of a balanced gut microbiota.

2.1. BENEFICIAL EFFECTS OF PROBIOTIC YEASTS AS DEMONSTRATED BY BIOLOGICAL AND CLINICAL TRIALS

Probiotics are live microorganisms that, when given in sufficient quantities, benefit the host's health (FAO/WHO, 2002). Pharmaceutical preparations and fermented dairy products contain these microorganisms in large quantities for the treatment or prevention of neurological, dermatological, gastrointestinal, and vaginal disorders. While many probiotic products for humans contain *lactobacilli* and bifidobacteria, only a few types of yeast, including *Saccharomyces boulardii*, are typically used.

Although *Saccharomyces cerevisiae* has been used for many years to make breads and alcoholic beverages, its probiotic qualities were not well understood until a few decades ago (McFarland, 2010). French microbiologist Henri Boulard traveled to Southeast Asia in the 1920s in search of a yeast strain that could withstand high temperatures and still make a good wine. During that period, the Indochina Peninsula was experiencing a cholera outbreak. Boulard noted that the locals were using an infusion made from the bark of a tropical fruit called lychee to treat and even prevent diarrhea. The fruit was later discovered to be coated in a yeast, and the presence of this microbe—which was subsequently identified and given the name *Saccharomyces boulardii*—was thought to be responsible for the fruit's ability to prevent diarrhea. The non-pathogenic yeast *Saccharomyces boulardii* grows at an exceptionally high temperature of 37°C (MacFarland, 2010). It is also known as *S. boulardii* lyo 17 because it is frequently sold as a probiotic in a lyophilized form. Several diarrheal illnesses, including antibiotic-associated diarrhea (AAD), have been treated with *Saccharomyces boulardii*.

The only yeast that has been commercialized globally as a probiotic for humans to date is *Saccharomyces boulardii*. Nonetheless, based primarily on *in vitro* tests and a few clinical trials, some writers have recommended using different yeast species or genera (Kumura et al., 2004; Kühle et al., 2005; Pedersen et al., 2012; Tiago et al., 2009). *Saccharomyces boulardii* is sometimes categorized as *S. cerevisiae* var. despite having a metabolic makeup that is significantly different from that of *S. cerevisiae* due to their nearly identical genetic makeup. *boulardii* (Fietto et al., 2004). According to earlier findings from research, *S. cerevisiae* strain 905, which was isolated during the production of sugar cane spirit, or "cachaça," was able to colonize and survive in the gastrointestinal tract (GIT) of conventional mice and germ-free mice, as well as shield the latter against oral challenges with *Clostridium difficile* or *Salmonella enterica* serovar Typhimurium (Martins et al., 2005).

Numerous mechanisms of action have been proposed to explain the actions of bacterial and yeast probiotics. These include the production of antagonistic compounds (organic acids, bacteriocins, H_2S), competition for nutrients, inhibition of pathogen adhesion to epithelium (competition for receptor and spatial exclusion, trapping by adhesion on probiotic surface), modification of the immune system (stimulation of IgA production or blood clearance, anti-inflammatory capacity during infection or inflammatory pathologies), interference on bacterial-induced signaling pathways, and actions on bacterial virulence factors (Vieira et al., 2013). The findings that follow illustrate which of these mechanisms might be at work in yeasts to account for the positive outcomes seen in biological and clinical tests, as well as to support the possibility of using them as probiotics.

2.2. MECHANISMS OF ACTION RESPONSIBLE FOR BENEFICIAL EFFECTS OF PROBIOTIC YEASTS

2.2.1. Antagonism

Initially, it was demonstrated that *S. boulardii* exhibited antagonistic activity against a variety of bacterial pathogens both *in vitro* (Brugier and Patte, 1975) and *in vivo* (Ducluzeau and Bensaada, 1982). *Saccharomyces boulardii* 17 inhibits Candida colonization as well as filamentous growth, adhesion, and biofilm formation in Candida albicans (Jawhara and Poulain, 2007). Krasowska et al. (2009) by means of capric acid secretion (Murzyn et al., 2010). However, when this yeast's ability to oppose *S. typhimurium* and *Shigella flexneri* was assessed in the digestive tract of regular or gnotobiotic mice, the *in vitro* results for bacterial pathogens were not verified. *S. boulardii* 17 was given in 10 mg doses to conventional animals every day, while germfree animals only received one 10 mg dose. Five days later, the pathogenic bacteria (108 or 102 viable cells, respectively) were given orally to both groups as a challenge. Instead of *S. boulardii* 17, saline was administered to the control groups. In mice treated with yeast, mortality, weight gain, and histopathological data demonstrated a protective effect against the pathogenic bacteria (Figure 2.1).

In these results showed that the protection against *S. typhimurium* and *S. flexneri* obtained in conventional and gnotobiotic mice previously associated with *S. boulardii* 17 is not due to the reduction of the bacterial populations in the intestines through the action of some antagonistic compound (Rodrigues et al., 1996; Martins et al., 2013). These results confirm that production of inhibitory compounds against pathogenic bacteria is typical and more frequently found in *lactobacilli* and bifidobacteria, and rarely found in yeast (Servin, 2004).

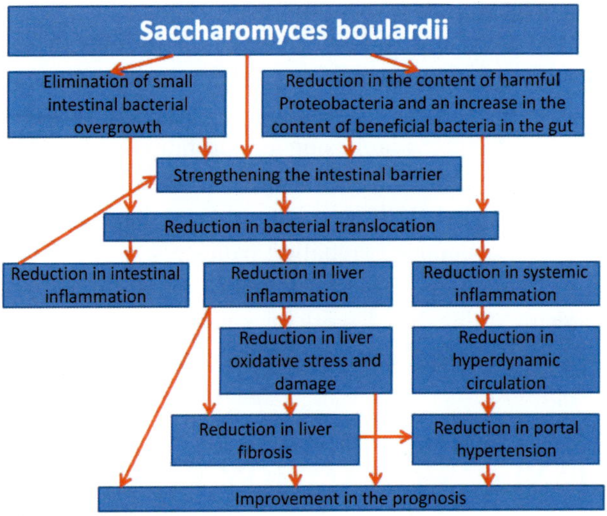

Figure 2.1. *Proposed mechanism of restoration of the gut–liver axis by Saccharomyces boulardii.*

Source: Maslennikov R, Benuni N, Levshina A, Adzhieva F, Demina T, Kucher A, Pervushova E, Yuryeva E, Poluektova E, Zolnikova O, et al. (2024). Effect of Saccharomyces boulardii on liver diseases: A systematic review. Microorganisms; 12(8):1678. https://doi.org/10.3390/microorganisms12081678.

2.2.2. Bacterial Motility

The ability of enteropathogenic bacteria to move into suitable niches, pass through the mucus layer, and infiltrate the intestinal epithelium is known as motility, and it plays a crucial role in their pathogenesis. The potential impact of *S. boulardii* 17 on *S. typhimurium* motility was assessed in a recent study. Human colonic T84 cells were used for the experiments. The cells were infected with *S. typhimurium* alone or in combination with yeast. Time-lapse video microscopy was used to record the *Salmonella* motility, and manual tracking was used to analyze the dynamics of the bacteria. When *S. boulardii* 17 was present, the median curvilinear velocity (CLV) of *Salmonella* cultured in isolation with T84 dropped from 43.3 μm/s to 31.2 μm/s. Track linearity measurements revealed that *S. boulardii* 17 increased the number of bacteria with rotator tract by 22% and decreased the number of bacteria with linear tract by 15%. By examining a non-motile, flagella-deficient *S. typhimurium* strain that moved with a significantly lower CLV of 0.5 μm/sec and a decrease in invasion properties, the correlation between *S. typhimurium* motility and invasion was further established. The motility of *S. typhimurium* and s was altered by this protective mechanism of *S. boulardii* 17.

2.2.3. Local and Systemic Immunomodulation (IgA, Anti-Inflammatory Cytokine, Clearance)

The gut epithelial barrier serves as the first line of defense against pathogen entry, and probiotics can stop pathogen-induced membrane damage by blocking pathogen adhesion and preserving the proper arrangement of the tight junction and cytoskeleton proteins. As a barrier against pathogen adhesion to the intestinal mucosa, IgA is an antibody that is essential to mucosal immunity. This immunoglobulin (Ig) can attach to molecules on the surface of bacteria that mediate adhesion to epithelial cells (preventing translocation) or to the surface of bacteria itself, trapping them in the mucous layer of the intestinal epithelia. Different microorganisms have varying capacities to stimulate the production of sIgA; in gnotobiotic mice, Gram-negative bacteria and yeast typically cause a greater stimulation than Gram-positive bacteria (Rodrigues et al., 2000). A recent investigation (Martins et al., 2009), using germ-free Swiss/NIH mice monoassociated with each probiotic, the impact of *S. boulardii* 17 on the immune system has been compared with some bacterial probiotics (*Lactobacillus* casei, *Escherichia coli* EM0, and *Bifidobacterium animalis* BB12).

Probiotics may help mice with experimental colitis, according to recent research, by increasing IL-10 and IL-10-dependent T regulatory cells (Mengheri, 2008). The main producers of the anti-inflammatory cytokine IL-10 are monocytes and, to a lesser extent, lymphocytes. It influences inflammation and immune regulation in pleiotropic ways. It suppresses the expression of co-stimulatory molecules, MHC class II antigens, and Th1 cytokines on macrophages. Moreover, it improves the survival, growth, and generation of antibodies by B cells. According to research utilizing knockout mice, this cytokine serves as a crucial immune-regulator in the intestines (Grimbaldeston et al., 2007). It has the ability to suppress the production of pro-inflammatory cytokines such as IFN-g, IL-2, IL-3, TNF, and GM-CSF by cells like macrophages and Th1 cells. It also has strong antigen-presenting cell (APC) suppression properties. Mice in the experimental intestinal obstructed (IO) group had higher serum levels of IL-10 (50.75 and 29.58 pg/ml, respectively) than the control (CTL) group ($P<0.05$) treatment with *S. cerevisiae* UFMG that has been both heat-killed and viable.

When translocation is successful, the reticuloendothelial system (RES) is crucial in clearing the circulatory system of microorganisms. Over 80% of the RES's function is attributed to Küpffer cells' hepatic phagocytosis. RES stimulators such as glucan and other *Saccharomyces* extracts are significant. Following glucan administration, it has been shown that pluripotent hematopoietic stem cells, granulocyte–macrophage colony-forming cells, and erythroid colony-forming cells proliferate in mouse bone marrow. Additionally, there is an increase in IL-1 production following glucan administration, which can be attributed to either a direct effect on macrophages or an indirect effect on the production of colony stimulating factor(s) from T-cells stimulated by glucan. Well-known to be among the largest populations of tissue macrophages, Küpffer cells have been linked to the removal of bacteria from the bloodstream along with neutrophils (Gregory and Wing, 2002). In yeast mono-associated mice (42.4 ± 4.4 and 52.9 ± 15.7 with *S. boulardii* 17 and *S. cerevisiae* UFMG 905, respectively), the number of Küpffer cells per 100 hepatocytes in the liver was significantly higher than in germ.

2.2.4. Inflammation and Permeability of Intestinal Epithelium

Numerous investigations demonstrate that enteropathogenic and enterohaemorrhagic *Escherichia coli*-induced inflammatory transduction pathways are modulated by *S. boulardii* 17 and *S. cerevisiae* UFMG 905 bacteria (Czerucka et al., 2000; Dahan et al., 2003), *S. typhimurium* (Martins et al., 2010, 2011, 2013), C. difficile (Chen et al., 2006), and Shigella (Mumy et al., 2008). In addition to its ability to prevent inflammation brought on by pathogenic bacteria, *S. boulardii* has an anti-inflammatory property that it demonstrates in inflammatory bowel disorders (Pothoulakis, 2009).

Using T84 human colorectal cancer (CRC) cells, it was shown that pre-infection incubation with *S. boulardii* completely eliminated *Salmonella* invasion at the molecular level. This is correlated with a reduction in Rac1 activation. *Saccharomyces boulardii* 17 reduced IL-8 synthesis induced by *S. typhimurium* while maintaining the T84 barrier function. This anti-inflammatory effect was linked to *S. boulardii* 17's inhibitory effect on the activation of NF-κB and the MAPKs ERK1/2, p38, and JNK induced by S. Typhimurium. Studies using electron and confocal microscopy revealed bacterial adhesion to yeast cells, which may be one of the ways that *S. boulardii* 17 provides protection. An additional explanation might be that the yeast is producing a substance that interferes with the cellular signaling pathways that are stimulated by bacteria. The anti-inflammatory effect of the *S. boulardii* 17 growth supernatant further suggests the involvement of soluble factor(s) produced by the yeast. LPS, IL-1β, and TNF-α-induced NF-κB activation was inhibited by a supernatant fraction containing a small (1 kDa) heat-stable and water-soluble anti-inflammatory molecule (Sougioultzis et al., 2006). Within the research of Martins et al. (2010), *S. boulardii* 17's heat treatment eliminated the inhibitory effect, but it remained in the supernatant that contained yeast cell walls. An *in vivo* protective effect against *S. typhimurium* infection verified this modulation of inflammation and signal transduction pathway in T84 cells infected by *S. typhimurium* (Martins et al., 2010). Regarding C. difficile, C is inhibited by *S. boulardii* 17. difficile toxin A-associated enteritis by preventing Erk1/2 MAP kinases from being activated (Chen et al., 2006).

Through collaboration with S. flexneri, Mumy et al. (2008) showed that *S. boulardii* 17 significantly improved barrier integrity and had a positive effect on the tight junction protein zonula occluden-2. It also reduced the activation of ERK, JNK, and NF-κB in response to S. flexneri, which was the cause of the observed decrease in IL-8 secretion. Additionally, IL-8 secretion could be decreased by a cell-free *S. boulardii* 17 culture supernatant, indicating once more that the yeast is secreting anti-inflammatory substances. The effects of oral treatment with viable or heat-killed cells of *S. boulardii* 17 and *S. cerevisiae* UFMG 905 on intestinal permeability and histological aspect of the ileum have been evaluated in a murine IO model described above (Generoso et al., 2010; 2011). Intestinal permeability is considered increased when permeation of molecules > 150 Daltons is observed. For this reason intestinal permeability can be evaluated by measuring blood radioactivity after the intake of 99mTc-diethylenetriaminopentaacetate (99mTc-DTPA). This molecule, used as a disodium complex, can be considered adequate for intestinal permeability determination due to its following characteristics: molecular weight of 549 Daltons, water soluble, non-charged, not destroyed in the gut, non-toxic, not metabolized or sequestered, quantitatively cleared by the kidneys, and easily detectable. The findings demonstrated that intestinal permeability was increased in mice subjected to experimental IO. Oral administration of viable

S. boulardii 17 or *S. cerevisiae* UFMG 905 maintained intestinal permeability at physiological levels comparable to those noted in the Sham group during all follow-up periods, as illustrated in the same figure. When compared to the lesions seen in the IO group, histological analysis revealed preservation of the ileum mucosa in mice that received both forms of the yeast (Generoso et al., 2010, 2011). Increase in intestinal permeability was also observed in *in vitro* infectious models, as demonstrated by transepithelial electrical resistance (TER) measure in T84 cell culture infected with enteropathogenic *E. coli* (EPEC) (Czerucka et al., 2000), enterohaemorrhagic *E. coli* (Dahan et al., 2003), *S. flexneri* (Mumy et al., 2008), or with *S. typhimurium* (Martins et al., 2010), and in all these studies treatment with *S. boulardii* 17 prevents the increase. In a recent clinical trial, Vilela et al. (2008) observed important alterations in the intestinal mucosa integrity in patients with Crohn's disease in remission and demonstrated that an oral treatment with *S. boulardii* 17 improved intestinal permeability. The mechanisms that could explain the maintenance of epithelial integrity by the yeasts are not well known. However, the ability of *S. boulardii* 17 and *S. cerevisiae* UFMG 905 to modulate the transduction pathways implicated in the control of tight-junction structure described above could be an explanation (Dahan et al., 2003).

2.2.5. Inhibition of Toxin Action

Protease, which is produced by *Saccharomyces boulardii*, breaks down *C. difficile* toxins A and B, as well as the colonic mucosa's corresponding receptors (Corthier et al., 1986; Castagliuolo et al., 1999; Pothoulakis et al., 1993; Qamar et al., 2001), and has an 'anti-toxin' action expressed by inhibition of the intestinal water/sodium hypersecretion and of mucosal lesions induced by inoculation of cholera (Czerucka et al., 1989; Czerucka et al., 1994; Dias et al., 1995) and *Escherichia coli* toxins (Massot et al., 1982).

The cholera toxin (CT) may also be prevented by specific receptors capturing it on the surface of yeast, as suggested by Brandão et al., 1998). Trehalase was activated in *S. boulardii* cells grown on a non-fermentable carbon source in a manner akin to that of *S. cerevisiae* cells upon the addition of glucose or dinitrophenol. Trehalase was also activated when CT was added to the same cells. Separate studies on the CT A and B subunits revealed that both are required for activation.

Similarly, in *S. boulardii*, the addition of CT produced a cyclic AMP (cAMP) signal but not that of its individual subunits. These findings imply that the cAMP-mediated protein phosphorylation cascade was most likely the mechanism by which CT stimulated trehalase. The necessity of CT subunit B for the cAMP signal and trehalase activation suggests that the yeasts have a particular receptor that can bind to the toxin, which is comparable to what is seen in mammalian cells. Experiments using 125I-labelled CT, which demonstrated the toxin's specific binding to yeast cells, supported this theory. Another mechanism that could explain the protection against the toxin observed in animals treated with yeasts is the adhesion of CT to a receptor on the yeast surface through the B subunit and internalization of the A subunit (necessary for the cAMP signal and trehalase activation) (Brandão et al., 1998).

2.2.6. Trapping onto Yeast Cell Surface

According to certain research, part of the probiotic effect may be explained by pathogenic bacteria adhering to the surface of probiotic yeasts rather than intestinal receptors (Gedek, 1999; Pérez-Sotelo et al., 2005; Tiago et al., 2012). The previous study assessed the pathogenic bacteria's adherence to the cell walls of *S. boulardii* 17 and the strains of *S. cerevisiae* UFMG 905, W303, and BY4741. Due to the difficulty of mutating polyploid yeast, such as *S. cerevisiae* UFMG 905 and *S. boulardii* 17, cell-wall mutants of the parental strain of *S. cerevisiae* BY4741 were also used to understand the mechanism of adhesion of pathogens to yeast. Adhesion tests revealed that only *E. coli*, *S. typhimurium*, and *Salmonella typhi* adhered to the surface of *S. boulardii* 17, *S. cerevisiae* UFMG 905, and *S. cerevisiae* BY4741 out of the 11 enteropathogenic bacteria tested. This adhesion was not dependent on the viability of the yeast and was inhibited by the presence of mannose and, to a lesser extent, bile salts. Of the 44 *S. cerevisiae* BY4741 cell-wall mutants, five were unable to fix the bacteria. The phenomenon of yeast-bacteria adhesion was demonstrated by electron microscopy to occur in the digestive tracts of dixenic mice both *in vitro* and *in vivo* (Tiago et al., 2012). Past research has examined the location and binding mechanisms to the epithelial surface of enteropathogenic bacteria, including *Salmonella* and *E. coli*, as well as type I fimbria.

2.2.7. Translocation

According to a number of studies, when tested in conventional mice, *S. boulardii* 17 and *S. cerevisiae* UFMG 905 both inhibited pathogenic bacterial translocation to the liver, mesenteric lymph nodes (MLN), and spleen. This advantageous effect was most likely the result of the various mechanisms mentioned above cooperating. The translocation of *S. typhimurium* to every organ examined in conventional mice was greatly decreased by the yeast treatment (Martins et al., 2007, 2010). A previous study (Martins et al.) found that mice treated with *S. cerevisiae* UFMG 905 had fewer hepatic inflammatory foci 28 days after the enteropathogenic bacteria challenge. This could be explained by the reduced translocation of *S. typhimurium* in these mice. Berg (1995) described two possible pathways for translocation to the liver: the extracellular passage (vascular route) or the intracellular (lymphatic route, via lymph nodes). The latter occurs when the intestinal epithelium is physically damaged. Lower translocation levels have also been seen when probiotic products based on bacteria are used in a variety of animal models, including burn injury, liver injury, experimental immunosuppression, and abdominal infection (Chiva et al., 2002; Gun et al., 2005) and yeast (Herek et al., 2004; Peret et al., 1998). It appears that bacterial translocation is regulated by every element of the host immune system, including humoral, cell-mediated, and mucosal immunity. It is believed that sIgA decreases bacterial penetration by preventing pathogenic bacteria from closely adhering to the mucosal epithelium. The lamina propria and lymphoid organs' cell-mediated immunity offers a second line of defense for the host against bacterial translocation. Serum immunoglobulins (Igs) most likely function as opsonins to boost macrophage and polymorphonuclear leucocyte phagocytosis efficacy, which in turn kills translocating bacteria.

2.2.8. Action on Bacterial Virulence Factors

According to one study, *S. boulardii* 17 reduces colitis caused by bacteria by acting on virulence factors of the bacteria (Wu et al., 2008). *S. boulardii* 17 treatment significantly reduced weight loss, improved crypt hyperplasia and histological damage score, decreased myeloperoxidase activity, and moderated increased mannitol flux in mice infected with *Citrobacter rodentium*. *S. boulardii's* mitigating actions were linked to a marked decrease in mucosal adherent *C. rodentium* counts, a notable decrease in the expression and secretion of EspB, a translocation apparatus protein, and a marked reduction in the secretion and translocation of Tir (the receptor for bacterial intimin) into mouse colonocytes. The authors deduce from the data that *S. boulardii* 17 reduced the adherence of bacteria to host epithelial cells by possibly affecting the type III secretion system (Table 2.1).

Table 2.1. *Characteristics of a Few Randomized Controlled Trials Included in the Systematic Review by Maslennikov et al. (2024)*

Study	CNCM I-745	Disease	Count of Participants	Groups	ALT/ AST	GGT/ ALP	Bilirubin	Serum Albumin	Serum Glucose/ Lipids	Cirrhosis Manifestations	Hemodynamics	Prognosis	Systemic Inflammation	Gut Micro-Biota	SIBO	Bacterial Translocation
Tang 2020 [21]	+	NHB	63 + 61	SB/C			+							+		
Suganthi 2016 [25]		NHB	86 + 95	SB/C			+									
Serce 2015 [28]	+	NHB	98 + 81	SB/C			+									
Hu 2022 [31]	+	NHB	55 + 45	SB/C			+									
Maslennikov 2022 [16]	+	Cirrhosis	24 + 16	SB/Pl	+	+	+	+	+	+	+		+			
Efremova 2024 [17]	+	Cirrhosis	20 + 13	SB/Pl								+			+	+

C—control (disease without tested interventions); NHB—neonate hyperbilirubinemia; Pl—placebo; SB—Saccharomyces boulardi; SIBO—small intestinal bacterial overgrowth.

Source: Maslennikov R, Benuni N, Levshina A, Adzhieva F, Demina T, Kucher A, Pervushova E, Yuryeva E, Poluektova E, Zolnikova O, et al. (2024). Effect of Saccharomyces boulardii on liver diseases: A systematic review. Microorganisms; 12(8):1678. https://doi.org/10.3390/microorganisms12081678.

2.3. PROBIOTICS INTERACT WITH THE GUT MICROBIOTA

Probiotics interact with the gut microbiota in a number of ways. They are defined as live microorganisms that, when consumed in sufficient amounts, confer health benefits. The diverse community of microorganisms that live in the GIT, including bacteria, viruses, fungi, and archaea, is referred to as the gut microbiota. Probiotics influence the diversity and composition of the gut microbiota, which is one of the main ways they interact with it. Probiotic supplements often contain strains of *Lactobacillus* and *Bifidobacterium*, two beneficial bacteria that can be introduced into the gut ecosystem by probiotics. In the gut, these probiotic strains can proliferate and establish themselves, supporting the general microbial balance. Probiotics contribute to the creation of an environment that is less conducive to the growth of potentially harmful microorganisms by increasing the abundance of beneficial bacteria.

Probiotics can also alter the gut microbiota by promoting the development and activity of particular bacterial species. They have the ability to create compounds that act as nutrients for specific good bacteria, encouraging their expansion. For instance, certain probiotics ferment dietary fibers to create short-chain fatty acids, or SCFAs. In addition to providing colonocytes with energy, SCFAs like acetate, propionate, and butyrate also encourage the growth of *Bifidobacterium* and other beneficial bacteria. By providing these growth-promoting conditions, probiotics contribute to shaping a healthy and diverse gut microbiota.

Probiotics can prevent the growth of potentially dangerous microorganisms in the gut in addition to encouraging the growth of helpful bacteria. They make it more difficult for pathogens and harmful bacteria to colonize and flourish in the gut by competing with them for nutrients and adhesion sites. Antimicrobial substances like bacteriocins, which are proteinaceous compounds with antimicrobial properties, can also be produced by probiotics. Bacteriocins offer an extra line of defense against GIT infections by directly preventing the growth of pathogens. Probiotics' metabolic processes also have an impact on the gut microbiota. They are able to generate enzymes that facilitate the digestion and breakdown of different food ingredients. For instance, lactase, an enzyme that aids in lactose digestion in those with lactose intolerance, is produced by certain probiotic strains. Probiotics help to optimize the metabolic processes of the gut microbiota by aiding in the digestion and utilization of nutrients. Additionally, probiotics can help correct microbial imbalances by modulating the gut microbiota elements like medication and dietary adjustments. It is noteworthy that there may exist strain-specific differences in the ways that probiotics and the gut microbiota interact. Probiotic strains and species may differ in their capacity to affect the makeup and functionality of the gut microbiota. Individual reactions to probiotics can also differ based on things like dietary habits, general health, and the composition of one's gut microbiota at baseline (Gou et al., 2022).

2.3.1. Mechanisms of Interaction

Probiotics and the gut microbiota interact through a variety of pathways that support the general equilibrium and smooth operation of the gut ecosystem. Live microorganisms known as probiotics have the potential to improve health through a variety of interactions with the gut microbiota that already exists there. Competition between organisms for adhesion sites and nutrients is a basic mechanism of interaction. Probiotics fight for the vital nutrients found in the gut environment with other microorganisms, such as pathogens and potentially dangerous bacteria. Probiotics impede the growth and colonization of other microorganisms by limiting their access to these nutrients. Probiotics can also fight for adhesion sites on the intestinal lining, which stops pathogens from attaching and lessens their capacity to spread infection. The synthesis of antimicrobial compounds is another crucial mechanism. Probiotics have the capacity to produce and release antimicrobial substances into the gut, including bacteriocins. Bacteriocins are proteinaceous compounds that directly hinder some pathogens' ability to proliferate and survive.

Probiotics can also alter the development and activity of particular bacterial species that are present in the gut microbiota. By producing metabolites that act as nutrients for *Lactobacillus* and *Bifidobacterium* species, among other beneficial bacteria, they can promote their growth. Additionally, probiotics have the ability to prevent the growth of potentially dangerous bacteria by either directly suppressing their proliferation or altering the environment in the gut to make it less conducive to their survival. Additionally, probiotics can influence the host immune system through influencing the gut microbiota. By encouraging the synthesis of anti-inflammatory cytokines and the growth of regulatory immune cells, they can affect the immune response. By limiting excessive inflammation and fostering immune tolerance, this immune modulation aids in maintaining the immune system's equilibrium. Additionally, probiotics can strengthen the lymphoid tissue associated with the gut, which is essential for immunological surveillance and pathogen defense. Competition for nutrients and adhesion sites, as well as production of certain nutrients, are the mechanisms through which probiotics and the gut microbiota interact (Gou et al., 2022).

2.3.2. Impact on Gut Microbiota Composition

Probiotics' influence on the composition of the gut microbiota is a crucial part of how they interact with the gastrointestinal ecosystem. The diversity, abundance, and general balance of the gut microbiota can be significantly impacted by probiotics, which are defined as live microorganisms that, when consumed in sufficient amounts, provide health benefits. Beneficial bacteria are one way probiotics affect the composition of the gut microbiota by entering the digestive system. Supplements containing probiotic strains like *Lactobacillus* and *Bifidobacterium* species are frequently used. These live microbes have the ability to colonize and establish themselves in the gut after ingestion. Probiotics enhance the overall microbial balance and maintain a healthy gut ecosystem by raising the number of beneficial bacteria.

Probiotics are also involved in preventing the growth of microorganisms that may be harmful. Probiotics increase the difficulty of pathogens and harmful bacteria colonizing and growing in the gut by competing with them for nutrients and adhesion sites. This competition helps to maintain a balanced composition of the gut microbiota and lessens the possibility of pathogenic overgrowth. Furthermore, it has been demonstrated that probiotics alter the gut microbiota's bacterial species' relative abundance. They have the ability to promote the development and activity of good bacteria, such as *Lactobacillus* and *Bifidobacterium*, which are linked to positive health outcomes. Probiotics support a more resilient and diverse gut microbiota by offering these advantageous microorganisms ideal growth conditions. Probiotics may aid in the restoration of microbial equilibrium in dysbiosis cases, which are characterized by an imbalance in the composition of the gut microbiota. Antibiotic use, dietary modifications, and specific medical conditions are some of the variables that can lead to dysbiosis. Probiotics can aid in the overgrowth's suppression through their colonization and competition mechanisms.

It is significant to remember that probiotics have strain-specific effects on the makeup of the gut microbiota. The capacity of various probiotic strains to alter particular bacterial populations may differ. For instance, certain strains might be better at boosting *Bifidobacterium* abundance, while other strains might have a stronger effect on *Lactobacillus* populations. Comprehending the effects that are unique to each strain is essential for customized and focused probiotic intervention strategies. Additionally, the response to probiotics can be influenced by individual factors, including host genetics, baseline gut microbiota composition, and dietary habits. Every person has a different gut microbiota profile, and probiotic response varies from person to person. This emphasizes the value of tailored strategies and the need for more study to fully comprehend how probiotics, the gut microbiota, and individual traits interact. Probiotics significantly alter the composition of the gut microbiota. They bring in good bacteria, stop the growth of bad bacteria, encourage the development of particular bacterial species, and, in certain situations, aid in the restoration of the microbial balance.

2.3.3. Modulation of Microbial Balance

A crucial component of the relationship between probiotics and the gut microbiota is the modification of microbial equilibrium. Restoring and preserving a balanced gut microbial community is greatly aided by probiotics, which are defined as live microorganisms that, when consumed in sufficient amounts, offer health advantages.

The complex ecosystem known as the gut microbiota is made up of a wide variety of microorganisms that live in the GIT, such as bacteria, viruses, fungi, and archaea. A balanced microbial population is a hallmark of a healthy gut microbiota, supporting both host health and gut health in general. An imbalance in the composition of the gut microbiota is known as dysbiosis and can be brought on by antibiotic use, dietary modifications, or specific medical conditions.

By encouraging the growth of helpful bacteria and preventing the overgrowth of potentially hazardous microorganisms, probiotics have the power to modify the microbial balance. Probiotics aid in microbial balance restoration by bringing good bacteria into the gut. Probiotic strains are frequently taken as supplements, including *Lactobacillus* and *Bifidobacterium* species. These live microbes have the ability to colonize and establish themselves in the gut after ingestion, which increases the number of helpful bacteria. The general balance of the gut microbiota is aided by this rise in helpful bacteria.

In addition to introducing beneficial bacteria, probiotics can inhibit the growth of harmful microorganisms. They compete with pathogens and detrimental bacteria for nutrients and adhesion sites within the gut, making it more difficult for these harmful microbes to colonize and thrive. Probiotics can also produce antimicrobial substances, such as bacteriocins, which have direct inhibitory effects on the growth and survival of certain pathogens. Through competition and the production of antimicrobial compounds, probiotics create an inhospitable environment for pathogenic overgrowth, helping to restore microbial balance.

The metabolic actions also allow probiotics to alter the gut microbiota. They have the ability to produce enzymes that help break down and use particular food ingredients. For those who are lactose intolerant, certain probiotic strains, for instance, generate enzymes that aid in lactose digestion. Probiotics support the metabolic processes of the gut microbiota by supplying these digestive enzymes. The gut microbial community's general equilibrium and functionality are aided by this metabolic interaction.

It is crucial to remember that probiotics' ability to modify microbial balance depends on a number of variables, including the precise probiotic strains utilized, the dosage, and the length of supplementation. The capacity of various probiotic strains to alter particular bacterial populations may differ. Optimizing probiotic interventions requires knowing the effects of individual strains as well as determining which strains work best under particular circumstances. Furthermore, the way in which probiotics work and how well the microbial balance is restored can be influenced by individual factors. Since every person has a different gut microbiota profile, probiotic responses can differ. Probiotics' ability to modify the microbial balance can be influenced by a number of factors, including host genetics, baseline gut microbiota composition, and dietary practices. To get the best results, probiotic interventions may require customized strategies.

Probiotics have a major impact on how the microbial balance in the gut is modulated. They contribute to the general equilibrium and smooth operation of the gut microbial community by introducing good bacteria and preventing the growth of pathogenic microorganisms. It is imperative to comprehend the effects of individual strains, individual variances in response, and the underlying mechanisms in order to utilize probiotics in a targeted and efficient manner in order to maintain and repair a healthy gut microbiota and enhance overall gut health. To better understand the complex interactions between probiotics and the gut microbiota and to create tailored probiotic intervention strategies, more research is required (Gou et al., 2022).

2.3.4. Metabolic Activities and Microbial Interactions

Probiotics' effects on gut health are largely determined by their metabolic processes and interactions with the gut microbiota. Probiotics participate in a variety of metabolic interactions within the gastrointestinal ecosystem. Probiotics are defined as live microorganisms that provide health benefits when consumed in adequate amounts. Certain dietary components can be broken down and utilized more easily with the help of the enzymes produced by probiotics. For instance, some probiotic strains generate the enzyme lactase, which aids in the digestion of lactose in those who are lactose intolerant. Probiotics help the gut microbiota's metabolic processes by supplying these digestive enzymes, which improve nutrient absorption and breakdown. Probiotics can also take part in the fermentation processes that occur in the gut. Food fibers that are indigestible to human enzymes can be fermented by them. Probiotics produce SCFAs like butyrate, propionate, and acetate during fermentation. The essential metabolic byproducts known as SCFAs provide the colonocytes—the cells lining the colon—with energy. These SCFAs also help regulate gut pH, promote the growth of beneficial bacteria.

The composition and overall functioning of the gut microbiota can be significantly influenced by the metabolic activities of probiotics. Probiotics have the ability to use particular nutrients and substrates selectively, which can affect how readily available these resources are to other microbial populations. Probiotics have the ability to produce a competitive environment that hinders the growth of potentially harmful microorganisms by consuming nutrients. This rivalry helps keep the gut microbiota in balance and stops harmful species from proliferating. Probiotics can also interact with both good and bad bacteria as well as other microorganisms in the gut microbiota. The gut ecology may be affected in a variety of ways by these microbial relationships. Probiotics, for example, have the ability to produce antimicrobial compounds called bacteriocins, which directly inhibit certain pathogens and harmful bacteria. The synthesis of antimicrobial compounds aids in probiotics' ability to outcompete and inhibit the growth of pathogenic microorganisms, thereby promoting a more beneficial composition of the gut microbiota.

Additionally, probiotics can affect how genes are expressed in the gut microbiota. They have the ability to alter the expression of microbial genes involved in different metabolic pathways, which has an impact on the microbiota's overall metabolic activity in the gut. These variations in gene expression may have an effect on the synthesis of metabolites and other compounds that support the gut ecosystem's general health. Moreover, probiotics have metabolic properties and interactions that go beyond their influence on gut microbiota. Probiotics have the ability to affect how the host uses its own metabolism. For instance, it has been demonstrated that some probiotics affect cholesterol levels and lipid metabolism, which improves cardiovascular health. It is significant to remember that the makeup of the gut microbiota and the particular probiotic strains can affect the metabolic processes and microbial interactions of probiotics. Various strains may have distinct substrate preferences and enzymatic capacities, which can result in a range of metabolic effects. Furthermore, the response can be influenced by host genetics, dietary factors, and individual differences in the composition of the gut microbiota.

The effects of probiotics on gut health are largely dependent on their metabolic processes and interactions with other bacteria. In addition to producing enzymes that facilitate the breakdown and absorption of nutrients, probiotics also take part in fermentation processes and interact both cooperatively and competitively with other gut flora. Gaining an understanding of these metabolic interactions and how they affect the composition and functioning of the gut microbiota is essential to maximizing the potential advantages of probiotics in enhancing gut health and general wellbeing. To fully understand the complex metabolic networks and mechanisms that underlie probiotic-microbiota interactions, more research is required (Gou et al., 2022).

2.3.5. Strain-Specific Effects

The term "strain-specific effects" describes the differences in how various probiotic strains affect the gut microbiota and general health outcomes. Probiotics include a variety of strains from various microbial species, including *Lactobacillus* and *Bifidobacterium*, and are defined as live microorganisms that provide health benefits when consumed in adequate amounts. Specific metabolic activities, substrate preferences, and interactions with the host's gut microbiota are just a few of the distinctive features that distinguish each probiotic strain. Because of this, distinct strains may have varied effects on the makeup of the gut microbiota, immune system regulation, metabolic processes, and general health outcomes. The capacity of probiotic strains to colonize and persist in the gut is one of the primary factors influencing strain-specific effects. Certain strains exhibit heightened ability to endure transit via the GIT, adhere to the intestinal lining, and establish a durable presence within the gut microbiome. The makeup and functionality of the gut microbiota are more likely to be permanently altered by these strains.

Furthermore, the capacity of different probiotic strains to alter particular bacterial populations within the gut microbiota varies. While some strains may be more effective at preventing the growth of harmful microorganisms, others may be more effective at boosting the abundance of beneficial bacteria, such as *Lactobacillus* and *Bifidobacterium* species. It is essential to comprehend how strains affect microbial populations in order to attain the intended health outcomes. Furthermore, the synthesis of bioactive compounds and metabolic processes may also be affected by strain-specific factors. Different probiotic strains may have different metabolic pathways and enzymatic capacities, which can affect how particular dietary components are metabolized and broken down. This fluctuation in metabolic activity can affect the synthesis of metabolites, which have varying effects on gut health and general well-being. Examples of these metabolites include SCFAs, neurotransmitters, and antimicrobial substances.

Furthermore, probiotics' ability to modulate immunity may be impacted by strain-specific effects. Different strains can influence immune cell responses, promote immune tolerance, and stimulate the production of particular cytokines in different ways. Conditions like autoimmune diseases, inflammatory bowel disease (IBD), and allergies may be affected by these immune modulation effects. It is significant to remember that individual factors also have an impact on strain-specific effects. Every person has a different host genetic makeup and gut microbiota profile, which can affect how they react to different probiotic strains. Individuals' dietary habits, host immune status, and baseline gut microbiota composition may all affect how they interact with different probiotic strains and how probiotic interventions turn out. The influence of probiotics on the composition of the gut microbiota, immune modulation, metabolic activities, and general health outcomes is significantly influenced by strain-specific effects. Probiotic strains differ in their traits and ways that they interact with the host's gut microbiota. It is critical to comprehend these strain-specific effects (Gou et al., 2022).

2.3.6. Prebiotics and Synbiotics

The terms "prebiotics" and "synbiotics" refer to the ways in which certain drugs and probiotics interact in the gut to support gut health and general wellbeing. Prebiotics are indigestible dietary fibers that specifically promote the development and function of good bacteria in the digestive system. Conversely, synbiotics are blends of probiotics and prebiotics that complement each other to improve the effects of both on the gut microbiota. Beneficial bacteria that are frequently used as probiotics, such as *Lactobacillus* species and *Bifidobacterium*, eat prebiotics. They pass through the upper GIT undigested, making their way to the colon, where the gut microbiota ferments them. SCFAs, like acetate, propionate, and butyrate, are produced during this fermentation process and are beneficial to human health in addition to providing colonocytes with energy. Prebiotics support the upkeep of a diverse and well-balanced gut microbiota by specifically encouraging the growth and activity of good bacteria.

Prebiotics like fructooligosaccharides (FOS), inulin, galactooligosaccharides (GOS), and resistant starch are a few examples of those that are frequently researched. Natural sources of these prebiotic fibers include chicory root, onions, garlic, bananas, and whole grains. In order to promote gut health, some prebiotics can also be extracted and added as components to functional meals, supplements, and pharmaceutical items. Prebiotics and probiotics work together synergistically when taken as synbiotics. The probiotic strains' growth and activity are supported by the prebiotic fibers, which also act as a substrate. Compared to probiotics or prebiotics alone, this combination increases the probiotics' ability to survive and colonize the gut, which has a greater effect on the makeup and function of the gut microbiota. To further support gut health, synbiotics can also increase the synthesis of SCFAs and other advantageous metabolites.

Prebiotic and synbiotic use has been linked to a number of health advantages. They have been demonstrated to enhance gastrointestinal health by encouraging frequent bowel movements, lessening constipation, and easing the signs and symptoms of a number of gastrointestinal conditions, including IBS. By affecting the lymphoid tissue associated with the gut and regulating the generation of immune-related molecules, prebiotics and synbiotics can also improve immune function. Furthermore, they might be beneficial for metabolic health, regulating blood sugar levels, improving lipid profiles, and possibly helping with weight loss. It is significant to remember that the effects of prebiotics and synbiotics can differ based on a person's genetic makeup and baseline composition of the gut microbiota. Furthermore, the effectiveness of particular prebiotics and synbiotics can differ based on the probiotic strains used, the dosage, and the length of time taken.

In order to maximize the benefits of probiotics and support gut health, prebiotics and synbiotics are crucial ideas. Prebiotics work to specifically promote the growth of good bacteria in the gut, whereas synbiotics work to maximize the benefits of both prebiotics and probiotics. These methods have been linked to a number of health advantages, such as stronger immune systems, better gastrointestinal health, and possible metabolic advantages. In order to create tailored approaches to their use based on individual gut microbiota profiles and health objectives, more research is required to investigate the precise mechanisms underlying the effects of prebiotics and synbiotics (Gou et al., 2022).

2.3.7. Impact of External Factors

Since different external factors can affect the effectiveness and results of probiotic interventions, it is important to take into account how these factors affect the interaction between probiotics and the gut microbiota. The term "external factors" refers to a variety of influences, such as lifestyle, food, drugs, and surroundings. The gut microbiota is greatly influenced by diet, which can also impact how the body reacts to probiotics. Prebiotics, dietary fibers, and polyphenols are a few examples of food ingredients that act as substrates for the development and activity of good bacteria in the stomach. These elements may aid in probiotics' colonization and efficacy. Conversely, diets heavy in processed foods, sugars, and saturated fats may harm the gut microbiota and consequently lessen the benefits of probiotics. Thus, the advantages of probiotics can be increased by eating a varied, well-balanced diet high in fiber and plant-based foods. The gut microbiota and the effectiveness of probiotics can also be influenced by lifestyle variables like stress and physical activity. Regular physical activity has been associated with a more diverse gut microbiota composition, which can enhance.

Antibiotics in particular have the potential to significantly affect the gut microbiota and change how the body reacts to probiotics. By getting rid of both good and bad bacteria, antibiotics can upset the microbial balance. The ability of probiotics to colonize and flourish in the gut may be impacted by this disturbance. The detrimental effects on the gut microbiota can be lessened by taking certain strains of probiotics that have been shown to be resistant to antibiotics or by timing the use of probiotics in conjunction with antibiotics. Environmental variables can potentially affect the gut microbiota and probiotic efficacy. These factors include exposure to pollutants, toxins, and specific chemicals. Environmental influences have the power to upset the gut microbiota's delicate balance and jeopardize its diversity and functionality. Optimizing the response to probiotics may involve minimizing exposure to toxins and pollutants in the environment, for example, by maintaining a clean-living environment and making healthy food choices.

It is significant to remember that, in addition to external factors, individual variances and characteristics like age, genetics, and general health status can also affect how the body reacts to probiotics. Every person has a different gut microbiota composition and set of health requirements, which can affect how well probiotics work and how they react to outside stimuli. The way that probiotics and the gut microbiota interact is greatly influenced by outside factors. The composition and function of the gut microbiota can be influenced by diet, lifestyle, medications, and environmental factors. This could have an impact on the effectiveness of probiotic interventions. Taking into account these outside variables, maintaining a healthy diet, getting regular exercise, controlling stress, and limiting exposure to dangerous substances can all help promote the best possible response to probiotics and increase their advantages. Personalized strategies that take into account unique traits and variations are also crucial for optimizing the effects of probiotic treatments (Gou et al., 2022).

2.4. IMMUNOMODULATORY EFFECTS OF PROBIOTICS

Probiotics' capacity to interact with and affect the immune system, fostering a well-balanced and responsive immune response, is known as their immunomodulatory effects. Probiotics are live microorganisms that, when taken in sufficient quantities, offer health benefits. Studies have indicated that probiotics significantly affect immune function and are essential for immune homeostasis maintenance. Probiotics have the ability to interact with immune cells, cytokines, and Igs, among other immune system components. Direct contact with immune cells or the release of bioactive molecules and metabolites can both result in these interactions.

Probiotics have the ability to both stimulate and suppress the production of pro- and anti-inflammatory cytokines, which is a crucial immunomodulatory mechanism. Probiotics have the ability to increase the secretion of cytokines that help reduce inflammation and regulate immune responses, such as interleukin-10 (IL-10), transforming growth factor-beta (TGF-β), and interleukin-4 (IL-4).

Probiotics have the ability to suppress the generation of pro-inflammatory cytokines, which play a role in the etiology of several inflammatory disorders. These cytokines include tumor necrosis factor-alpha (TNF-α), interleukin-6 (IL-6), and interleukin-1 beta (IL-1β). The immune system's balance is preserved and excessive inflammation is avoided thanks to this regulation of cytokine production.

Additionally, probiotics can affect how immune cells develop and function. They have the ability to promote the development and maturation of regulatory immune cells, which are essential for immune tolerance and control. Examples of these cells are dendritic cells and regulatory T cells (Tregs). Dendritic cells are essential for both triggering and regulating immune responses, while tregs assist in preventing excessive immune responses and preserving immune homeostasis. Probiotics support immune balance and help prevent immune-related disorders by encouraging the growth and activity of these regulatory immune cells.

Probiotics can also strengthen the gut-associated lymphoid tissue (GALT), a significant part of the gut's immune system. The GALT is made up of lymphoid tissues and a variety of immune cells that support the regulation of immune responses in the GIT. Probiotics can help the GALT's immune cells mature and become activated, which will help the immune system respond to pathogens and antigens in a coordinated manner. The interaction of probiotics with mucosal surfaces is another immunomodulatory mechanism. As a physical barrier separating the gut lumen from the underlying immune tissue, intestinal epithelial cells (IECs) are a surface on which probiotics can adhere and interact. Probiotics can affect the activation and responsiveness of the immune system by modifying the interaction between immune cells and IECs. This can lead to a balanced immune response and better gut health.

Numerous health benefits have been linked to probiotics' immunomodulatory effects. It has been demonstrated that they enhance immune function in a number of populations, such as young children, the elderly, and those with weakened immune systems. Probiotics have shown promise in preventing allergies, enhancing immunization-induced immune responses, and lessening the intensity and duration of respiratory tract infections. It is significant to remember that probiotics may have strain-specific immune-modulatory effects. The capacity of various probiotic strains to engage with the immune system and produce immunomodulatory effects may differ. The best probiotic strains for a given immune-related condition depend on your understanding of the effects of each strain. Probiotics exhibit noteworthy effects on immunomodulation through their ability to interact with immune cells, modulate cytokine production, influence the growth and function of regulatory immune cells, augment the lymphoid tissue associated with the gut, and impact mucosal surfaces. By reducing excessive inflammation and bolstering the immune system, these immunomodulatory effects help maintain a responsive and balanced immune system.

2.4.1. Cancer Immunotherapy and Probiotics

A promising method of treating cancer is cancer immunotherapy, which tries to strengthen the body's immune response against cancer cells. Due to their immunomodulatory properties, probiotics have drawn interest as possible adjuvants in cancer immunotherapy. Here are some key aspects to consider when exploring the relationship between cancer immunotherapy and probiotics:

- **Enhancing Immune Responses**: The potential of probiotics to strengthen immune responses against cancer cells has been investigated. They have the ability to increase the development and function of immune cells that are essential for tumor surveillance and eradication, such as dendritic cells, cytotoxic T lymphocytes (CTLs), and natural killer (NK) cells. By enhancing the immune system's ability to combat tumors, probiotics may increase the effectiveness of cancer immunotherapy.
- **Modulating the Tumor Microenvironment**: Tumor microenvironment response to immunotherapy and tumor progression are closely related. Probiotics have been demonstrated to alter the ratio of pro-to anti-inflammatory factors, thus modulating the tumor microenvironment. By fostering an environment that is immune-favorable, they can lessen immunosuppression and increase the effectiveness of immunotherapeutic treatments.
- **Improving Treatment Outcomes**: Research indicates that probiotics could enhance the effectiveness of cancer immunotherapy treatments. Probiotic supplements can improve tumor control, response rates, and overall survival in some cancer types when used in conjunction with immunotherapies like immune checkpoint inhibitors or cancer vaccines, according to preclinical and clinical research.
- **Gut Microbiota and Systemic Immune Responses**: An important factor in determining systemic immune responses, such as anti-tumor immunity, is the gut microbiota. Probiotics can affect the gut microbiota's composition and function, which can impact the activation of the systemic immune system and the body's reaction to cancer immunotherapy. Optimizing the use of probiotics in cancer treatment requires an understanding of the complex interactions that occur between them, the gut microbiota, and systemic immune responses.

- **Combination Therapies**: Probiotics can be incorporated into immunotherapy combination treatments for cancer. Probiotics have been studied in conjunction with other therapeutic modalities like radiation, chemotherapy, or targeted therapies to improve treatment outcomes and reduce side effects. More research is needed to determine how probiotics work in concert with other cancer treatments.
- **Safety considerations:** Probiotics are generally regarded as safe, but when using them as adjuvants in cancer immunotherapy, it's important to take their safety into account. To reduce risks and guarantee optimal efficacy, factors like strain selection, dosage, and unique patient characteristics should be carefully taken into account.
- **Personalized Approaches**: Individual variations in the gut microbiota composition, immune status, and genetic factors can influence the response to probiotics and cancer immunotherapy. Personalized approaches, considering the specific needs and characteristics of each patient, are essential for maximizing the benefits of probiotics in cancer treatment.

One interesting area of research is the use of probiotics as adjuvants in cancer immunotherapy. Probiotics may improve treatment outcomes, alter the tumor microenvironment, and strengthen immune responses. To fully understand the mechanisms underlying the interactions between probiotics, the gut microbiota, and cancer immunotherapy, more research is required. Comprehending these interplays can enhance the application of probiotics as supplemental treatments to augment the efficacy of cancer treatment approaches.

2.4.2. Immune Function in Specific Populations

Comprehending the immunomodulatory impacts of probiotics within particular demographics is imperative in customizing probiotic interventions to address the distinct immune requirements of diverse groups. Here are some key populations to consider when exploring the impact of probiotics on immune function:

- **Infants and Children**: Long-term health is dependent on the immune system's development and maturation during the early years of life. Probiotics have been investigated for their ability to improve immunological development, avoid allergies, and lower the risk of respiratory tract infections in infants and young children. Examining how probiotics affect immune responses, immune cell populations, and immune markers in pediatric populations can shed light on the potential for immune modulation of these supplements.
- **Elderly Individuals**: Immune system alterations, such as immune

senescence and heightened susceptibility to infections, are linked to aging. Probiotics have demonstrated promise in boosting vaccination responses and lessening the frequency and severity of respiratory infections, among other immune function improvements in the elderly. Investigating how probiotics affect immune parameters and immune-related health outcomes in the elderly can yield important insights into the promotion of healthy aging.

- **Athletes and Active Individuals**: Athletes and other physically active people are more prone to infections because intense physical activity has the temporary effect of suppressing immune function. The potential of probiotics to enhance immune function and lower the frequency of respiratory tract infections in athletes has been studied. Probiotics' function in preserving immune resilience in this population can be better understood by looking into how they affect immune parameters, inflammatory markers, and athletic performance.
- **Individuals with Compromised Immune Systems**: The immune system may be weakened by specific medical disorders and treatments, which raises the possibility of infection. Probiotics have been studied in patients with weakened immune systems, including those receiving organ transplants, chemotherapy, and HIV/AIDS patients. Optimizing the use of probiotics as supportive therapies can be achieved by having a thorough understanding of their effects on immune function, infection rates, and immune-related complications in these populations.
- **Individuals with Autoimmune Disorders**: Immune system dysregulation in autoimmune disorders causes healthy tissues to be attacked. Probiotics have been studied for their ability to control immune dysregulation and lessen inflammation in diseases like multiple sclerosis, IBD, and rheumatoid arthritis. Probiotics' potential as adjuvant therapies can be clarified by investigating their effects on immune markers, disease activity, and quality of life in people with autoimmune disorders.
- **Pregnant Women**: Immune system changes occur during pregnancy in order to support fetal growth and preserve mother health. Pregnant women who take probiotics may be able to modify their immune responses, avoid complications during pregnancy, and lower their child's risk of allergic diseases. Probiotic use during pregnancy can be better understood by looking into how it affects immune parameters, pregnancy outcomes, and the immune health of the offspring.

When investigating the immunomodulatory effects of probiotics, it is important to take into account individual variations within these populations, such as baseline immune status, composition of the gut microbiota, and genetic factors. The most effective way to use probiotics to support immune function and enhance health outcomes is through personalized approaches that consider unique immune requirements and characteristics of each population (Gou et al., 2022).

2.4.3. Probiotics and Vaccination Responses

Probiotics have drawn interest because of their capacity to improve immunization responses and alter the immune system. By inducing an immune response against particular pathogens or antigens, vaccinations are crucial for preventing infectious diseases. When examining the connection between probiotics and vaccination responses, keep the following important factors in mind:

- **Immune-Enhancing Effects**: The potential of probiotics to improve the immune system's reaction to vaccinations has been investigated. They have the ability to change immune signaling pathways, improve immune cell function, and increase the production of antibodies. Probiotics may increase protection against specific diseases and increase the effectiveness of vaccinations by boosting immune responses.
- **Modulation of Immune Pathways**: Probiotics can influence immune pathways involved in vaccine responses. They can enhance the production of specific antibodies, such as IgA and IgG, which play a crucial role in neutralizing pathogens. Probiotics may also affect the activation and maturation of APCs, such as dendritic cells, which are responsible for initiating immune responses to vaccines.
- **Gut Microbiota and Vaccine Efficacy**: Systemic immune responses, including those resulting from vaccinations, are significantly influenced by the gut microbiota. The composition and function of the gut microbiota can be altered by probiotics, potentially affecting the effectiveness of vaccinations. Certain gut bacteria can affect the generation of antibodies, control immune signaling, and affect the growth and functionality of immune cells. Probiotics may improve the composition of the gut microbiota, which would improve vaccination responses.
- **Specific Vaccine-Targeted Studies**: Scholars have examined the impact of probiotics on the immune response to particular vaccinations, including those against influenza, pneumococcal disease, and rotavirus. These investigations have looked at things

like T-cell responses, antibody production, and vaccination effectiveness. Examining how probiotics affect various vaccines offers important insights into how they might be used as adjuvants to improve vaccine efficacy.

- **Timing and Duration of Probiotic Supplementation**: The immune response may be influenced by the probiotic supplementation's timing and duration in relation to vaccination. Research has looked into the effects of probiotics given prior to, during, or following immunization. To optimize the effects of probiotics on vaccination responses, more research is required to determine the ideal timing and length of probiotic supplementation.
- **Strain Specificity**: The effects of various probiotic strains on vaccination responses might differ. The immunomodulatory effects of particular probiotic strains, such as *Lactobacillus* and *Bifidobacterium* species, have been researched. Targeted probiotic interventions can be developed with the help of knowledge about the effects of particular strains and the strains that work best for particular vaccinations.
- **Safety Considerations**: Although probiotics are generally thought to be safe, safety considerations should be carefully considered, especially when using them in immunocompromised or specific medical conditions individuals. To protect people's health, it is critical to evaluate the advantages and disadvantages of probiotic supplementation in relation to immunization.

Probiotics have the potential to improve vaccination responses through immune pathway modulation, gut microbiota influence, and immune cell stimulation. To better understand how probiotics affect vaccination responses, determine the best strains and dosage schedules, and maximize their use as adjuvants, more research is required. Enhancing vaccination efficacy and public health outcomes can be achieved by having a better understanding of the interactions between probiotics, the immune system, and vaccines.

2.4.4. Personalized Approaches to Probiotic Interventions

Probiotic use can be customized to meet specific health goals and take into account individual characteristics such as immune system function, genetic makeup, gut microbiota composition, and immune status. Personalized probiotic interventions take these variables into account in an effort to maximize each person's benefits and results. The following are important factors to take into account when investigating customized probiotic intervention strategies:

- **Gut Microbiota Profiling**: Analyzing the composition of a person's gut microbiota can reveal important details about their unique microbial ecology. Metagenomic sequencing and targeted gene sequencing are two methods that can be used to determine the variety and abundance of various bacterial species in the gut. For the best possible health results, this information can help select probiotic strains that either alter or enhance the gut microbiota already present.
- **Strain Selection**: Probiotic strains differ in their properties and impacts on the immune system and gut microbiota. It is critical to comprehend strain-specific characteristics, such as immunomodulatory potential, adhesion to the gut epithelium, and survivability, when choosing the best strains for a given set of circumstances or objectives. Personalized methods entail choosing probiotic strains that meet a person's unique requirements, such as strains that have been shown to be effective for a given medical condition or strains that are in harmony with the person's gut microbiota.
- **Dosage and Duration**: To achieve desired results, it is crucial to determine the ideal probiotic supplementation dosage and duration. Individual tolerance, the seriousness of the ailment being treated, and the intervention's objectives are all taken into consideration in personalized approaches. For the desired effects, some people might need to take probiotics at higher doses or for longer periods of time, while others might benefit from taking them at lower doses or for shorter periods of time.
- **Immune Status and Health Conditions**: Customizing probiotic interventions requires taking into account each person's immune status as well as their unique medical conditions. For people with weakened immune systems or particular medical conditions, such as IBD, allergies, or autoimmune disorders, some strains or combinations of strains may be more helpful. Knowing the underlying immune dysregulation or health issues can help select the right probiotic strains and define the intervention's objectives.
- **Genetic Factors**: Genetic differences can affect how probiotics affect a person's health and how they react to them. Genetic markers linked to immunological response, gut health, or probiotic metabolism can be found through genetic testing, such as single nucleotide polymorphism (SNP) analysis. Personalized methods can make use of this genetic data to choose probiotic strains or create interventions that are specific to a person's genetic needs and predispositions.

- **Monitoring and Adjustment**: Regular monitoring of a person's immune system, gut microbiota composition, and health markers may be part of personalized probiotic interventions. By keeping an eye on things, you can evaluate how probiotics are working, spot any changes or advancements, and, if needed, modify the intervention. Probiotic interventions must be continuously assessed and modified to maintain their efficacy and accommodate shifting patient requirements.
- **Combination Therapies**: Probiotics may be combined with other interventions, such as particular diets, prebiotics, synbiotics, or other therapeutic modalities, in personalized approaches. In addition to improving the synergistic effects, combination therapies offer a more specialized method of addressing particular health issues or objectives.

2.5. INTESTINAL PROBIOTICS RESTORE THE INTESTINAL BARRIER

The mechanical, chemical, immune, and microbial barriers that make up the intestinal barrier are the various layers of defense that work together to keep the intestinal microenvironment in a state of homeostasis. An integral component of the intestinal barrier, intestinal microbes support healthy intestinal barrier function. These bacteria are widespread and colonize the host intestine in a symbiotic fashion. Certain gut microorganisms function as probiotics, meaning they improve the health of the host. "Live microorganisms, which when administered in adequate amounts confer a health benefit on the host" is how the World Health Organization (WHO) and the Food and Agriculture Organization of the United Nations define probiotics. Single or compound probiotics can be used to restore the intestinal barrier, offering a new clinical treatment option for diseases related to intestinal barrier damage. This is made possible by advancements in molecular biology, genetic engineering, fermentation culture, microfluidics, and experimental models in recent years.

2.5.1. Intestinal Mechanical Barrier

The intestinal mechanical barrier is composed of different types of IECs and ISC-differentiated IJCs that are found at the base of crypts. Probiotics have the ability to regulate IECs apoptosis, stimulate IEC proliferation, and increase the expression of genes and proteins involved in tight junction (TJ) signaling, all of which contribute to the restoration of the intestinal mechanical barrier.

2.5.1.1. Probiotics Affect the TJs Between IECs

In addition to maintaining normal intestinal barrier function, the TJs between IECs can selectively transport substances and keep harmful substances and pathogenic bacteria out of the intestinal lumen. Intracellular and membrane proteins make up the TJ-related complex proteins of IECs. There are several intracellular proteins, including cingulin, zonula occludens (ZO)1, ZO2, and ZO3. Membrane proteins comprise a range of transmembrane proteins found in TJs, including junction adhesion molecules, occlusive protein (occludin), and synaptic connexin (claudin). By controlling the expression of genes and proteins involved in TJ signaling in IECs, probiotics have an impact on the function of the intestinal barrier (Gou et al., 2022) (Table 2.2).

Table 2.2. List of Probiotics that Affect the Tight Junction to Restore the Intestinal Mechanical Barrier

Probiotic	**Experimental Subject**	**Gene/Protein Expression Increased (↑) or Decreased (↓)**
Escherichia coli Nissle 1917	Germ-free mice	ZO1↑ gene and protein expression
Lactobacillus reuteri (LR1 R2LC, 4,659, I5007 DSM 17938)	IPEC-1 cells DSS colitis mice OVA-sensitized rat	ZO1↑ ZO1 redistribution Occludin ↑claudin1, 3, 5, 7, 8, 9↑ gene and protein expression
Lactobacillus rhamnosus GG	T84 cell line mice	ZO1↑Occludin↑ protein expression ZO1 redistribution Claudin1 redistribution
Lactobacillus plantarum MB452	Caco-2 cells	Occludin↑ ZO1↑ Claudin2↑ gene and protein expression
L. plantarum WCFS1	Healthy volunteers Caco-2 cells	Occludin↑ ZO1↑ gene and protein expression
Bifidobacterium infantis ATCC No. 15697 *Lactobacillus acidophilus* ATCC No. 53103	Caco-2 cells	Occludin ↑Claudin1↑ protein expression

B. infantis strain BB-02	NEC mice	Claudin4 and occludin redistribution
L. plantarum CGMCC No.1258	Caco-2 cells	Occludin↑ ZO1 ↑ Claudin1↑ JAM1↑ protein expression
Bacillus subtilis CW14	Caco-2 cells	ZO1↑ Claudin1↑ protein expression
L. acidophilus LA1	Caco-2 cells DSS colitis mice	Occludin↑ TLR2↑ protein expression
Probiotic mixture (*Bifidobacterium infants, L. acidophilus, Enterococcus, Bacillus cereus*)	NEC mice	Occludin↑ Claudin1↑ protein expression
S. boulardii CNCM I-745	T84 cell line	ZO1 distribution

Source: Gou HZ, Zhang YL, Ren LF, Li ZJ, Zhang L. (2022). How do intestinal probiotics restore the intestinal barrier? Front Microbiol.;13:929346. doi: 10.3389/fmicb.2022.929346. PMID: 35910620; PMCID: PMC9330398.

This does not, however, imply that EcN 1917 has the same effect in the intestinal environment of humans, and it is possible that bacterial colonization and adhesion require the development and upkeep of the intestinal defense system. By boosting TJ protein expression in IECs, *Lactobacillus reuteri* preserves the integrity of the intestinal barrier. But *L. reuteriLR1* may mitigate the damage to the integrity of the intestinal epithelial barrier caused by ETEC K88 infection by upregulating the gene and protein expression of the TJ molecules ZO1 and occludin via the myosin light chain kinase signaling pathway.

Lactobacillus rhamnosus affects TJ protein expression to restore the intestinal mechanical barrier. *E. coli* O157:H7 caused abnormal distribution of ZO1 and claudin-1 in polarized epithelial cells, as well as a decrease in ZO1 expression, resulting in increased permeability and decreased barrier function in an *in vitro* cell model. Cultured *L. rhamnosus* GG with polarized epithelial cells to induce the redistribution of ZO1 and claudin-1 and increased expression of ZO1, improving barrier function. The postbiotic HM0539 from *L. rhamnosus* GG can enhance the resistance of mice to *E. coli* O157:H7 infection by attenuating the destruction of TJ proteins.

In order to stabilize TJs and enhance intestinal barrier function, *Lactobacillus plantarum* MB452 increases the gene and protein expression of ZO1, ZO2, occludin, and cingulin and controls the expression of TJ protein-degrading genes (such as itchy E3 ubiquitin protein ligase and snail family transcriptional regressor 1). Moreover, *L. plantarum* Mb452-induced reduction in proteasome gene expression could be a further way to improve TJ integrity.

In a model of IECs *in vitro*, *Bifidobacterium infantis* and *Lactobacillus acidophilus* normalize the expression of TJ proteins, occludin and claudin1, preventing barrier damage brought on by IL-1 stimulation. *B. infantis* also modulates the correct localization of occludin and claudin 4 in TJs, attenuates intestinal permeability, preserves intestinal barrier function, and lowers the incidence of NEC in a neonatal mouse model.

The intestinal barrier can be restored by a combination of probiotics in addition to a single strain that can strengthen the intestinal barrier and restore its function. By controlling the pregnane X receptor-c-Jun N-terminal kinase signaling pathway, the probiotic mixture (*B. infantis*, *L. acidophilus*, *Enterococcus*, and *Bacillus cereus*) may improve the expression of claudin-1 and occludin and lessen intestinal barrier damage in neonatal NEC. Furthermore, it was discovered that *Saccharomyces boulardii* CNM I-745 modified the distribution of ZO1 while maintaining the barrier function when co-cultured *in vitro* with T84 cells (Gou et al., 2022).

2.5.1.2. Probiotics affect IEC Apoptosis and Proliferation

Probiotics found in the gut have the ability to control IECs' apoptosis, encourage their growth, lessen intestinal injury, and rebuild the intestinal mechanical barrier. *L. rhamnosus* GG secretes the protein p40, which stimulates the growth of IECs. In order to phosphorylate the epidermal growth factor receptor and activate the PI3K/PKB signaling pathway, p40 upregulates the catalytic activity of disintegrin and metalloproteinase domain-containing protein 17, which in turn decreases apoptosis and maintains barrier function. Through the aryl hydrocarbon receptor (AhR), *L reuteri* D8 stimulates the lamina propria lymphocytes to secrete IL-22. This, in turn, activates the phosphorylation of STAT3, which promotes the proliferation of IECs and increases the growth of intestinal organs, thereby recovering the intestinal mucosal barrier and repairing the structural damage to the intestinal epithelium caused by TNF treatment. In a model of IBD in mice, constant *B. subtilis* during the remission phase can lessen IBD and preserve the integrity of the intestinal barrier by controlling the growth of IECs. Further, *Lactobacillus casei* and *Clostridium butyricum* can also promote the proliferation of IECs and restore the intestinal barrier.

2.5.2. Intestinal Chemical Barrier

Mucin (MUC), digestive juices, lysozymes, mucopolysaccharides, antimicrobial peptides (AMPs), and other chemicals make up the intestinal chemical barrier.

Transmembrane mucins and gel-forming mucins are the two types of mucins. MUC2 is the main secretory mucin in the GIT and a significant component of intestinal mucus, whereas MUC13 and MUC17 are the predominant membrane mucins in the intestines of humans and mice. The primary element of the intestinal chemical barrier, MUC2 covers IECs to create an intestinal mucus layer that enhances food absorption, offers intestinal symbiotic bacteria attachment sites, and restricts the combination of pathogens and IECs. Paneth cells secrete AMPs, a class of polypeptide with anti-inflammatory, bactericidal, immune-boosting, and tissue-repair properties. Probiotics have been demonstrated in numerous studies to control mucin expression, influence mucus layer formation, and preserve the integrity of the intestinal barrier.

The increase in MUC2 expression by *L. acidophilus* A4 and its cell extracts significantly inhibited the attachment of *E. coli* O157:H7 to HT-29 IECs. Similar results were obtained when treating colorectal adenocarcinoma (Caco-2) cells with *L. acidophilus* LA1. In an *in vitro* model, *L. casei* GG increased the expression of MUC2 gene and protein to inhibit the translocation of specific pathogenic bacteria.

2.5.3. Intestinal Immune Barrier

Gut-associated lymphoid tissue (GALT) and intestinal immune cells make up the intestinal immune barrier. Peyer's node, immune cells from the innate and adaptive immune systems, such as thymus (T) and bone marrow-or bursa-derived (B) cells, dendritic cells, and macrophages, are found in the propria of the intestinal epithelial mucosa. Paneth cells secrete antibacterial peptides, and plasma cells secretory immunoglobulin A (IgA). These take part in the intestinal barrier's immune defense system.

2.5.3.1. Probiotics and Immune Cells

Probiotics can control immunological and anti-inflammatory processes either directly or indirectly. Mice used in *in vivo* studies showed that *Lactobacillus* could stimulate the maturation of dendritic cells by upregulating the expression of CD40, CD80, and CD86, as well as MHC II. *B. infantis* encourages dendritic cell maturation and tolerogenic CD103 dendritic cell accumulation in GALT, which further controls T-cell differentiation, increases the expression of anti-inflammatory factors, and enhances the immune response of the intestinal mucosa. In order to preserve intestinal immune homeostasis, EcN 1917 controls T cell activation by influencing the T cell cycle and apoptosis. *In vitro, L. plantarum* 8,826 from the National Collection of Industrial Food and Marine Bacteria boosts the activity of cytokine IL-10 as well as T cell subsets and natural killer cells. To boost immunity, *L acidophilus* and *B bifidum* expanded the phagocytic potential of aged mouse macrophages. According to a recent probiotic study, *L. rhamnosus* activates the release of IL-10 from macrophages, which in turn reduces intestinal ischemia–reperfusion injury in mice.

2.5.3.2. Probiotics Regulate IgA Secretion

Probiotics stimulate plasma cells to secrete IgA, but probiotic-induced IgA secretion may be strain-specific. Another study on probiotics found that after gavage of mice with *E. faecalis* CECT7121 increased intestinal mucosal IgA levels and enhanced local mucosal immune responses.

2.5.3.3. Probiotic Metabolites and Intestinal Immune Factors

Probiotic metabolites like SCFAs have the ability to influence immune cells like lymphocytes and mononuclear phagocytes. They can also prevent the growth of immune effector cells and influence the release of inflammatory mediators and immune chemotaxis. Additionally, they may take part in the intestinal regulation of immunity. By blocking histone deacetylase (HDAC), SCFAs can lower the activity of the nuclear transcription factor NF-κB. They can also stop neutrophils and macrophages from releasing IL-8, tumor necrosis factor-α, and other inflammatory factors. Finally, they can stop neutrophils from migrating to areas of inflammation, which lowers intestinal inflammation. SCFAs stimulate the production of immune factors, improve intestinal immune function in mice, and improve intestinal barrier function. They also activate mitogen kinase signaling through G protein-coupled receptors on IECs (Gou et al., 2022).

2.5.4. Intestinal Microbial Barrier

The commensal gut microbiota firmly adheres to the surface of the intestinal epithelial mucosa to form the intestinal microbial barrier, a bacterial membrane barrier. The intestinal microbial barrier, which aids in maintaining the proper balance between the quantity and composition of intestinal microflora, is significantly influenced by intestinal probiotics. In addition to competing with pathogens for nutrients, probiotics can inhibit the attachment sites of targeted cells or the spread of microcolonies to prevent pathogen invasion by creating space barriers.

L. rhamnosus and *L. acidophilus* can competitively attach to the adhesion sites of HEp-2 cells and T84 cells, reducing the adhesion sites on the cell surface and preventing the invasion of pathogenic microorganisms, such as EPEC. EcN can secrete a non-bacteriocin component to act on pathogenic microorganisms or host cells to weaken the adhesion of pathogenic microorganisms. *Bacillus mesentericus*, *C. butyricum* and *E. faecalis* increase the diversity and abundance of intestinal microbes and maintain the balance of intestinal flora.

In the cecum of mice given cyclophosphamide, *Lactobacillus plantarum* boosted the number of *Bifidobacterium* and *Lactobacillus* and decreased the number of *Enterococcus* and *E. coli*. Furthermore, the combination of *Lactobacillus fermentum* GOS57 and *Lactobacillus plantarum* GOS42 was able to improve the function of the intestinal barrier, decrease the amount of Enterobacteriaceae, and increase the abundance of *Lactobacillus*. It also adjusted the balance of intestinal flora. *Lactobacillus rhamnosus* and *Lactobacillus bulgaricus* have the ability to suppress the growth of pathogenic bacteria, boost the number of good bacteria in the intestines, and aid in correcting the intestinal flora imbalance brought on by antibiotics. Children who experience recurrent respiratory infections have an unbalanced intestinal flora, with an increase in *E. coli* and a marked decrease in *Bifidobacterium* and *Lactobacillus*. Giving children with infections more *Bifidobacterium* and *Lactobacillus* can help preserve the balance of their intestinal flora and lower their risk of infection. This is achieved by giving them *Bifidobacterium* quadruple live bacterial tablets (Gou et al., 2022).

REFERENCES

1. Berg, R. D. (1995). Bacterial translocation from the gastrointestinal tract. *Trends in Microbiology, 3*(4), 149–154.
2. Brandão, R. L., Castro, I. M., Bambirra, E. A., Amaral, S. C., Fietto, L. O., Tropia, M. J. M., Neves, M. J., Santos, R. G., Gomes, N. C. M., & Nicoli, J. R. (1998). Intracellular signal triggered by cholera toxin in *Saccharomyces boulardii* and *Saccharomyces cerevisiae. Applied and Environmental Microbiology, 64*(2), 564–568.
3. Brugier, S., & Patte, F. (1975). Antagonisme *in vitro* entre l'ultralevure et différents germes bactériens. *Médecin Paris, 45*(1), 61–66.
4. Camilleri, M. (2019). Leaky gut: Mechanisms, measurement, and clinical implications in humans. *Gut, 68*(8), 1516–1526.
5. Castagliuolo, I., Riegler, M. F., Valenick, L., Lamont, J. T., & Pothoulakis, C. (1999). *Saccharomyces boulardii* protease inhibits the effects of *Clostridium difficile* toxins A and B in human colonic mucosa. *Infection and Immunity, 67*(1), 302–307.
6. Chen, X., Kokkotou, E. G., Mustafa, N., Bhaskar, K. R., Sougioultzis, S., O'Brien, M., Pothoulakis, C., & Kelly, C. P. (2006). *Saccharomyces boulardii* inhibits ERK1/2 mitogen-activated protein kinase activation both *in vitro* and *in vivo* and protects against *Clostridium difficile* toxin A-induced enteritis. *Journal of Biological Chemistry, 281*(35), 24449–24459.
7. Chiva, M., Soriano, G., Rochat, I., Peralta, C., Rochat, F., Llovet, T., Mirelis, B., Schiffrin, E. J., Guarner, C., & Balanzó, J. (2002). Effect of *Lactobacillus johnsonii* La1 and antioxidants on intestinal flora and bacterial translocation in rats with experimental cirrhosis. *Journal of Hepatology, 37*(4), 456–462.
8. Chopyk, D. M., & Grakoui, A. (2020). Contribution of the intestinal microbiome and gut barrier to hepatic disorders. *Gastroenterology, 159*(3), 849–863.
9. Corthier, G., Dubos, P., & Ducluzeau, R. (1986). Prevention of *Clostridium difficile*-induced mortality in gnotobiotic mice by *Saccharomyces boulardii. Canadian Journal of Microbiology, 32*(11), 894–896.
10. Dahan, S., Dalmasso, G., Imbert, V., Peyron, J. F., Rampal, P., & Czerucka, D. (2003). *Saccharomyces boulardii* interferes with enterohemorrhagic *Escherichia coli*-induced signaling pathways in T84 cells. *Infection and Immunity, 71*(2), 766–773.
11. Dias, R. S., Bambirra, E. A., Silva, M. E., & Nicoli, J. R. (1995). Protective effect of *Saccharomyces boulardii* against cholera toxin in rats. *Brazilian Journal of Medical and Biological Research, 28*(3), 323–325.

12. Ducluzeau, R., & Bensaada, M. (1982). Effet comparé de l'administration unique ou en continu de *Saccharomyces boulardii* sur l'établissement de diverses souches de *Candida* dans le tractus digestif de souris gnotoxéniques. *Annales de Microbiologie, 133B*(4), 491–501.
13. Fietto, J. L. R., Araújo, R. S., Valadão, F. N., Fietto, L. G., Brandão, R. L., Neves, M. J., Gomes, C. O., Nicoli, J. R., & Castro, I. M. (2004). Molecular and physiological comparisons between *Saccharomyces cerevisiae* and *Saccharomyces boulardii*. *Canadian Journal of Microbiology, 50*(8), 615–621.
14. Gedek, B. R. (1999). Adherence of *Escherichia coli* serogroup O157 and *Salmonella typhimurium* mutant DT 104 to the surface of *Saccharomyces boulardii*. *Mycoses, 42*(7, 8), 261–264.
15. Generoso, S. V., Viana, M. V., Santos, R., Arantes, R. M. E., Martins, F. S., Nicoli, J. R., Machado, J. A. N., Correia, M. I. T. D., & Cardoso, V. N. (2011). Protection against increased intestinal permeability and bacterial translocation induced by intestinal obstruction in mice treated with viable and heat-killed *Saccharomyces boulardii*. *European Journal of Nutrition, 50*(4), 261–269.
16. Generoso, S. V., Viana, M. V., Santos, R., Martins, F. S., Machado, J. A. N., Arantes, R. M. E., Nicoli, J. R., Correia, M. I. T. D., & Cardoso, V. N. (2010). *Saccharomyces cerevisiae* strain UFMG 905 protects against bacterial translocation, preserves gut barrier integrity, and stimulates the immune system in a murine intestinal obstruction model. *Archives of Microbiology, 192*(7), 477–484.
17. Gou, H. Z., Zhang, Y. L., Ren, L. F., Li, Z. J., & Zhang, L. (2022). How do intestinal probiotics restore the intestinal barrier? *Frontiers in Microbiology, 13*, 929346.
18. Grimbaldeston, M., Nakae, S., Kalesnikov, J., Tsai, M., & Galli, S. J. (2007). Mast cell-derived interleukin 10 limits skin pathology in contact dermatitis and chronic irradiation with ultraviolet B. *Nature Immunology, 8*(10), 1095–1104.
19. Herek, O., Kara, I. G., & Kaleli, I. (2004). Effects of antibiotics and *Saccharomyces boulardii* on bacterial translocation in burn injury. *Surgical Today, 34*(3), 256–260.
20. Jawhara, S., & Poulain, D. (2007). *Saccharomyces boulardii* decreases inflammation and intestinal colonization by *Candida albicans* in a mouse model of chemically induced colitis. *Medical Mycology, 45*(8), 691–700.
21. Krasowska, A., Murzyn, A., Dyjankiewicz, A., Łukaszewicz, M., & Dziadkowiec, D. (2009). The antagonistic effect of *Saccharomyces*

boulardii on *Candida albicans* filamentation, adhesion, and biofilm formation. *FEMS Yeast Research, 9*(8), 1312–1321.

22. MacFarland, L. V. (2010). Systematic review and meta-analysis of *Saccharomyces boulardii* in adult patients. *World Journal of Gastroenterology, 16*(18), 2202–2222.
23. Martins, F. S., Nardi, R. M. D., Arantes, R. M. E., Rosa, C. A., Neves, M. J., & Nicoli, J. R. (2005). Screening of yeasts as probiotics based on capacities to colonize the gastrointestinal tract and to protect against enteropathogen challenge in mice. *Journal of General and Applied Microbiology, 51*(2), 83–92.
24. Martins, F. S., Silva, A., Vieira, A., Barbosa, F. H. F., Arantes, R. M. E., Teixeira, M. M., & Nicoli, J. R. (2009). Comparative study of *Bifidobacterium animalis, Escherichia coli, Lactobacillus casei*, and *Saccharomyces boulardii* probiotic properties. *Archives of Microbiology, 191*(8), 623–630.
25. Moher, D., Liberati, A., Tetzlaff, J., & Altman, D. G. (2009). Preferred reporting items for systematic reviews and meta-analyses: The PRISMA statement. *BMJ, 339*, b2535. https://doi.org/10.1136/bmj.b2535.
26. Odenwald, M. A., & Turner, J. R. (2017). The intestinal epithelial barrier: A therapeutic target? *Nature Reviews Gastroenterology & Hepatology, 14*(1), 9–21. https://doi.org/10.1038/nrgastro.2016.169.
27. Pedersen, L. L., Owusu-Kwarteng, J., Thorsen, L., & Jespersen, L. (2012). Biodiversity and probiotic potential of yeasts isolated from Fura, a West African spontaneously fermented cereal. *International Journal of Food Microbiology, 159*(2), 144–151.
28. Peret, L. A., Penna, F. J., Bambirra, E. A., & Nicoli, J. R. (1998). Dose effect of oral *Saccharomyces boulardii* treatments on morbidity and mortality in immunosuppressed mice. *Journal of Medical Microbiology, 47*(2), 111–116.
29. Pontier-Bres, R., Prodon, F., Munro, P., Rampal, P., Lemichez, E., Peyron, J. F., & Czerucka, D. (2012). Modification of *Salmonella Typhimurium* motility by the probiotic yeast strain *Saccharomyces boulardii. PLoS ONE, 7*(2), e33796.
30. Qamar, A., Aboudola, S., Warny, M., Michetti, P., Pothoulakis, C., LaMont, J. T., & Kelly, C. P. (2001). *Saccharomyces boulardii* stimulates intestinal immunoglobulin A immune response to *Clostridium difficile* toxin A in mice. *Infection and Immunity, 69*(5), 2762–2765.
31. Ratajczak, W., Rył, A., Mizerski, A., Walczakiewicz, K., Sipak, O., & Laszczyńska, M. (2019). Immunomodulatory potential of gut microbiome-

derived short-chain fatty acids (SCFAs). *Acta Biochimica Polonica, 66*(1), 1–12.

32. Rodrigues, A. C. P., Cara, D. C., Fretez, S. H. G. G., Cunha, F. Q., Vieira, E. C., Nicoli, J. R., & Vieira, L. Q. (2000). *Saccharomyces boulardii* stimulates sIgA production and the phagocytic system of gnotobiotic mice. *Journal of Applied Microbiology, 89*(3), 404–414.
33. Rodrigues, A. C. P., Nardi, R. M., Bambirra, E. A., Vieira, E. C., & Nicoli, J. R. (1996). Protective effect of *Saccharomyces boulardii* against experimental oral infection with *Salmonella Typhimurium* and *Shigella flexneri* in gnotoxenic mice. *Journal of Applied Bacteriology, 81*(3), 251–256.
34. Servin, A. L. (2004). Antagonistic activities of lactobacilli and bifidobacteria against microbial pathogens. *FEMS Microbiology Reviews, 28*(4), 405–440.
35. Sharma, R., Young, C., & Neu, J. (2010). Molecular modulation of intestinal epithelial barrier: Contribution of microbiota. *Journal of Biomedicine and Biotechnology, 2010*, 305879.
36. Sugahara, H., Odamaki, T., Fukuda, S., Kato, T., Xiao, J. Z., Abe, F., et al. (2015). Probiotic *Bifidobacterium longum* alters gut luminal metabolism through modification of the gut microbial community. *Scientific Reports, 5*, 13548.
37. Thomas, S., Metzke, D., Schmitz, J., Dörffel, Y., & Baumgart, D. C. (2011). Anti-inflammatory effects of *Saccharomyces boulardii* mediated by myeloid dendritic cells from patients with Crohn's disease and ulcerative colitis. *American Journal of Physiology-Gastrointestinal and Liver Physiology, 301*(6), G1083–G1092.
38. Tiago, F. C. P., Martins, F. S., Souza, E. L. S., Pimenta, P. F. P., Araujo, H. R. C., Castro, I. M., Brandão, R. L., & Nicoli, J. R. (2012). Adhesion to the yeast cell surface as mechanism for trapping pathogenic bacteria by *Saccharomyces* probiotics. *Journal of Medical Microbiology, 61*(8), 1194–1207.
39. Van der Aa Kühle, A., Skovgaard, K., & Jespersen, L. (2005). *In vitro* screening of probiotic properties of *Saccharomyces cerevisiae* var. *boulardii* and food-borne *Saccharomyces cerevisiae* strains. *International Journal of Food Microbiology, 101*(1), 29–39.
40. Vilela, E. G., Abreu Ferrari, M. L., Gama Torres, H. O., Martins, F. P., Goulart, E. M., Lima, A. S., & Cunha, A. S. (2008). Influence of *Saccharomyces boulardii* on the intestinal permeability of patients with Crohn's disease in remission. *Scandinavian Journal of Gastroenterology, 43*(7), 842–848.

41. Wu, X., Vallance, B. A., Boyer, L., Bergstrom, K. S., Walker, J., Madsen, K., O'Kusky, J. R., Buchan, A. M., & Jacobson, K. (2008). *Saccharomyces boulardii* ameliorates *Citrobacter rodentium*-induced colitis through actions on bacterial virulence factors. *American Journal of Physiology-Gastrointestinal and Liver Physiology, 294*(2), G295–G306.

CHAPTER 3

THE ROLE OF PROBIOTICS IN DIGESTIVE HEALTH

Contents

INTRODUCTION

Probiotics are microscopic organisms that benefit their host's health. Probiotics are involved in the myriad of microorganisms that our bodies depend on to function and maintain our health. These are the beneficial bacteria that your body should genuinely have. Probiotics also aid in immune system function. They interact with the gut-associated lymphoid tissue (GALT), which is essential to the gastrointestinal tract's (GITs) immune defense. Probiotics aid in the management of inflammatory responses and the prevention of infections by increasing the generation of helpful cytokines and boosting immune cell activity.

Short-chain fatty acids (SCFAs), which have anti-inflammatory qualities and promote the health of the gut lining, are among the advantageous metabolites that probiotics produce, and they have an additional impact on digestive processes. Probiotic bacteria use dietary fibers to ferment them into SCFAs, which are then used by intestinal cells as an energy source to support gut health. Additionally, probiotics can aid in the prevention and treatment of gastrointestinal illnesses. *Lactobacillus* and *Bifidobacterium* species, for instance, are frequently used to modulate gut flora and reduce inflammation, thereby alleviating symptoms of inflammatory bowel disease (IBD) and irritable bowel syndrome (IBS). Additionally, they help prevent and treat diarrhea, especially infectious diarrhea and diarrhea brought on by antibiotics. They help regulate dangerous microorganisms like germs and support healthy bodily systems from your mouth to your stomach. Probiotics facilitate digestion and enhance nutrient absorption when taken in the proper amounts. In addition to being found in cultured milk and fermented foods, probiotic supplements are also produced (Borchers et al., 2009; Brenner et al., 2009; Moayyedi et al., 2010).

3.1. THE ROLE OF PROBIOTICS IN DIGESTIVE HEALTH

Digestion health is still the main advantage of probiotics, despite the fact that they have been shown to have health benefits outside of the gastrointestinal tract (GIT). Technological developments have allowed for even more thorough examinations of the gut microbiota. However, because everyone's microbiota is unique and varies with age, it is impossible to define a healthy microbiota. Probiotics generally work to maintain health and lower the risk of illness rather than curing or preventing it, though probiotic medications may also be an option. Probiotics have been shown to have health benefits in a number of areas, including constipation, inflammatory bowel diseases (IBD), and specific types of diarrhea. These findings are supported by meta-analyses. Additionally, our knowledge of the mechanisms underlying probiotic action is constantly growing.

It is still difficult to identify particular strains or combinations that have clear health benefits. It is likely that in the future we will see new probiotics from genera other than *Lactobacillus*, *Bifidobacterium*, or *Saccharomyces*; these may focus on new or existing health targets. This is because of our growing understanding of the composition and activity of the intestinal microbiota.

The most commonly used definition of probiotics is that of live microorganisms that, when given in sufficient quantities, benefit the host's health. This definition was put forth by a working group of the Food and Agriculture Organization and the World Health Organization (WHO) and was recently confirmed by the International Scientific Association for Probiotics and Prebiotics. Thus, the definition emphasizes three key elements: (i) the microbes' viability; (ii) the dose; and (iii) the health benefit that has been proven.

3.1.1. Viability

The microbes' viability presents some technological difficulties. Although strains and species from other genera are also used, the majority of probiotics are found in the genera *Lactobacillus* and *Bifidobacterium*. In order to do this, the microbes must continue to exist until the end of the shelf life. Probiotics are usually taken as a dietary supplement or as part of a meal.

Foods generally have a high-water activity, and foods that have undergone fermentation have a low pH. Such foods often have a short shelf life—weeks as opposed to months or years—and must be stored in a refrigerator. Although dietary supplements typically have low water activity, they can have up to two-year shelf lives and are typically stored at room temperature. There must be an adequate dose available until the end of the shelf life. The amount of viable microbes at the time of manufacture is less interesting because it does not indicate what the minimal dose is at the time of consumption; ideally, this is stated on the product.

3.1.2. Dose

Hence, dosage is crucial. Nevertheless, it's not always obvious what this dosage ought to be. Health Canada suggests at least 10^7 colony-forming units (CFU) for a general probiotic claim, but as a general guideline, the dose should be at least 10^9 CFU. The dosage, however, will probably vary depending on the specific probiotic strain (combination), the health benefit, and the consumer's subpopulation. As a result, the dosage that has been shown to have health benefits should be taken in accordance with the recommendations; this is also necessary, according to Health Canada's requirements, for certain health claims.

3.1.3. Health Benefit

Lastly, there is the health benefit; what is a health benefit and what is not? Probiotic health claims have not been approved by the European Food Safety Authority as of October 2015, when this article was written. That being said, this should not be taken to mean that probiotics have no health benefits. It is outside the purview of this work to analyze in detail the regulatory obstacles to probiotic health claims in different jurisdictions. To keep things simple, an improvement in a clinical outcome will be considered a health benefit in this context. This implies that modifications to the makeup of the microbiota or specific immune markers will not be taken into account. The general consensus is not on the strength of these markers.

3.1.4. Safety

Probiotics are clearly commercialized and should be clearly marked and safe, even though this is not stated in the definition. It is best to identify using molecular methods. It's not necessary, but genome sequencing is now so affordable that asking for commercial probiotic strains would not be out of the ordinary. The general public is not concerned about safety when it comes to members of the genera *Bifidobacterium* and *Lactobacillus*. A risk–benefit analysis should be done for consumers (patients) who are not well, both generally and specifically in terms of their immune system. Probiotics tend to be most beneficial to this group of people, and the risk (usually infection) is minimal. In the improbable event that an infection develops, it is easily treatable. Lastly, transferable antibiotic resistance should not exist in commercialized probiotics for the benefit of all consumers. It is possible to test for this using both phenotypic and genome analysis methods.

3.1.5. Strain Specificity Versus Commonality

The majority of commercial probiotics that fall into the specific category of lactic acid bacteria (LAB), such as *lactobacilli* and bifidobacteria, are typically selected to withstand low pH and bile acids after being isolated from similar sources like fermented dairy products or the intestine. Thus, it makes sense to presume that while some traits may be unique to a given strain, other traits may be shared by several strains; Figure 3.1 provides a schematic representation of this.

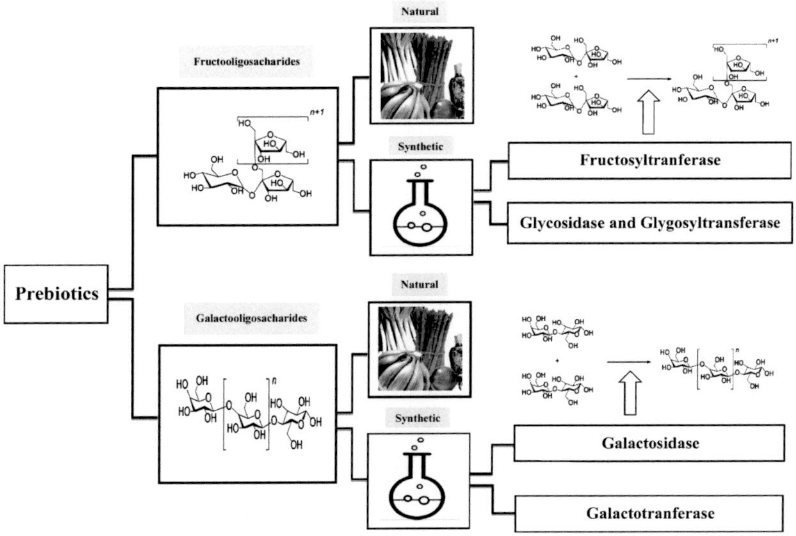

Figure 3.1. *Sources and production of major prebiotics, including fructo-oligosaccharides (FOS) and galacto-oligosaccharides (GOS).*

Source: Davani-Davari D, Negahdaripour M, Karimzadeh I, Seifan M, Mohkam M, Masoumi SJ, Berenjian A, Ghasemi Y. (2019). Prebiotics: Definition, types, sources, mechanisms, and clinical applications. Foods, 8(3):92. https://doi.org/10.3390/foods8030092.

3.1.6. Intestinal Microbiota

The ability to analyze the composition and activity of the intestinal microbiota is constantly evolving, making analyses previously thought to be impossible possible. A comprehensive examination of the intestinal microbiological analyses is outside the purview of this work, and readers are directed to other resources, such as Lankelma et al. (2015) for future information. All agree, however, that the intestinal microbiota—the right term for what was formerly called the intestinal microflora—plays a significant role in both health and disease and in a much wider range of contexts than previously believed. Remember, we are talking about small life, not small plants. Since fecal microbiota and intestinal microbiota are frequently used interchangeably, it is critical to understand that distinct microbiota is present in various parts of the GIT.

Finding particular organisms or groups of organisms that are causally linked to particular health conditions has proven challenging, despite advancements in analytical techniques. Because it is unclear what change is desired—not even an increase in the amounts of bifidobacteria and/or *lactobacilli*—changing the composition of the microbiota is frequently not regarded as beneficial to health. Nonetheless, one thing becomes evident: diversity is generally beneficial. Reduced microbial diversity seems to be linked to a number of disease conditions. This is not shocking because, according to ecology, monocultures are vulnerable to disruptions and diversity in an ecosystem is linked to resilience. While most people agree that the intestinal microbiota is important for both health and disease, our understanding of the precise mechanisms and roles played by each member is still lacking. This is further complicated by the fact that each person has a unique microbiota; there is no such thing as a "standard" healthy microbiota because they are all unique and vary throughout life.

It has been demonstrated that some probiotics alter the intestinal microbiota's makeup. But this does not always end up like this. This may be because a healthy microbiota, in whatever form it exists, is resistant to changes brought about by external agents, both beneficial (probiotics) and harmful (pathogens). Probiotics may aid in the recovery of a disrupted microbiota, allowing it to revert to its initial state. The use of antibiotics is a glaring illustration of this. Although it has been shown that the majority of commercial probiotics can survive gastrointestinal transit, this does not necessarily indicate a health benefit and does not explain how probiotics can aid in the recovery of the microbiota because they are more than just supplements for species that have been depleted. Instead, probiotics seem to be helping the local environment to improve, which permits the microbiota to heal (Borchers et al., 2009; Brenner et al., 2009; Moayyedi et al., 2010).

3.1.7. Digestive Health

In formal terms, digestive health is not defined. The absence of intestinal complaints should serve as the practical definition. Although this is obviously unsatisfactory, it might be feasible in practice. If there are no symptoms, or if they are not even reminded of their intestines, then the majority of healthy consumers will be content with their digestive health. Thus, maintaining digestive health is primarily concerned with lowering the chance of gastrointestinal problems and/or treating them when they do arise. Probiotics are mostly sold as dietary supplements or as ingredients in food, though they can also be used as drugs. Probiotics may lower the chance of illness, but they are not expected to prevent it. Probiotics may lessen the symptoms of illness, but they neither cure nor lessen disease. Therefore, the primary goal of most probiotics is to maintain health rather than to be treated like drugs. Furthermore, unless they contain particular strains with proven health benefits, like some probiotic yogurts, traditional fermented foods like yogurt, sauerkraut, and kimchi are not probiotics (Figure 3.2).

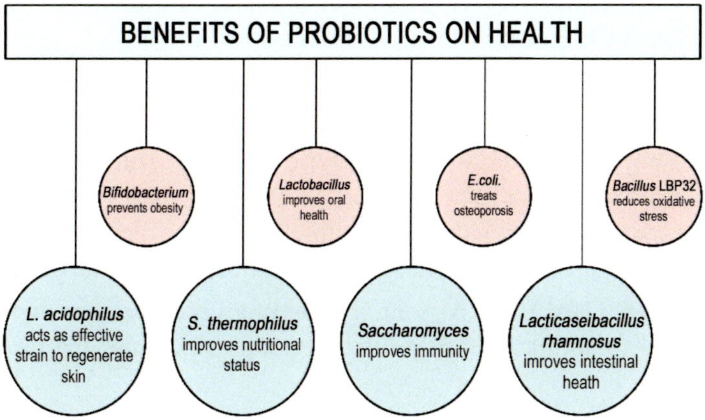

Figure 3.2. *Schematic representation of probiotic products.*

Source: Ranjha MMAN, Shafique B, Batool M, Kowalczewski PŁ, Shehzad Q, Usman M, Manzoor MF, Zahra SM, Yaqub S, Aadil RM. (2021). Nutritional and health potential of probiotics: A review. Applied Sciences; 11(23):11204. https://doi.org/10.3390/app112311204.

3.1.8. Probiotics and Digestion

Probiotics are believed to help with digestive issues such as:

- colic
- constipation
- Crohn's disease
- inflammatory bowel disease (IBD)
- irritable bowel syndrome (IBS)
- lactose intolerance
- ulcerative colitis

However, there is still much to learn about their effectiveness and how they function across different populations. The majority of probiotic research has been limited, and it is unclear how precisely probiotics improve digestion. Additionally, they do not explain how to take probiotic supplements or how to dose them consistently.

Probiotics are understood to be safe for most people, and there are some general guidelines. Here is what we *do* know:

- **Digestive Ecosystems Differ**: The microbiomes in the intestines of constipated individuals differ from those of non-constipated individuals. We are unsure of whether constipation results from these various ecosystems or is a cause of it.

- **They Lower pH Levels**: Probiotics may aid in the faster passage of stool through the colon by lowering its pH level.
- **They may Relieve Antibiotic-Related Diarrhea**: Probiotics may be particularly beneficial in treating diarrhea brought on by *Clostridium difficile* and antibiotics. Probiotics are supposed to restore the beneficial bacteria that antibiotics may have destroyed.
- **They Can Help Absorb Protein**: Probiotics can improve your body's ability to absorb other vitamins, minerals, and protein from your diet.

Not all probiotics are equal, and more research is needed to understand the benefits of each type. Most research covers the *Lactobacillus* and the *Bifidobacterium* strains. These strains include:

- L. acidophilus
- L. casei
- L. plantarum
- B. lactis
- B. longum
- B. bifidum

L. casei and *B. lactis* were found to be useful in treating constipation in a 2010 study. Five previously published studies involving adults and children were reviewed for this study. Before using any probiotics, more research is necessary to determine their safety as they may not belong to the *Lactobacillus* and *Bifidobacterium* genera. Probiotics must be able to withstand the extremely acidic environment of the stomach and GIT in order to be effective because they are taken orally. Probiotics of the highest caliber will enter your stomach undamaged and proceed to the intestines, where nutrients are absorbed. This is the main area of healing and beneficial effects of probiotics. Since probiotics are living organisms, a variety of issues with storage and packaging can compromise their viability. There is still more to discover about the optimum way to take probiotics. Furthermore, we need to find out more about how long probiotics can stay in your gut after they enter it. As of right now, it is thought that no probiotic dosage will remain in your system permanently. Nowadays, probiotics are taken on a daily basis.

3.1.9. Constipation, Slow Intestinal Transit

Constipation is among the most prevalent digestive issues. Constipation affects overall health significantly and is linked to serious intestinal conditions like diverticulitis, even though it is benign. The definition of constipation is having three or fewer spontaneous bowel movements in a week. It tends to get worse with age and affects women more frequently than men. Constipation is a clearly defined condition, but it is somewhat subjective and challenging to quantify. The intestinal transit time and the Bristol Stool Scale are substitute measurements. Numerous methods, including the use of dyes like carmine red, radiopaque markers, and so-called smart pills, can be used to measure intestinal transit time. Since most subjects only defecate once a day or less, the former, which calculates oroanal transit time, is not very sensitive. Radiation exposure is involved with radiopaque markers, which are sensitive but need subjects to consume the markers consistently for six days in order to establish a steady state. Though the method is costly and has not been regularly used for measuring transit for research purposes, the so-called smart pill offers more data than just colonic transit time.

Many probiotics have been investigated for their impact on slow intestinal transit; the majority are strains of *Bifidobacterium lactis*, including *B. lactis* DN-173010, *B. lactis* HN019, or *Bifidobacterium longum* BB536 plus *B. lactis* 420. It has been discovered that the probiotics under test generally shorten colonic transit. It is crucial to note that probiotics do not induce diarrhea in subjects with "normal" initial transit times; in other words, they do not shorten the transit time (Figure 3.3). Enhancement of stool consistency and frequency has also been demonstrated by probiotics. Meta-analyses have not detected a dose-response effect despite the use of a wide range of doses in studies.

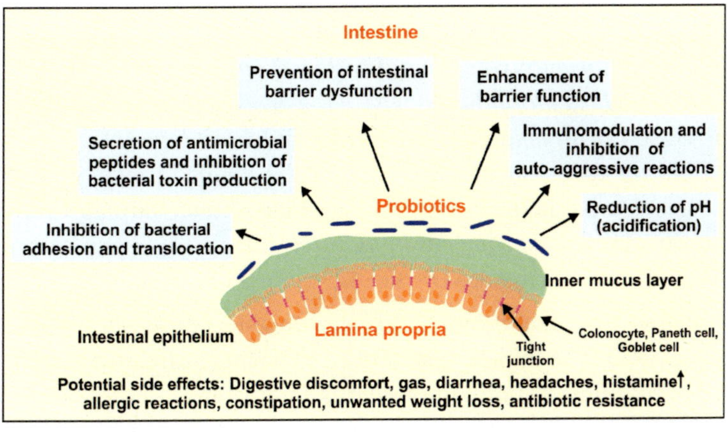

Figure 3.3. *Principle by which probiotics can provide benefits for the gut.*

***Source**: Chandrasekaran P, Weiskirchen S, Weiskirchen R. (2024). Effects of probiotics on gut microbiota: An overview. International Journal of Molecular Sciences; 25(11):6022. https://doi.org/10.3390/ijms25116022.*

3.1.10. Antibiotic-Associated Diarrhea

Antibiotic-associated diarrhea (AAD) has been documented to affect 5% to 39% of patients receiving antibiotics. Nevertheless, the source data used to calculate these figures is somewhat outdated and focuses on hospitalized patients. According to a more recent epidemiological study, 96% of hospital patients who received antibiotic prescriptions went on to develop AAD; these patients were primarily 65 years of age and older. It is less known how common AAD is in patients who can walk. However, AAD lowers quality of life and has a significant financial impact on the healthcare system. Complementary treatments that can lower its incidence are therefore very welcome. It is believed that *Staphylococcus aureus*, *Clostridium difficile*, and *Clostridium perfringens* are the most frequent pathogens causing AAD. Since these pathogens do not explain all cases of AAD, there are still many unidentified causes, some of which might not be related to microbiology and instead be related to physiological alterations in the host (such as altered bile acid resorption).

Probiotics have been demonstrated to be highly effective in lowering the risk of AAD. When examining the settings, the majority of studies have been carried out and shown to be successful in a clinical or hospital setting. Information about ambulatory patients and community settings is far less abundant. According to calculations, using probiotics can cut the cost of treating diarrhea by 25%; this translates to a number needed to treat (NNT) of 13.

In general, probiotic combinations, such as *Bifidobacterium bifidum* W23 + *B. lactis* W18 + *B. longum* W51 + *Enterococcus faecium* W54 + *Lactobacillus acidophilus* W37 + *L. acidophilus* W55 + *Lactobacillus paracasei* W72 + *Lactobacillus plantarum* W62 + *Lactobacillus rhamnosus* W71 + *Lactobacillus salivarius* W24 or *L. acidophilus* NCFM + *L. paracasei*Lpc-37 + *B. lactis* Bi-07 + B. lactis BL-04 or *L. acidophilus* CL1285 and *Lactobacillus casei* LBC80R have been shown to be successful. Among the single strains, *L. rhamnosus* GG has been shown to help; *E. faecium* SF68 has shown varying results. Also, yeasts, in particular *Saccharomyces boulardii* (nom. inval.), have been reported to be efficacious in reducing AAD risk.

Remarkably, two uncommon probiotic dose-response studies have been carried out in AAD. The higher dose proved to be more effective than the lower tested dose, according to both studies. It is interesting to note that probiotics have been shown to have dose-response effects, though this should not be generalized (Borchers et al., 2009; Brenner et al., 2009; Moayyedi et al., 2010).

3.1.11. Community-Acquired Diarrhea

Many pathogens, toxins, and other factors can cause diarrhea that is acquired in the community. Rotavirus and traveler's diarrhea are two of the most thoroughly studied community-acquired diarrhea targets for probiotics. Probiotics like *L. rhamnosus* GG have been demonstrated to help treat rotavirus diarrhea and lower the risk of nosocomial infections, or diarrhea caused by the virus, in children receiving medical care. For every tested strain, there is little to no evidence that rotavirus infection poses a risk to children in daycare centers. It is unclear how important this target is left to pursue in hospital and community settings now that a rotavirus vaccine is available. Probiotics like *B. lactis* HN019 and *L. paracasei*Lpc-37 have been shown to be effective in lowering risk, especially in young children, when it comes to community-acquired diarrhea, even though the etiology is frequently unknown.

With an estimated (nonsignificant) reduction of 8%, probiotics are less effective in preventing traveler's diarrhea, which is not surprising given the variety of causes. Furthermore, there is a high attrition rate in the studies that have been used to assess the efficacy. Nonetheless, it has been estimated that using probiotics can cut the cost of treating diarrhea by 25% when they are effective. The cost of the probiotic was higher at first, but later on, extra medication and consultations were less expensive. The fact that a probiotic can lengthen transit times in cases of diarrhea and shorten them in cases of constipation may seem contradictory. Probiotics normalize the transit time, which is a more accurate viewpoint (Figure 3.3).

3.1.12. Irritable Bowel Syndrome

IBS is a syndrome rather than an illness; it should not be confused with IBD. Further, IBS is defined according to the Rome III criteria as: recurrent abdominal pain or discomfort at least 3 days per month in the last 3 months associated with two or more of the following:

- Improvement with defecation
- Onset associated with a change in frequency of stool
- Onset associated with a change in form (appearance) of stool.

Furthermore, the subject's symptoms are not explained by any indication of an inflammatory, anatomical, metabolic, or neoplastic process. Using the Bristol Stool Scale, IBS can be classified as undefined, mixed, constipation, or diarrheal. Even though IBS-like symptoms are among the most common reasons that patients visit gastroenterologists, treatment is difficult because the etiology is unknown. Studies involving IBS must be sufficiently long (several months) and have a sufficient number of volunteers (hundreds of people/treatment group) because the condition appears to have a strong psychosomatic component and is therefore susceptible to a placebo effect. To ensure a sufficiently homogeneous study population, it might also be necessary to concentrate on a single IBS subtype. Probiotics appear to have the potential to relieve specific symptoms of IBS, such as abdominal pain (e.g., VSL#3 or *Bacillus coagulans* GBI-30, 6086), bloating (e.g., *L. acidophilus* NCFM + *B. lactis* Bi-07 or *B. bifidum* MIMBb75), and flatulence (e.g., *S. boulardii*). However, it seems difficult to improve IBS quality of life overall. As a result, determining which particular strains or combinations of strains would work best is also challenging. However, as previously mentioned, some strains have been shown to alleviate specific IBS symptoms.

3.1.13. Necrotizing Enterocolitis

Necrotizing enterocolitis (NEC) is an extremely dangerous illness that has a high death rate. Those babies born weighing less than 1500 grams are the primary risk group. Thankfully, NEC rarely occurs in live births overall (<0.1%). Although the pathophysiology of NEC is not entirely understood, it is most likely complex and involves a significant microbial component. Compared to term infants, preterm infants' intestinal microbiota has been shown to be less diverse and to be more frequently colonized by potentially pathogenic species from genera like Klebsiella, Enterobacter, and *Clostridium*. As of right now, there is no known cure for NEC prevention. However, one of the best-documented benefits of probiotics is that they have been shown to be highly effective in lowering the risk of NEC in the at-risk population. Probiotic administration should begin as soon as possible and continue until at least the corrected gestational age of 36 or 37 weeks due to the potential pathogens' involvement in the etiology of NEC. However, given that the majority of studies are small and that different doses and study designs are used, it is challenging to suggest particular strains for this use (Borchers et al., 2009; Brenner et al., 2009; Moayyedi et al., 2010).

3.1.14. Inflammatory Bowel Disease

IBD is categorized into three distinct conditions: pouchitis, an inflammation of an ileoanal pouch, ulcerative colitis (UC), an inflammatory condition restricted to the colon, and Crohn's disease (CD), an inflammatory condition that can occur anywhere in the intestinal tract. IBD is typified by remission intervals in between bouts of actively manifesting illness that is managed. Although the cause of IBD is unknown, the microbiota seems to be involved. It follows that the idea of using probiotics for IBD is not shocking. *Escherichia coli* Nissle 1917 and the probiotic combination VSL3 are the two main probiotics under investigation. While the latter appears to be effective in maintaining remission in pouchitis, inducing remission in combination with conventional therapy, and maintaining remission in UC, the former appears to be effective in maintaining remission in CD.

3.1.14.1. Helicobacter Pylori Eradication

Gastric ulcers and gastric cancers have been linked to *Helicobacter pylori*, a common resident of the gastric mucosa. *H. pylori*, however, plays two roles: it has been discovered that exposure to the organism is associated with a lower risk of atopic dermatitis in children, and that eliminating *H. pylori* is associated with a higher risk of obesity. Its pathogenic nature is the reason it is eliminated from carriers. Although research conducted on animals and *in vitro* has suggested that *Lactobacillus* probiotics in particular have antagonistic properties toward *H. pylori*, human studies have not supported this. Thus, probiotics cannot be regarded as a replacement for conventional treatment. Given the possible benefits the organism may have, it may be preferable for certain strains, like *Lactobacillus johnsonii* La1, to decrease *H. pylori* counts and activity.

Probiotics have been found to be promising as adjuvants to standard therapy, reducing side effects, improving compliance, and increasing success rate (72% vs. 82%) even though they are not eradicating. The Maastricht IV consensus report on *H. pylori* treatment also makes this recommendation. The use of *S. boulardii* is advised by the Maastricht IV report, with a recommendation grade of D. Although other factors appear to be involved as well, the beneficial effect of certain probiotics in the elimination of *H. pylori* can be explained by mechanisms similar to those for reducing the risk of AAD.

3.1.15. How to Use Probiotics for Digestive Health

Increase your daily intake of probiotics to begin reaping their benefits. Supplements or food sources can help you achieve this. Probiotics are found in abundance in many traditional foods from around the world. These foods include:

- sauerkraut
- kimchi
- kefir
- Greek yogurt
- kombucha

Increase your probiotic consumption gradually and be mindful of any side effects, both good and bad. Because probiotics come in a wide variety of strains, pay attention to how each one affects you uniquely. For this initial phase, a food and supplement journal would be very beneficial.

Here are some additional tips for taking probiotics:

- **Slowly Increase CFUs**: Doses between 1 billion and 10 billion colony forming units (CFUs) are advised daily. Gradually raise the CFU from 1 while monitoring your body's reaction. Each cup of kefir has 15–20 CFUs. Supplements ought to have a CFU list on the label.

- **Take Advice If It's Your First Time**: If fermented foods are new to you, find a friend or highly-rated restaurant to introduce you to them and how to best enjoy.
- **Take at Specific Times in Regards to Meals:** Take probiotics just before eating or with your meal, but *not* after you eat.
- **Try Natural Sources Before Supplements**: Probiotics are available as supplements, but it is usually better to consume them in their natural form. Probiotics can be found in a wide range of food sources, but supplements should identify each strain specifically.
- **Pay Attention to Labels and Quality**: The U.S. Food and Drug Administration (FDA) does not regulate supplements, so quality is not always a guarantee. Read labels closely and follow storage instructions. Try to pick out a trustworthy brand. If you can, find a doctor who can help you learn more.

3.2. USE OF PROBIOTICS IN GASTROINTESTINAL DISORDERS

Probiotics are being used more and more to treat a variety of illnesses, including respiratory infections, bacterial vaginosis, urinary tract infections, allergic diseases (atopic dermatitis, possibly allergic rhinitis), and dental caries prevention. Numerous gastrointestinal (GI) conditions are treated with probiotics.

3.2.1. Historical Perspectives

Because probiotics improve digestive health and longevity, they have been used medicinally for many centuries in various parts of the world. Probiotics are described as "live organisms which, when administered in adequate amounts, confer a health benefit on the host" by the WHO. Today's probiotics fall into two categories: bacteria, such as strains of *Escherichia coli* and LAB; yeast, including *Saccharomyces boulardii*, which is the most well-known among them (Table 3.1). Prebiotics are non-digestible food ingredients that encourage the growth or activity of bacteria in the GI tract that are good for the body. These bacteria can be found in onions, garlic, asparagus, leeks, artichokes, bananas, tomatoes, wheat, oats, soybeans, and other plants. Prebiotics include lactulose, inulin, psyllium, and other oligosaccharides. Synbiotics, like inulin and *Lactobacillus rhamnosus* GG or *Bifidobacter longum*, are a blend of probiotic and prebiotic. On the other hand, substances known as antibiotics either destroy or stop the growth of bacteria.

Table 3.1. *Sources of Probiotic Metabolites and heir Effects*

Metabolites	**Mechanism/Source of Production**	**Effects**
SCFA-acetate	Diet and endogenous production through acetyl-CoA	Increased satiety, weight loss, improved insulin sensitivity, and decreased pro-inflammatory cytokines
SCFA-propionate	Dietary fiber fermentation	Decreased weight gain, intestinal and hepatic gluconeogenesis, and decreased pro-inflammatory cytokine levels
SCFA-butyrate	Dietary fiber fermentation	The maintenance of mucosal integrity, the regulation of local and systemic immunity, anti-obesity effects, the stimulation of leptin synthesis, and the release of anorexigenic hormones
TMAO	Egg, milk red meat, and fish	Increased levels are associated with adverse cardiovascular disease
Tryptophan metabolites, indole derivatives, and tryptamine	Gut microbiota-derived	Anti-microbial effects, anti-obesity properties, appetite suppression, and slow gastric emptying
Primary bile acids (e.g., cholic acid)	Liver	The facilitation of fat digestion and nutrient absorption, and protection of the mucosal barrier
Secondary bile acids (e.g., deoxycholic acid and lithocholic acid)	Produced in colon	The inhibition of *Clostridioides difficile* spore germination and associated colorectal cancer and HCC
Polyamines (e.g., putrescine, spermidine, and spermine)	Lower GI—synthesized by the gut microbiome Upper GI—food-derived	The regulation of stress, antioxidant effects, and impact on cell proliferation and differentiation

*Abbreviations used are as follows: GI, gastrointestinal tract; HCC, hepatocellular carcinoma; SCFA, short-chain fatty acid(s); TMAO, trimethylamine N-oxide.

Source: Chandrasekaran P, Weiskirchen S, Weiskirchen R. (2024). Effects of probiotics on gut microbiota: An overview. International Journal of Molecular Sciences; 25(11):6022. https://doi.org/10.3390/ijms25116022.

The Russian Nobel Prize winner Ilya Metchnikoff likely published the first article on probiotics in Western medicine during the early 1900s. He reported on the long life spans of people in Eastern Europe who mostly consumed *Lactobacillus acidophilus* (LAB) fermented milk. According to his theory, the aging process is caused by toxic substances produced by proteolytic microbes in the colon. He suggested that consuming fermented milk would coat the colon with LABs, which lowers intestinal pH, suppresses proteolytic bacteria, and slows down the aging process. Consuming milk fermented by this "Bulgarian *Bacillus*," Metchnikoff and his adherents claimed health advantages.

During a severe outbreak of shigellosis in 1917, Alfred Nissle isolated a strain of *E. coli* from the feces of a soldier who did not develop enterocolitis. When treating acute cases of infectious intestinal diseases like salmonellosis and shigellosis, Nissle employed the E coli strain with notable success. *E. coli* Nissle 1917 is one of the few non-LAB probiotics that is still in use today. Probiotics have been researched and used in a range of medical conditions by scientists and medical professionals. More than 5,000 articles have been published in the medical literature in the past 10 years. Furthermore, because probiotics are unregulated, the use of them has increased significantly as a result of direct-to-consumer marketing (Borchers et al., 2009; Brenner et al., 2009; Moayyedi et al., 2010).

3.2.2. Mechanisms of Action

As a conduit between the host and the environment, the GI tract plays crucial roles. Approximately 10 trillion microbes of various species, weighing between 1–2 kg, are present in it. Of these species, only 300–500 can be studied and cultured *in vitro*. Because they are more invasive and have flagella, intestinal epithelial cells (IECs) can discriminate between pathogenic and nonpathogenic bacteria; however, the precise mechanisms underlying this ability remain unclear. Probiotics' specific mode(s) of action remain unclear at this time. Potential mechanisms to consider include: (i) modulation of GI immunity by altering inflammatory cytokine profiles and downregulating proinflammatory cascades or inducing regulatory mechanisms in a strain-specific manner; (ii) displacement of gas-producing, bile salt-deconjugating bacterial species and thus possibly inhibiting pathogenic bacterial adherence; (iii) alteration of bacterial flora by acidification of the colon by nutrient fermentation; (iv) enhancement of epithelial barrier function; (v) induction of μ-opioid and cannabinoid receptors in IECs; and (vi) reduction of visceral hypersensitivity, spinal afferent traffic, and stress response.

3.2.3. Probiotic Formulations and Delivery Systems

Probiotics come in a broad range of forms, including yogurts, milk, juices, and tablets and powders. Doctors typically suggest tablets and powders, but direct-to-consumer marketing actively promotes other formulations. A bacterial strain must meet specific requirements in order to be considered a probiotic, including being completely identified, safe for ingestion, adhering to the luminal mucosa, colonizing the gut, and having benefits for health that have been scientifically proven (Table 3.2). Delivering a probiotic in a formulation that holds up well in storage is important. Before entering the small intestine and colon, the bacteria must survive the bilious and acidic environment of the upper GI tract, and their colony size and viability must be consistent. It is challenging to determine probiotics' safety and efficacy with precision because their composition and quality are unregulated.

Table 3.2. Functions of Probiotics

Type	Strain	Function
Lactobacillus	*Lactobacillus acidophilus*	Used in the treatment of diarrhea and for relief from the symptoms of irritable bowel syndrome; can reduce cytokines to relieve inflammatory bowel disease, modulate immunity, lower cholesterol, relieve diarrhea, and alleviate cancer.
	Lacticaseibacillus casei	Prevention or treatment of diseases that disrupt the intestinal microbiota: diarrhea and constipation, relief from the symptom of irritable bowel syndrome, and gingivitis, and the anti-inflammatory response.
	Lactobacillus rhamnosus	Has anti-diabetic and anti-viral activity; used to treat obesity; has the ability to fight against pathogenic bacteria and fungi in the genitourinary tract, preventing the recurrence of urinary tract infections in postmenopausal women; prevents enteric colonization by *Candida* species and treats recurrent *Clostridium difficile*-induced colitis in children; and has an effect on symptoms of maternal depression and anxiety during the postpartum period.

	Lactiplantibacillus plantarum	Has the ability to prevent the production of endotoxin; improves of the symptom of irritable bowel; and has antimicrobial activity and cholesterol lowering activity.
	Lactobacillus reuteri	Treatment of gingivitis in pregnant women and chronic periodontitis.
Bifidobacterium	*Bifidobacterium infantis*	Can improve the symptoms of irritable bowel syndrome and inhibits the secretion of allergen induced IgE.
	Bifidobacterium adolescentis	Can reduce inflammation of the spleen and brain and changes the microbiota of cecum and colon.
	Bifidobacterium bifidum	Can reduce cholesterol; is used in the treatment of infant diarrhea; and has greater cytokine (IL-6) production and active phagocytic properties.
	Bifidobacterium longum	Used in treating diarrhea and provides relief from the symptoms of irritable bowel syndrome; modulates the immune system through IL-10 production.
Other Lactic Acid Bacteria	*Streptococcus thermophilus*	Can produce antioxidant compounds and mitigate the risk of some types of cancer; has anti-inflammatory, antimutagenic effects and stimulates the gut immune system; and is useful in inflammatory bowel disease.
	Enterococcus faecium	Modulates the Th2-mediated pathologic response.
Other Microorganisms	*Bacillus subtilis*	Used in treating diarrhea, in the eradication of *H. pylori*, and in the production of vitamin K.
	Bacillus coagulans	Can regulate the balance of intestinal microbiota and improve immunity; promotes the metabolism and utilization of nutrients; and is able to resist high temperatures and has acid and bile resistance.

	Saccharomyces boulardi	Used in treating diarrhea, ulcerative colitis, and a symptom of irritable bowel syndrome.
	Escherichia coli NISSLE 1917	Used in the treatment of intestinal diseases (diarrhea, inflammatory bowel disease, ulcerative colitis); can exhibit antagonistic effects on a variety of intestinal pathogenic bacteria; regulates the secretion of immune factors *in vivo*; and enhances the ability of host immunity.

Source: Maftei N-M, Raileanu CR, Balta AA, Ambrose L, Boev M, Marin DB, Lisa EL. (2024). The potential impact of probiotics on human health: An update on their health-promoting properties. Microorganisms; 12(2):234. https://doi.org/10.3390/microorganisms12020234.

3.2.4. Efficacy

Many questions related to efficacy, viability, most optimal dose, and method of delivery remain:

- It is unknown how many colony forming units (CFUs) of each bacterial strain should be supplied. Despite the variations in intestinal surface area, the doses utilized in human trials are based on those from animal studies. Studies on dose-response are generally deficient. Probiotic formulations that are sold commercially usually contain at least 10^6 CFUs, but they can contain as many as 10^{12} CFUs.
- By using fecal recovery studies, very few studies have actually shown that a probiotic that has been administered survives as it passes through the gut. Different *Bifidobacterium* species have varying growth requirements and acid tolerances, which can affect their ability to transfer to other probiotics and affect the rate of fecal recovery.
- The delivery method, for example yogurt as opposed to milk, could affect the number and health of bacterial colonies. Additionally, for two weeks, only one strain of *B. longum* could endure in fermented milk.
- Probiotics may have their effects on both living and nonliving bacteria, indicating that their immunomodulatory actions may be mediated by metabolic, secreted, or structural/cellular components. Furthermore, multiple experiments show that cell wall components play a major role in mediating the ability to induce the secretion of different cytokines.

- In various disease states, different probiotic species and genuses may have different physiological and immunological effects. Wagner et al. demonstrated that the effectiveness of various *Lactobacillus* species in preventing fungal sepsis in mice varies.
- Aging (>60 years) appears to alter the makeup of the colonic bacterial microflora. It's unclear if probiotic treatments for elderly patients should differ from those for younger patients.
- Combination probiotics may interact and have an impact on host intestinal flora differently than single probiotic preparations.
- The ideal length of time for probiotic therapy and the persistence of the effect are unknown. There is uncertainty regarding the duration required for a particular probiotic to colonize, modify the microflora, and affect immune function. The length of the treatments varied significantly in the human trials, which probably accounts for the discrepancies in the results that have been reported.

Therefore, it is not possible to generalize the findings of a single study using a single probiotic species, dosage, and formulation in a single disease state to other probiotic(s) as a whole. These are the causes of the highly erratic and challenging to understand data from probiotic clinical trials (Borchers et al., 2009; Brenner et al., 2009; Moayyedi et al., 2010).

3.2.5. Safety

Humans have been consuming probiotics in one form or another for more than a century, and they typically have a good safety record. According to a Finnish epidemiological study, there is no rise in *Lactobacillus* infections among healthy people in areas where the use of products containing *Lactobacillus* has been shown to have increased significantly. Research on probiotic supplementation in healthy individuals indicates that some probiotic strains may improve nonspecific immune responses, but there does not seem to be much of an impact on humoral and adaptive cellular immune responses.

However, there have been doubts expressed concerning the safety of probiotic administration when a serious illness is present. The most dreaded side effect associated with probiotic therapy is probiotic sepsis. One uncommon but well-researched cause of adult endocarditis is *lactobacillus*. Numerous reports of bacteremia in adults and children when probiotics are administered have been published in the literature. Furthermore, the literature has documented multiple instances of *Saccharomyces boulardii* fungemia, including two series in which the fungi spread to nearby patients who weren't taking the probiotic. It was previously believed that patients with immunocompromised conditions or those who had undergone intestinal surgery (jejunostomy) or chronic illnesses (valvular heart disease) were the source of this spread. It was believed that only one instance of probiotic sepsis resulted in death. A randomized, double-blind, placebo-controlled trial was performed using probiotic prophylaxis (six different strains of viable bacteria: *L. acidophilus, L. casei, L. salivarius, L. lactis, B. bifidum*, and *B. lactis*) in a total daily dose of 10^{10} bacteria orally twice daily for 28 days in patients hospitalized with severe acute pancreatitis. Compared to the placebo group (6%, relative risk (RR) 2.53, 95% confidence interval (CI) 1.22–5.25), the probiotics group had a 16% higher death rate but no decrease in infectious complications. Compared to none in the placebo group, nine out of the 152 patients in the probiotics group experienced bowel ischemia, which resulted in eight deaths.

Additional safety concerns are related to the unpredictable way that certain disease states alter the intestinal flora, which in turn affects immune modulation. For instance, a probiotic formulation's ability to exacerbate Crohn's disease (CD) in patients or exacerbated indomethacin-induced enteropathy by *Lactobacillus* GG in animal models. Even though these side effects seem uncommon, the safety profile of probiotics has to be thoroughly investigated, especially in hospitalized patients. Just as there are safety data on regulated medications, there are no official clinical trials evaluating the safety of probiotics. We currently have no other option than to rely on case reports, which is obviously not ideal.

3.2.6. Gastrointestinal Disorders for Which Studies of Probiotics Have Shown Some Benefit

It is believed that ingested probiotics contribute to the pathophysiology of numerous GI disorders by altering the deranged bowel flora or altering the patient's tolerance to their own commensal flora. The following conditions have been researched in relation to probiotic treatments the most frequently.

AAD may be prevented by coadministration of probiotics, as suggested by several randomized controlled trials (RCTs). Recent large-scale meta-analyses have all demonstrated that probiotics considerably reduced the incidence of AAD (RR 0.39–0.43). All probiotic categories, formulations, and treatment lengths had comparable effects. Probiotics that were most frequently used included *S. boulardii*, LABs, and various combinations of LABs. These were administered in doses ranging from 10^7 to 10^{11} for durations of 5–49 days, which typically corresponded with the course of antibiotic therapy. *S. boulardii*, *L. rhamnosus*, and various combinations of two distinct probiotics were shown to be the most effective in preventing AAD in one of the meta-analyses. Some specific preparations have received less research attention, which could account for the less significant findings regarding their efficacy. One randomized, double-blind, placebo-controlled trial in individuals over the age of 50 using combination *L. casei*, *L. bulgaricus*, and *S. thermophilus* twice daily during a course of antibiotics and for 1 week after the completion of antibiotic therapy showed reduction in the incidence of AAD.

Clostridium difficile colitis (CDC) has not consistently been shown to be prevented by probiotic cotreatment in a number of studies. The largest study conducted to date found that *S. boulardii*, at a dose of 2×10^{10} per day, in combination with metronidazole and vancomycin, significantly reduced the risk of CDC recurrence. Furthermore, Hickson and colleagues' above study demonstrated effectiveness in preventing CDC. Nevertheless, this benefit has not been validated by other research. A recent Cochrane review found that there was not enough data, even when probiotics were used in conjunction with vancomycin and metronidazole, due to the paucity of high-quality studies. But in reality, a lot of doctors advise probiotics following antibiotic therapy, especially for patients who have experienced CDC relapse.

Infectious diarrhea in both adults and children may be shortened by the use of probiotics. A systematic review of trials in active infectious diarrhea suggests a 30-hour reduction in the duration of symptoms. With a risk reduction of 0.66 (95% CI 0.55–0.77) at 3 days, probiotics were found to lower the risk of persistent diarrhea when compared to placebo or no probiotics, based on a Cochrane review of 23 studies involving nearly 2000 participants, 352 of whom were adults. Of the 23 studies that tested probiotics, 18 of them used LABs, while two of them used *S. boulardii*.

IBD treatment with antibiotics is fraught by variable efficacy. The pathophysiology and maintenance of IBD may be aided by the recent discovery of genetic polymorphisms (NOD2/CARD15) in CD that are involved in bacterial peptidoglycan recognition and may be responsive to changes in enteric flora.

- **Ulcerative Colitis (UC)**: Few randomized controlled trials (RCTs) have been conducted on probiotics in the induction and remission of UC. The majority have been carried out in a comparatively small number of patients and use various probiotic formulations (Table 3.3). The first and largest controlled trial published by Remnacken to date demonstrated that steroids, mesalazine, and antibiotics were the most effective treatments for inducing remission, with no additional benefit from *E. coli* Nissle 1917. Three more trials with a small patient population, brief treatment durations, and different care standards demonstrated improvements in cytokine profiles as well as other disease activity metrics. Mallon and colleagues conducted a systematic review using the Cochrane database; however, because of variations in probiotics, outcomes, and methodology, a formal meta-analysis was not feasible. Nevertheless, they came to the conclusion that probiotics alone did not increase the rate of remission. However, this analysis showed a reduction in disease activity in mild to moderately severe UC. A second systematic review published recently also suggested a similar efficacy profile between probiotics and anti-inflammatory agents. With regard to maintenance of UC remission, probiotics have been tested in a larger number of patients (Table 3.3). One trial by Kruis and colleagues tested *E. coli* Nissle 1917 and found no difference in relapse rates in patients on a probiotic *versus* mesalamine. When Zocco et al. compared *Lactobacillus* GG with mesalamine versus a combination of the two, they also found no difference in relapse rates at 6 or 12 months. It does seem that the probiotic-taking patients took longer to relapse. In the short-term trials, probiotics may be just as effective at maintaining remission as mesalamine, according to all of these studies.

Table 3.3. *Prevention and Treatment of Pediatric Gastrointestinal Infections with Probiotics (The Numbers in Brackets Are the Back References in the Cited Source)*

Participants	Probiotic Strain/ Treatment	Time	Primary Outcome	Reference
986 children	*L. casei rhamnosus*, *L. rhamnosus* T cell-1, and a mixture of strains	7 months	*L. casei rhamnosus* reduced respiratory infections, whereas multiple probiotic supplementation reduced the gastrointestinal disease. *L. rhamnosus* T cell-1 decreased the incidence of bacterial infection at 7 months	Song-Lin et al. (2009)
281 children	*L. rhamnosus* GG	3 months	Only prevention of upper respiratory tract infections	Hojsak et al. (2010)

742 children	*L. rhamnosus* GG	1 week	*L. rhamnosus* GG treatment significantly reduced the risk for gastrointestinal infections, vomiting, and episodes of gastrointestinal infections	Hojsak et al. (2010)
124 children	*L. rhamnosus* GG	1 month	*L. rhamnosus* GG decreased diarrheal episodes and restored normal intestinal permeability	Kulandaipalayam et al. (2014)
90 children	*L. rhamnosus* GG plus vitamins B, C and zinc	2 weeks	Treatment reduced incidence of gastrointestinal infections and length of hospitalization	Bruzzese et al. (2016)
80 children	*L. salivarius* CECT5713	6 months	*L. salivarius* CECT5713 decreased incidence of diarrhea and respiratory infections	Maldonado et al. (2010)
215 children	*L. fermentum* CECT5716 plus GOS	6 months	Synbiotic administration prevented community-acquired gastrointestinal infections in infants	Maldonado et al. (2012)
91 children	*L. fermentum* CECT5716 plus GOS	3 years follow-up	All variables measured were similar compared with placebo	Maldonado et al. (2015)
109 children	*L. rhamnosus* GG	5 years follow-up	A decrease in the incidence of acute gastroenteritis was not detected	Scalabrin et al. (2017)
494 children	RCC plus *L. casei* CRL431, or RCC plus *L. reuteri* DSM17938	6 months	Incidence of all reported diarrhea and diarrhea incidence in children with a lower nutritional status were significantly lower in the *L. reuteri* group	Agustina et al. (2012)

126 children	*L. paracasei* CBA L74	3 months	Probiotic treatment decreased the number of episodes of acute gastroenteritis	Corsello et al. (2017)
638 children	*L. casei* DN-114 001	3 months	*L. casei* DN-114 001 decreased gastrointestinal infections	Merenstein et al. (2010)
106 children	*L. reuteri* DSM 17938	1 week	No effects	Wanke et al. (2012)
599 children	*L. casei* CNCM I-1518	3 months	No effects	Prodeus et al. (2016)
156 children	*B. lactis* B94 plus inulin	1 week	Synbiotic treatment decreased the duration of diarrhea	Islek et al. (2014)
727 children	*B. animalis* subsp. lactis BB-12	1 week	No effects	Hoksak et al. (2015)
67 children	*B. animalis* subsp. lactis BB-12	2 years follow-up	No effects	Taipale et al., (2016)
290 children	*B. animalis* subsp. lactis and *L. rhamnosus* GG	6 months	No effects	Laursen et al. (2017)

Abbreviations: GOS, galactooligassacharides; mixture of strains, three bifidobacteria, seven lactobacilli, S. thermophilus, and E. faecium; RCC, regular calcium content.

Source: Plaza-Díaz J, Ruiz-Ojeda FJ, Gil-Campos M, Gil A. (2018). Immune-mediated mechanisms of action of probiotics and synbiotics in treating pediatric intestinal diseases. Nutrients; 10(1):42. https://doi.org/10.3390/nu10010042.

- **Crohn's Disease**: The research on remission induction and maintenance in CD is inconsistent and challenging to understand (Table 3.4). This is caused in part by the small number of patients included in the trials and the imprecise definition of the degree of inflammatory involvement in the patients who were studied. Moreover, a paucity of research has explored the potential additive benefit of probiotics for active CD. Probiotics had no additional benefit over steroids and antibiotics in inducing remission, according to a study involving just 11 patients. In an open-label trial, probiotics (*B. breve*, *B. longum*, and *L. casei*) and prebiotics (psyllium) were given concurrently to 10 patients who were resistant to prednisolone and aminosalicylates. 6 out of 10 patients showed a full response with no negative side effects. More controlled research has been done on keeping CD patients in remission in adults (Table 3.4), but the majority of these studies have not demonstrated any advantages to taking probiotics.

Data are even more robust on the prevention of relapse following surgical intervention, but again probiotics fail to prevent endoscopic or clinical recurrence (Table 3.4). Several meta-analyses and systematic reviews have shown that probiotics were ineffective in maintenance of remission in CD.

Table 3.4. The Effectiveness of Probiotics in the Treatment of IBD (The Numbers in Brackets Are the Back References in the Cited Source)

Strain	**IBD**	**Research Model and Study Scheme**	**Results**	**Reference**
Bifidobacterium infantis (Baby Life, Solaray)	CD-like	Rat model (TNBS) 10 days supplementation with 0.205 g of *B. infantis* dissolved in 1.0 mL distilled water prior colitis. Colitis was induced by enema of 1.0 mL 5% (w/v) TNBS which lasted for 7 days.	Reduction in the symptoms, weaker damage to the mucosal architecture, protective function for mucus goblet cells and the epithelial cell layer.	Vanderpool et al. (2008)
Summary: a beneficial effect was observed				

Bifidobacterium bifidum PRL2010	CD-like	Murine model (TNBS) 7 days oral supplementation of 10^9 of *B. bifidum* PRL 2010. Colitis (TNBS 2.5 mg/mice) was induced in the 5th day of probiotic strain feeding.	Reduction in the edema, reduction in the macroscopic damage and histological scores, reduction in weight-loss, anti-inflammatory effect.	Vaughan (1965)
Summary: a beneficial effect was observed				
Bifidobacterium bifidum 231	CD-like	Rat model (TNBS) Colitis was induced by 31 mg/kg of TNBS. 14 days supplementation of 1.4×10^{11} CFU/rat/day *B. bifidum* (in saline) after colitis induction	Reduction in the edema, reduction in the macroscopic damage and histological scores, reduction in weight-loss, anti-inflammatory effect.	Wagner et al. (1997)
Summary: a beneficial effect was observed				
Lactobacillus plantarum species 299	CD-like	Rat model (TBSN) Colitis was induced by 30 mg (0.6 mL of 5% aqueous solution) of TNBS. 7 days supplementation of 10^9 colony forming units (CFU) of *Lactobacillus plantarum* (in oat fiber) after colitis induction.	No beneficial effects on the rat's gut permeability, weight changes, colon microscopic scores, and the level of blood albumins.	Waller et al. (2011)
Summary: Lack of positive effect				

Bifidobacterium animalis subsp. *lactis* BB12	UC-like	Murine model (DSS) 7 days supplementation (twice a day) with 1.2×10^{10} CFU *Bifidobacterium animalis* subsp. *lactis* BB12 by oral gavage prior to colitis. Colitis was induced by 3% DSS added to drinking water for 6 days.	Protection against a reduction in colon length, better picture of the colon histology, reduction in apoptosis in the epithelial layer, decrease in the level of TNF-α.	Yan et al. (2007)
Summary: a beneficial effect was observed				
Bifidobacterium longum subsp. *infantis* BB-02	UC-like	Murine model (DSS) 10 days supplementation (by oral gavage) once a day with 0.1 mL of a suspension containing 9.0 $\log_{10}$ CFU/mL in phosphate-buffered saline prior to colitis induction and continued during its development for next 7 days. Colitis was induced by DSS 3.5% (w/v) in drinking water ad libitum for 7 days.	Reduction in the clinical symptoms of the disease, protection of the colonic structure, reduction in edema.	Zigra et al. (2007)
Summary: a beneficial effect was observed				
Bifidobacterium lactis from fermented milk (Activia; Danone)	UC-like	Murine model (T-bet$^{-/-}$Rag2$^{-/-}$) 100 mg of diary product was orally instilled daily, and additional 100 mg per mouse in its cage was provided for consumption.	Reduction in severity of colitis and inflammation in an early stage, decrease in the level of *Enterobacteriaceae* (colitogenic).	Zocco et al. (2006)

Summary: a beneficial effect was observed				
Lactobacillus delbrueckii	UC-like	Murine model (DSS) Colitis was induced by the addition of 3% (w/v) DSS in the drinking water for 7 days. Administration of probiotic (5 × 10^9 bacteria/mouse/day started 1 day) before colitis induction and lasted until sacrifice.	Modulation of the NF-kB pathway, reduction in the inflammatory state.	Santos Rocha, C. et al 2012
Summary: a beneficial effect was observed				
VSL#3 (*L. paracasei*, *L. plantarum*, *L. acidophilus*, *L. delbrueckii* subspecies *bulgaricus*, *B. longum*, *B. breve*, and *B. infantis, Streptococcus thermophilus*)	UC-like	Murine model (DSS) 8 days supplementation (by oral gavage) with 0.1 mL of a suspension containing 5 × 10^{10} probiotic CFU/kg of body weight dissolved in saline solution after colitis induction. Colitis was induced by 5% DSS (w/v) in drinking water for 8 days.	Discrepancy in results (two different batches of VSL#3, contrary data); VSL#3 batch A: reduction in macroscopic scores, intestinal permeability, -reduction in expression of TNFα, IL-1β, IL-6 mRNAs, increase in the expression of TGFβ, IL-10, occludin, zonula occludens-1 (ZO-1) mRNAs, shift of colonic macrophages from a M1 to M2 phenotype-lack of effect in VSL#3 batch B.	Biagioli, M. et al 2017

	CD-like	Murine model (TBNS) Eight days supplementation (by oral gavage) with 0.1 mL of a suspension containing 5×10^{10} probiotic CFU/kg of body weight dissolved in saline solution after colitis induction. Colitis was induced by 1 mg of TNBS fasted for 12 h.		Biagioli, M. et al 2017
Summary: positive effect only one batch of the same product				
Bifidobacterium animalis spp. *lactis* Bl 5764	CD-like	Murine model (TNBS) 5 days supplementation (by oral gavage) of a 5×10^8 CFU/day/mice prior to colitis induction and continued during its development for next 2 days. Colitis was induced by TNBS (110 mg/kg, dissolved in 0.9% NaCl/ethanol (50/50 v/v)).	Lower body weight-loss, better macroscopic indicators of inflammation (Wallace scores), histopathological analysis (Ameho score), and level of lipocalin-2, promotion in the bone marrow-derived dendritic cell maturation and IL-17A secretion.	Hrdý, J. et al 2020

Lactobacillus reuteri Lr 5454	CD-like	Murine model (TNBS) 5 days supplementation (by oral gavage) of a 5×10^8 CFU/day/mice prior to colitis induction and continued during its development for next 2 days. Colitis was induced by TNBS (110 mg/kg, dissolved in 0.9% NaCl/ethanol (50/50 v/v)).	Lower body weight-loss, better macroscopic indicators of inflammation (Wallace scores), histopathological analysis (Ameho score), and level of lipocalin-2; greater involvement in the development of tolerogenic DC, induce Tregs population and expression of Reg3b in a NOD2-independent manner.	Hrdý, J. et al 2020
Summary: a beneficial effect was observed				
Saccharomyces boulardii (Floratil®)	CD	Human model Patients with CD in remission (based on Crohn's disease activity index) were supplemented with *S. boulardii* about 4×10^8 cells every 8 h as an oral capsule formulation during 3 months.	Help in maintaining remission and bowel sealing.	Garcia Vilela, E. et al 2008
Summary: a beneficial effect was observed				

Bifidobacterium breve strain Yakult, *Bifidobacterium bifidum* strain Yakult, *Lactobacillus acidophilus* strain Yakult (Yakult Co. Ltd. Japan)	UC	Human model Patients with mild to moderate active UC were supplemented with 100 mL/day (10 billion cells) of bifidobacterial-fermented milk for 3 months.	Promising usefulness in sustaining the remission phase, improvement in clinical activity index score and histological scores.	Kato, K. et al 2004
Summary: a beneficial effect was observed				
Escherichia coli Nissle1917 (Mutaflor 100 mg; Ardeypharm GmbH, Herdecke, Germany)	UC	Human model Randomized, double blind, double dummy trial, patients with UC remission were supplemented with 2.5–25 × 10^9 viable bacteria for 12 months.	Promising behavior in sustaining the remission phase, prevention from inflammatory state.	Kruis, W. et al 2004
Summary: the beneficial effect was demonstrated				
Lactobacillus fermentum, Lactobacillus delbruekii. (Lacteol Fort; Rameda, Egypt)	UC	Human model Patients with mild to moderate UC assessed by Mayo score were supplemented with 10 billion CFU of probiotic cells (powder to dissolve in 50 mL fresh water) for 8 weeks.	Decrease in the IL-6 and TNF-alpha levels and lowering regulation of NF-kB.	Hegazy, S.K. et al 2010
Summary: a beneficial effect was observed				
Bifidobacterium infantis 35624	UC	Human model Randomized, double-blind placebo-controlled studies, patients with mild to moderate active UC (based on a clinical activity index) were supplemented with 1 × 10^{10} CFU viable probiotic cells for 8 weeks.	Reduction in the levels CRP and TNF-α in both gastrointestinal and non-gastrointestinal inflammatory disorders, but did not particularly affect UC disease.	Groeger, D. et al 2015

Summary: Lack of effect for UC group, positive effect for gastrointestinal, and non-gastrointestinal inflammatory disorders group				
Bifidobacterium breve strain Yakult	UC	Human model Patients with UC (active and inactive) were supplemented with 1 g of the freeze-dried powder containing probiotic (10^9 CFU/g) for 1 year.	Better endoscopic scores, decrease of MPO level, modulation of luminal environmental factors such as intestinal microflora and pH.	Ishikawa, H. et al 2011
Summary: a beneficial effect was observed				
Bifidobacterium breve strain Yakult, *Lactobacillus acidophilus* strain Yakult (Yakult fermented milk (Mil–Mil))	UC	Human model Multileft, randomized, placebo-controlled, double-blind parallel-group study; patients with UC in remission were supplemented with 100 mL/day (10 billion cells of *Bifidobacterium* and 1 billion of cells of *Lactobacillus*) of fermented milk for 12 months.	No beneficial effect.	Matsuoka, K. et al 2018
Summary: no beneficial effect was observed, the study was discontinued				
Lactobacillus acidophilus strain LA-5 and *Bifidobacterium animalis* subsp. *lactis* BB12 (Probio-Tec AB25)	UC	Human model Randomized, double blind, placebo-controlled study; patients with UC in remission were supplemented with 1.5×10^{11} CFU daily (2 capsules 3 times daily) for 52 weeks.	Maintaining remission in colitis.	Wildt, S. et al 2011
Summary: a beneficial effect was observed				

Bifidobacterium longum 536 (Morinaga Milk Industry Co. Ltd, Tokyo, Japan)	UC	Human model A randomized, double-blinded, placebo-controlled multileft trial study; patients with mild to moderate UC (based on disease activity index) were supplemented with 2–3 × 10^{11} freeze-dried viable probiotic capsule 3 times daily for 8 weeks.	Decrease in the disease activity index and downscale the rectal bleeding, clinical remission.	Tamaki, H. et al 2016
Summary: a beneficial effect was observed				
Lactobacillus salivarius, Lactobacillus acidophilus, Bifidobacterium bifidum strain BGN4; Acronelle®, Bromatech SRL, Milan, Italy)	UC	Human model Patients with moderate to severe UC (based on disease activity index) were supplemented with probiotic blend for 24 months.	Reduction in recovery time, weaker activity of the disease, better endoscopic picture.	Palumbo, V.D. et al 2016
Summary: a beneficial effect was observed				
VSL#3	UC	Murine model UC associated carcinogenesis model was based on a single injection of 12.5 mg/kg body weight azoxymethane intraperitoneally, 1 week later 2.5% DSS was added to drinking water for 5 days, followed by 10 weeks and 2 days of regular drinking water. Probiotic mixture (1.5 × 10^9 CFU/mice) was supplemented alone or together with mesalazine.	Decrease in the level of TNF-α and IL-6, reduction of number of pathogenic microbiota, increase in the population of *Bifidobacterium* and other non-pathogenic species in the intestinal mucosa.	Wang, C.S.E et al 2018

Summary: preventive effect for UC associated carcinogenesis				
VSL#3	UC	Human model Patients with mild to moderate, active UC (based on Activity Index) were supplemented with the probiotic mixture twice daily for 12 weeks.	Improvement in rectal bleeding and stool frequency, mucosal appearance and overall physician's evaluation.	Sood, A. et al 2009
Summary: a beneficial effect was observed				
VSL#3	UC	Human model A multileft, double-blind, randomized, placebo-controlled, parallel study, patients with mild to moderately active UC (based on Activity Index) were supplemented with the probiotic mixture for 8 weeks in addition to standard therapy.	Reduction in UCDAI scores and frequency of rectal bleeding; – no differences in parameters such as the physician's rate of disease activity, or endoscopic scores.	Tursi, A. et al 2010
Summary: a beneficial effect was observed				
VSL#3	UC	Human model, children population Patients with mild to moderately active UC (based on activity index) were supplemented with the probiotic twice daily for 8 weeks with a dose of probiotic based on their age (from one-half sachet to two and one-half).	Remission in colitis, improvement in microbiota composition, decrease in level of IFN-γ, TNF-α, CRP, ESR.	Huynh, H.Q. et al 2009
Summary: a beneficial effect was observed				

Faecalibacterium prausnitzii	CD n.a	Murine model (TNBS) A double-blind controlled trial study, 5 days before colitis induction mice were supplemented with 10^9–10^{10} CFU bacterial suspension or bacterial medium. Colitis was induced by TNBS (100 mg/kg body weight), which was administrated intrarectally. The observation took 20 days. Cell line CaCo-2	Anti-inflammatory effect, blocking of NF-kB pathway and IL-8 production, anti-inflammatory effect.	Sokol, H. et al 2008
Summary: a beneficial effect was observed				
Bifidobacterium lactis	Cancer model	Murine model 6 days of supplementation with probiotic (in different dose) prior to colitis. An acute colitis was induced by 3.5% DSS in drinking water for 7 days. Colitis associated cancer was induced by azoxymethane (10 mg/kg) prior to 5 days DSS challenge. Cell line model (HT-29) Cell line HT-29 was incubated with the different concentration of *Bifidobacterium* lactis.	Inhibition of NF-kB and NF-kB-regulated genes in epithelial cells and prevention meaning for the acute colitis and cancer model, reduction in number and size of the tumors.	Kim, S.W. et al 2010

Summary: a beneficial effect was observed				
Lactobacillus acidophilus Lactobacillus fermentum	Cancer model	Murine cancer model 12 weeks supplementation with probiotic (0.5 × 10^{10} CFU of each strain) among $Apc^{Min/+}$ mice. Cell line model (CaCo-2) Bacteria (alone and as a mixture) prepared in simulated artificial intestinal juice were incubated with Caco-2 (up to 72 h).	Reduction in cancer cells proliferation, increase in apoptosis level, protection of normal colon cell growth from toxic treatment.	Kahouli, I. et al 2017
Summary: a beneficial effect was observed				
VSL#3	Cancer model	Murine cancer model ($IL10^{-/-}$) Supplementation with 1.2 billion bacteria per mouse/ day) or conjugated linoleic acid. Cancer was induced by azoxymethane, DSS (the first step) and a single dose of 5 × 10^7 CFU *Helicobacter typhlonius* by oral gavage (the second step cancer induction). Observation took 68 days.	Shorter recovery time, lower disease severity in an active phase of cancer; VSL#3 treatment- higher mRNA expression of TNF-α, increase in the angiostatin level and T-cells subpopulations; CLA treatment- decreased the level of COX-2.	Bassaganya-Riera, J. et al 2012
Summary: a beneficial effect was observed				

VSL#3	Cancer model	Murine cancer model/ IL-10$^{-/-}$ Mice with IL-10 deficient were supplemented by VSL# from the day of cancer induction for 17 weeks (once a day without the weekends). Cell line model (CaCo-2) Bacteria (alone and as a mixture) prepared in simulated artificial intestinal juice were incubated with Caco-2 (up to 72 h).	No protection against inflammatory processes and tumor development, increase in the tumor penetrance, worsening histologic dysplasia scores, extension of *Clostridium* population.	Arthur, J.C et al 2013
Summary: lack of beneficial effect, worsening of the neoplastic processes				
VSL#3	Cancer model	Rat cancer model (TNBS) Supplementation with VSL#3 1 week before the colorectal cancer induction by TNBS. Then rats were supplemented with probiotic in drinking water until end of experiment (17 weeks).	Delay in the carcinoma growing processes, improvement in the histological picture of the colon, increase in the level of angiostatin vitamin D receptor (VDR).	Appleyard, C.B. et al 2011
Summary: a beneficial effect was observed				
Propionibacterium freudenreichii, Isolated: SlpB, SlpE, two proteins with SLH domains, HsdM3	n.a	Proteomic and transcriptomic study	Potential connection with IL-10 increase and anti-inflammatory value.	Deutsch, S.M et al 2017

Summary: anti-inflammatory effect as a results of combination cytoplasmatic and surface protein				
Lactobacillus rhamnosus GG, isolated antigens: surface layer protein, genomic DNA and unmethylated cytosine-phosphate-guanine-containing oligodeoxynucleotides, alone or in combination.	n.a	Cell line RAW 264.7 Cell line was incubated with bacterial cells, components, or their combination (2 h), then was challenged with lipopolysaccharide (0.5 h).	Suppression in the inflammatory paths, inhibition of TLR, MAPK and NF-κB signaling pathways.	Qi, S.R et.al 2020
Summary: anti-inflammatory properties were observed				
bacteria's cells-free growth medium, *Lactobacillus acidophilus, Lactobacillus casei, Lactococcus lactis, Lactobacillus reuteri*, and *Saccharomyces boulardii*	n.a	Cell line HT-29 After LPS challenge (4 h), cell line was incubated with cell free bacterial medium (18 h.)	Anti-inflammatory properties, modulation of the level of IL-10, IL-1β, TNF-α, PGE-2, IL-8.	De Marco, S. 2018
Summary: anti-inflammatory properties were observed				
Bifidobacterium, Lactobacillus acidophilus, Enterococcus (Bifico, Shanghai Sine Pharmaceutical)	CD-like	Murine model (IL-10-deficient, DSS) Colitis was induced by 4% DSS in drinking water by 14 days. Mice were supplemented with 3×10^7 CFU probiotic in all study groups (with colitis or not) for 14 days.	Increase in the number of Treg, decrease in the total number of T cells in the colon and the peripheral blood, positive influence on the tight junctions.	Zhang, Y.et al 2018

Summary: a beneficial effect was observed				
Lactobacillus rhamnosus GG.	UC	Case report A female patient affected by UC with severe active pan-colonic involvement.	Symptoms of bacteremia.	Meini, S et al 2015
Summary: probiotic strain as a cause of bacteremia				
VSL#3 (Seaford Pharmaceuticals)	n.a	Rat model 7 days intragastrical supplementation with 3×10^9 bacteria.	Influence on the mucus structure, stimulation of expression of the Muc2 gene, increase in secretion of the non-mucin glycoprotein, improvement in the permeability of the intestinal barrier.	Caballero-Franco et al 2007
Summary: a beneficial effect on secretion and mucus within epithelial barrier				
VSL#3	UC-like	Murine model (*$Muc2^{-/-}$* and *$Muc2^{+/+}$; DSS)* 15 days of supplementation with VSL#3. Colitis among $Muc2^{-/-}$ group was induced by 1% DSS added to drinking water for 3 days.	Induction of mucus secretion in the crypts' goblet cells in the colon, reduction in the wall thickness and MPO level – lack of protection from colitis severity.	Kumar, M. et al 2017
Summary: a beneficial effect on mucus secretion and oblet cells, lack of usage in colitis severity protection				

VSL#3	n.a	Cell line Caco-2 Cell line monolayer incubated with 10^8 CFU/ mL (4 h).	Discrepancy in results (two different manufacturers of VSL#3, contrary data) – Italy-made product: increased the permeability of the barrier, and decreased the ZO-1/occludin expression – US-made product: increased the occludin level, pretreatment with VSL#3 prevented the heat-induced epithelial barrier integrity loss.	Palumbo, P.et. al 2018
Summary: different usefulness of the same product but produced by two different manufacturers				
Lactobacillus casei DN-114 001, lysate	CD-like	Murine model (DSS) Mice were supplemented with 6 x 10^8 CFU of heat-killed bacteria, by gavage. The administration of bacteria was repeated every 7 days (4 doses together) prior to colitis. Colitis was induced 7 days later by 3% DSS dissolved in tap water for 7 days.	Protective behavior of probiotic only on the BALB/c mice, increase in the barrier function by the upregulation of ZO-1, increase in the amount of Treg, decrease in the level of proinfla-mmatory factors, TNF-α, INF-γ, IL-10, influence on microbiota composition;	Zakostelska, Z et al 2011
Summary: a beneficial effect was observed				

Lactobacillus rhamnosus GG, surface layer protein HM0539	n.a	Cell line CaCo-2 Caco-2 cell line was incubated with HM0539 (50 ng/mL) for 12 h, then cell line was stimulated using TNF-a, (10 ng/mL) or lipopolysaccharide (1 mg/mL) for 6 h.	Increase in the level of the tight junction, increase in mucin secretion.	Gao, J. et al 2019
Summary: a beneficial effect was observed				
Lactobacillus plantarum, isolated MIMP	n.a	Murine model (DSS) Mice were supplemented with MIMP (0.1 μg/20 g) for 7 days prior colitis. Colitis was induced by DSS administered in drinking water for 7 days. Cell line Caco-2 Cell line was incubated with MIMP in different concentrations for 48 h.	Reduction in the permeability, increase in expression of JAM-1, occludin, ZO-1.	Yin, M 2018
Summary: a beneficial effect was observed.				

Source: Jakubczyk D, Leszczyńska K, Górska S. (2020). The effectiveness of probiotics in the treatment of inflammatory bowel disease (IBD)—A critical review. Nutrients; 12(7):1973. https://doi.org/10.3390/nu12071973.

- **Pouchitis:** The prevention and treatment of pouchitis provides the best evidence for the use of probiotics in IBD. Pouchitis, or acute and chronic inflammation of the ileal reservoir, is the most common long-term complication of proctocolectomy with ileal pouch–anal anastomosis, occurring in up to 20% of patients at one year. Research on the pouch's microflora has shown a lack of Streptococcal species. This has prompted several prospective controlled clinical trials involving the use of the probiotic VSL#3 for the treatment and prevention of pouchitis over a 9–12-month period. These studies consistently demonstrate a decline in the frequency of inflammatory response relapses. A 69% remission rate was observed in one uncontrolled trial involving patients treated with VSL#3 for mildly active pouchitis. In contrast, a three-month trial on a single species of *Lactobacillus* GG revealed no efficacy.

IBS is a multisymptom GI disorder with unclear etiology and pathogenesis. Changes in GI microflora in IBS patients have been reported by a number of investigators. Small intestinal bacterial overgrowth (SIBO) prevalences in IBS have been reported to vary recently. Probiotics have been empirically used to treat the challenging symptoms of IBS, such as flatulence and bloating, which have been linked to potential changes in the intestinal microbiota. Acute gastroenteritis may precede a bout of postinfectious IBS, indicating that changed gut microbiota or the induction of a changed inflammatory or immune state may result in altered bowel function and IBS symptoms. There have been reports of an increase in pro-inflammatory cytokines and lymphocytes.

Brenner et al. conducted a recent analysis of 16 RCTs involving IBS patients who met the Rome II or Manning Criteria and who were given either a single probiotic or a combination of probiotics in place of a placebo. *Bifidobacterium infantis* 35624 showed promise in two well planned RCTs. There was no indication of a rise in unfavorable events, and both overall and specific IBS symptoms (abdominal pain, bloating, incomplete evacuation, intestinal gas, straining, and bowel function) considerably improved. In suitably constructed RCTs, no other probiotic—including isolated *Lactobacillus* species—exhibited a discernible improvement in the symptoms of IBS. Probiotics were shown to have a statistically significant effect on reducing IBS symptoms in another systematic review of the literature, with a NNT of 4 (95% CI 3–12.5). *Lactobacilli* was found in almost all probiotic combinations, but a continuous data meta-analysis indicated that it had no effect. Bifidobacteria was also present. This suggests that in probiotic combinations, Bifidobacter might be the active ingredient.

IBS patients have been found to have higher levels of interleukin 6 (IL-6), IL-6R, IL-1Beta, and tumor necrosis factor alpha (TNF-α), as well as a lower IL-10/IL-12 ratio when compared to controls. These findings could indicate a possible link between IBS and increased secretion of pro-inflammatory cytokines. Plasma cytokine levels, however, might originate from activated immune cells in the spleen or liver rather than reflecting the expression or amounts of cytokines in the mucosa of the colon wall. A recent study by Macsharry and colleagues confirmed the imbalance between IL-10 and IL-12 found in peripheral blood mononuclear cells at the mucosal level. This finding raises the possibility that the imbalance is an underlying phenotype in IBS and a potential biomarker for a subset of IBS patients. *B infantis* has been demonstrated to raise the IL-10/IL-12 ratio in IBS patients, pointing to a potential mode of action for this probiotic. Studies have also looked at the impact of probiotics on other gastrointestinal disorders, such as lactose intolerance, *Helicobacter pylori* infection, microscopic colitis, diverticulitis prevention and treatment, and even colon cancer prevention (Borchers et al., 2009; Brenner et al., 2009; Moayyedi et al., 2010).

3.3. BENEFITS OF PROBIOTICS IN DIGESTIVE HEALTH

Probiotics are good bacteria that may aid in promoting a healthy digestive system and provide additional health advantages. Live microorganisms known as probiotics are ingested through fermented foods or supplements. An increasing number of studies demonstrate the connection between general health and illness and the balance or imbalance of bacteria in your digestive system. Probiotics have been connected to numerous health advantages and help maintain a balanced population of good bacteria in the gut. Benefits for immunological system performance, digestive health, and weight loss are among them.

Probiotics may help manage:

- IBS
- high cholesterol
- nutrient absorption
- diarrhea
- constipation
- lactose intolerance
- weight gain

Your quality of life and general sense of wellbeing can be enhanced by lessening the negative effects of any of the aforementioned problems. Probiotics in your gut have advantages beyond regular, healthy bowel movements. Enhancements in immune response and mental well-being have been connected to a healthy digestive system. Their advantages for dermatological and oral health are currently being researched. Probiotics can lessen muscle strain and help your body recover from exercise by assisting in the absorption of proteins. As worries about infections resistant to antibiotics grow, probiotics may be effective defense and healing tools that complement our bodies' natural processes.

3.3.1. Probiotics Help Balance the Friendly Bacteria in Your Digestive System

Probiotics contain beneficial bacteria. These are live microorganisms that, when eaten, may have health benefits. These advantages are assumed to arise from probiotics' capacity to rectify the innate equilibrium of intestinal microbiota. An imbalance is defined as an excess of pathogenic bacteria relative to beneficial bacteria. It can be brought on by sickness, drugs like antibiotics, an unhealthy diet, and other factors. Digestive disorders, allergies, mental health concerns, obesity, and other conditions are examples of consequences. Probiotics are typically consumed as supplements or found in fermented foods. Furthermore, the majority of people seem to find them safe.

3.3.2. Probiotics Can Help Prevent and Treat Diarrhea

The ability of probiotics to either prevent or lessen diarrhea is well known. One of the most frequent side effects of antibiotic use is diarrhea. It happens because the balance of good and bad bacteria in the gut can be adversely affected by antibiotics.

Probiotic use may be linked to a lower incidence of diarrhea brought on by antibiotics, according to several studies. In one study, researchers discovered that probiotic use resulted in a 60% reduction in C difficile infection (CDAD). Other types of diarrhea that are not brought on by antibiotics can also be helped by probiotics. Certain probiotic strains have been shown to shorten the duration of diarrhea by an average of 25 hours, according to a comprehensive review of 35 studies. Additionally, probiotics lessened the chance of traveler's diarrhea. Depending on the kind and dosage of the probiotic taken, effectiveness varies. Most frequently, strains like *Saccharomyces boulardii*, *Lactobacillus rhamnosus*, and *Lactobacillus casei* are linked to a lower risk of diarrhea.

3.3.3. Probiotics Can Help Reduce Symptoms of Certain Digestive Disorders

IBDs, such as UC and Crohn's disease, affect more than a million people in the United States. Probiotics derived from specific strains of *Lactobacillus* and *Bifidobacterium* have been shown to alleviate mild UC symptoms. Probiotics do not seem to have much of an impact on Crohn's disease symptoms, though. Probiotics, however, might be advantageous for other gastrointestinal issues. According to preliminary research, they might be beneficial for IBS symptoms (Borchers et al., 2009; Brenner et al., 2009; Moayyedi et al., 2010).

REFERENCES

1. Albers, R., Bourdet-Sicard, R., Braun, D., et al. (2013). Monitoring immune modulation by nutrition in the general population: Identifying and substantiating effects on human health. *British Journal of Nutrition, 110*(Suppl 2), S1–S30.
2. Allen, S. J., Okoko, B., Martinez, E., Gregorio, G., & Dans, L. F. (2004). Probiotics for treating infectious diarrhoea. *Cochrane Database of Systematic Reviews, 2*, CD003048.
3. Bassetti, S., Frei, R., & Zimmerli, W. (1998). Fungemia with *Saccharomyces cerevisiae* after treatment with *Saccharomyces boulardii*. *American Journal of Medicine, 105*(1), 71–72.
4. Besselink, M. G., van Santvoort, H. C., Buskens, E., Boermeester, M. A., van Goor, H., Timmerman, H. M., et al. (2008). Probiotic prophylaxis in predicted severe acute pancreatitis: A randomised, double-blind, placebo-controlled trial. *Lancet, 371*(9613), 651–659.
5. Borchers, A. T., Selmi, C., Meyers, F. J., Keen, C. L., & Gershwin, M. E. (2009). Probiotics and immunity. *Journal of Gastroenterology, 44*(1), 26–46.
6. Borruel, N., Carol, M., Casellas, F., Antolin, M., de Lara, F., Espin, E., et al. (2002). Increased mucosal tumour necrosis factor alpha production in Crohn's disease can be downregulated *ex vivo* by probiotic bacteria. *Gut, 51*(5), 659–664.
7. Boyle, R. J., Robins-Browne, R. M., & Tang, M. L. (2006). Probiotic use in clinical practice: What are the risks? *American Journal of Clinical Nutrition, 83*(6), 1256–1264.
8. Brenner, D. M., Moeller, M. J., Chey, W. D., & Schoenfeld, P. S. (2009). The utility of probiotics in the treatment of irritable bowel syndrome: A systematic review. *American Journal of Gastroenterology, 104*(4), 1033–1049.
9. Candela, M., Seibold, G., Vitali, B., Lachenmaier, S., Eikmanns, B. J., & Brigidi, P. (2005). Real-time PCR quantification of bacterial adhesion to Caco-2 cells: Competition between bifidobacteria and enteropathogens. *Research in Microbiology, 156*(8), 887–895.
10. Cesaro, S., Chinello, P., Rossi, L., & Zanesco, L. (2000). *Saccharomyces cerevisiae* fungemia in a neutropenic patient treated with *Saccharomyces boulardii*. *Supportive Care in Cancer, 8*(6), 504–505.
11. Cherifi, S., Robberecht, J., & Miendje, Y. (2004). *Saccharomyces cerevisiae* fungemia in an elderly patient with *Clostridium difficile* colitis. *Acta Clinica Belgica, 59*(4), 223–224.

12. Chermesh, I., Tamir, A., Reshef, R., Chowers, Y., Suissa, A., Katz, D., et al. (2007). Failure of Synbiotic 2000 to prevent postoperative recurrence of Crohn's disease. *Digestive Diseases and Sciences, 52*(2), 385–389.

13. Collado, M. C., Hernandez, M., & Sanz, Y. (2005). Production of bacteriocin-like inhibitory compounds by human fecal *Bifidobacterium* strains. *Journal of Food Protection, 68*(5), 1034–1040.

14. Cremonini, F., Di Caro, S., Nista, E. C., Bartolozzi, F., Capelli, G., Gasbarrini, G., et al. (2002). Meta-analysis: The effect of probiotic administration on antibiotic-associated diarrhoea. *Alimentary Pharmacology & Therapeutics, 16*(8), 1461–1467.

15. Cremonini, F., Mullan, B. P., Camilleri, M., Burton, D. D., & Rank, M. R. (2002). Performance characteristics of scintigraphic transit measurements for studies of experimental therapies. *Alimentary Pharmacology & Therapeutics, 16*(10), 1781–1790.

16. D'Souza, A. L., Rajkumar, C., Cooke, J., & Bulpitt, C. J. (2002). Probiotics in prevention of antibiotic-associated diarrhoea: Meta-analysis. *BMJ, 324*(7347), 1361.

17. De Groote, M. A., Frank, D. N., Dowell, E., Glode, M. P., & Pace, N. R. (2005). *Lactobacillus rhamnosus* GG bacteremia associated with probiotic use in a child with short gut syndrome. *Pediatric Infectious Disease Journal, 24*(3), 278–280.

18. Derrien, M., & van Hylckama Vlieg, J. E. (2015). Fate, activity, and impact of ingested bacteria within the human gut microbiota. *Trends in Microbiology, 23*(6), 354–366.

19. Dinan, T. G., Quigley, E. M., Ahmed, S. M., Scully, P., O'Brien, S., O'Mahony, L., et al. (2006). Hypothalamic-pituitary-gut axis dysregulation in irritable bowel syndrome: Plasma cytokines as a potential biomarker? *Gastroenterology, 130*(1), 304–311.

20. Enck, P., Zimmermann, K., Rusch, K., Schwiertz, A., Klosterhalfen, S., & Frick, J. S. (2009). The effects of ageing on the colonic bacterial microflora in adults. *Zeitschrift für Gastroenterologie, 47*(7), 653–658.

21. Engelbrektson, A., Korzenik, J. R., Pittler, A., et al. (2009). Probiotics to minimize the disruption of faecal microbiota in healthy subjects undergoing antibiotic therapy. *Journal of Medical Microbiology, 58*(Pt 5), 663–670.

22. Focareta, A., Paton, J. C., Morona, R., Cook, J., & Paton, A. W. (2006). A recombinant probiotic for treatment and prevention of cholera. *Gastroenterology, 130*(6), 1688–1695.

23. Fredenucci, I., Chomarat, M., Boucaud, C., & Flandrois, J. P. (1998). *Saccharomyces boulardii* fungemia in a patient receiving Ultra-levure therapy. *Clinical Infectious Diseases, 27*(2), 222–223.

24. Fujimori, S., Tatsuguchi, A., Gudis, K., Kishida, T., Mitsui, K., Ehara, A., et al. (2007). High dose probiotic and prebiotic cotherapy for remission induction of active Crohn's disease. *Journal of Gastroenterology and Hepatology, 22*(7), 1199–1204.

25. Furrie, E., Macfarlane, S., Kennedy, A., Cummings, J. H., Walsh, S. V., O'Neil, D. A., et al. (2005). Synbiotic therapy (Bifidobacterium longum/Synergy 1) initiates resolution of inflammation in patients with active ulcerative colitis: A randomised controlled pilot trial. *Gut, 54*(2), 242–249.

26. Gionchetti, P., Rizzello, F., Helwig, U., Venturi, A., Lammers, K. M., Brigidi, P., et al. (2003). Prophylaxis of pouchitis onset with probiotic therapy: A double-blind, placebo-controlled trial. *Gastroenterology, 124*(6), 1202–1209.

27. Gionchetti, P., Rizzello, F., Morselli, C., Poggioli, G., Tambasco, R., Calabrese, C., et al. (2007). High-dose probiotics for the treatment of active pouchitis. *Diseases of the Colon & Rectum, 50*(12), 2075–2082.

28. Glanville, J., King, S., Guarner, F., Hill, C., & Sanders, M. E. (2015). A review of the systematic review process and its applicability for use in evaluating evidence for health claims on probiotic foods in the European Union. *Nutrients, 14*, 16.

29. Hart, A. L., Lammers, K., Brigidi, P., Vitali, B., Rizzello, F., Gionchetti, P., et al. (2004). Modulation of human dendritic cell phenotype and function by probiotic bacteria. *Gut, 53*(11), 1602–1609.

30. Hengst, C., Ptok, S., Roessler, A., Fechner, A., & Jahreis, G. (2009). Effects of polydextrose supplementation on different faecal parameters in healthy volunteers. *International Journal of Food Science & Nutrition, 60*(Suppl 5), 96–105.

31. Hennequin, C., Kauffmann-Lacroix, C., Jobert, A., Viard, J. P., Ricour, C., Jacquemin, J. L., et al. (2000). Possible role of catheters in *Saccharomyces boulardii* fungemia. *European Journal of Clinical Microbiology & Infectious Diseases, 19*(1), 16–20.

32. Henry, S., D'Hondt, L., Andre, M., Holemans, X., & Canon, J. L. (2004). *Saccharomyces cerevisiae* fungemia in a head and neck cancer patient: A case report and review of the literature. *Acta Clinica Belgica, 59*(4), 220–222.

33. Hickson, M., D'Souza, A. L., Muthu, N., Rogers, T. R., Want, S., Rajkumar, C., et al. (2007). Use of probiotic *Lactobacillus* preparation to prevent

diarrhoea associated with antibiotics: Randomised double blind placebo controlled trial. *BMJ, 335*(7611), 80.

34. Hill, C., Guarner, F., Reid, G., et al. (2014). Expert consensus document: The International Scientific Association for Probiotics and Prebiotics consensus statement on the scope and appropriate use of the term probiotic. *Nature Reviews Gastroenterology & Hepatology, 11*(8), 506–514.
35. Inohara, N., Ogura, Y., Fontalba, A., Gutierrez, O., Pons, F., Crespo, J., et al. (2003). Host recognition of bacterial muramyl dipeptide mediated through NOD2. Implications for Crohn's disease. *Journal of Biological Chemistry, 278*(7), 5509–5512.
36. Ishikawa, H., Akedo, I., Umesaki, Y., Tanaka, R., Imaoka, A., & Otani, T. (2003). Randomized controlled trial of the effect of bifidobacteria-fermented milk on ulcerative colitis. *Journal of the American College of Nutrition, 22*(1), 56–63.
37. Kamil, R., Geier, M. S., Butler, R. N., & Howarth, G. S. (2007). *Lactobacillus rhamnosus* GG exacerbates intestinal ulceration in a model of indomethacin-induced enteropathy. *Digestive Diseases and Sciences, 52*(5), 1247–1252.
38. Kassinen, A., Krogius-Kurikka, L., Makivuokko, H., Rinttila, T., Paulin, L., Corander, J., et al. (2007). The fecal microbiota of irritable bowel syndrome patients differs significantly from that of healthy subjects. *Gastroenterology, 133*(1), 24–33.
39. Kato, K., Mizuno, S., Umesaki, Y., Ishii, Y., Sugitani, M., Imaoka, A., et al. (2004). Randomized placebo-controlled trial assessing the effect of bifidobacteria-fermented milk on active ulcerative colitis. *Alimentary Pharmacology & Therapeutics, 20*(10), 1133–1141.
40. Kim, H. J., Camilleri, M., McKinzie, S., Lempke, M. B., Burton, D. D., Thomforde, G. M., et al. (2003). A randomized controlled trial of a probiotic, VSL#3, on gut transit and symptoms in diarrhoea-predominant irritable bowel syndrome. *Alimentary Pharmacology & Therapeutics, 17*(7), 895–904.
41. Kim, H. J., Vazquez Roque, M. I., Camilleri, M., Stephens, D., Burton, D. D., Baxter, K., et al. (2005). A randomized controlled trial of a probiotic combination VSL#3 and placebo in irritable bowel syndrome with bloating. *Neurogastroenterology & Motility, 17*(5), 687–696.
42. Kobayashi, K. S., Chamaillard, M., Ogura, Y., Henegariu, O., Inohara, N., Nunez, G., et al. (2005). Nod2-dependent regulation of innate and adaptive immunity in the intestinal tract. *Science, 307*(5710), 731–734.
43. Komanduri, S., Gillevet, P. M., Sikaroodi, M., Mutlu, E., & Keshavarzian, A. (2007). Dysbiosis in pouchitis: Evidence of unique microfloral patterns in pouch inflammation. *Clinical Gastroenterology and Hepatology, 5*(3), 352–360.

44. Kruis, W., Fric, P., Pokrotnieks, J., Lukas, M., Fixa, B., Kascak, M., et al. (2004). Maintaining remission of ulcerative colitis with the probiotic *Escherichia coli* Nissle 1917 is as effective as with standard mesalazine. *Gut, 53*(11), 1617–1623.
45. Kruis, W., Schutz, E., Fric, P., Fixa, B., Judmaier, G., & Stolte, M. (1997). Double-blind comparison of an oral *Escherichia coli* preparation and mesalazine in maintaining remission of ulcerative colitis. *Alimentary Pharmacology & Therapeutics, 11*(5), 853–858.
46. Lankelma, J. M., Nieuwdorp, M., de Vos, W. M., & Wiersinga, W. J. (2015). The gut microbiota in internal medicine: Implications for health and disease. *Netherlands Journal of Medicine, 73*(2), 61–68.
47. Lewis, S. J., & Heaton, K. W. (1997). Stool form scale as a useful guide to intestinal transit time. *Scandinavian Journal of Gastroenterology, 32*(9), 920–924.
48. Liebregts, T., Adam, B., Bredack, C., Roth, A., Heinzel, S., Lester, S., et al. (2007). Immune activation in patients with irritable bowel syndrome. *Gastroenterology, 132*(3), 913–920.
49. Lin, H. C. (2004). Small intestinal bacterial overgrowth: A framework for understanding irritable bowel syndrome. *JAMA, 292*(7), 852–858.
50. Lin, Y. P., Thibodeaux, C. H., Pena, J. A., Ferry, G. D., & Versalovic, J. (2008). Probiotic *Lactobacillus reuteri* suppress proinflammatory cytokines via c-Jun. *Inflammatory Bowel Diseases, 14*(8), 1068–1083.
51. Mackay, A. D., Taylor, M. B., Kibbler, C. C., & Hamilton-Miller, J. M. (1999). *Lactobacillus* endocarditis caused by a probiotic organism. *Clinical Microbiology and Infection, 5*(5), 290–292.
52. Macsharry, J., O'Mahony, L., Fanning, A., Bairead, E., Sherlock, G., Tiesman, J., et al. (2008). Mucosal cytokine imbalance in irritable bowel syndrome. *Scandinavian Journal of Gastroenterology, 43*(12), 1467–1476.
53. Makras, L., Triantafyllou, V., Fayol-Messaoudi, D., Adriany, T., Zoumpopoulou, G., Tsakalidou, E., et al. (2006). Kinetic analysis of the antibacterial activity of probiotic lactobacilli towards *Salmonella enterica* serovar Typhimurium reveals a role for lactic acid and other inhibitory compounds. *Research in Microbiology, 157*(3), 241–247.
54. Malchow, H. A. (1997). Crohn's disease and *Escherichia coli*. A new approach in therapy to maintain remission of colonic Crohn's disease? *Journal of Clinical Gastroenterology, 25*(6), 653–658.
55. Mallon, P., McKay, D., Kirk, S., & Gardiner, K. (2007). Probiotics for induction of remission in ulcerative colitis. *Cochrane Database of Systematic Reviews, 4*, CD005573.

56. Matto, J., Malinen, E., Suihko, M. L., Alander, M., Palva, A., & Saarela, M. (2004). Genetic heterogeneity and functional properties of intestinal bifidobacteria. *Journal of Applied Microbiology, 97*(3), 459–470.

57. McCarthy, J., O'Mahony, L., O'Callaghan, L., Sheil, B., Vaughan, E. E., Fitzsimons, N., et al. (2003). Double blind, placebo controlled trial of two probiotic strains in interleukin 10 knockout mice and mechanistic link with cytokine balance. *Gut, 52*(7), 975–980.

58. McFarland, L. V. (2006). Meta-analysis of probiotics for the prevention of antibiotic-associated diarrhea and the treatment of *Clostridium difficile* disease. *American Journal of Gastroenterology, 101*(7), 812–822.

59. McFarland, L. V., Surawicz, C. M., Greenberg, R. N., Fekety, R., Elmer, G. W., Moyer, K. A., et al. (1994). A randomized placebo-controlled trial of *Saccharomyces boulardii* in combination with standard antibiotics for *Clostridium difficile* disease. *JAMA, 271*(24), 1913–1918.

60. Miller, L. E., & Ouwehand, A. C. (2013). Probiotic supplementation decreases intestinal transit time: Meta-analysis of randomized controlled trials. *World Journal of Gastroenterology, 19*(29), 4718–4725.

61. Mimura, T., Rizzello, F., Helwig, U., Poggioli, G., Schreiber, S., Talbot, I. C., et al. (2004). Once daily high dose probiotic therapy (VSL#3) for maintaining remission in recurrent or refractory pouchitis. *Gut, 53*(1), 108–114.

62. Moayyedi, P., Ford, A. C., Talley, N. J., Cremonini, F., Foxx-Orenstein, A. E., Brandt, L. J., et al. (2010). The efficacy of probiotics in the treatment of irritable bowel syndrome: A systematic review. *Gut, 59*(3), 325–332.

63. Mukai, T., Kaneko, S., Matsumoto, M., & Ohori, H. (2004). Binding of *Bifidobacterium bifidum* and *Lactobacillus reuteri* to the carbohydrate moieties of intestinal glycolipids recognized by peanut agglutinin. *International Journal of Food Microbiology, 90*(3), 357–362.

64. Nance, D. M., & Sanders, V. M. (2007). Autonomic innervation and regulation of the immune system (1987–2007). *Brain, Behavior, and Immunity, 21*(5), 736–745.

65. Niault, M., Thomas, F., Prost, J., Ansari, F. H., & Kalfon, P. (1999). Fungemia due to *Saccharomyces* species in a patient treated with enteral *Saccharomyces boulardii*. *Clinical Infectious Diseases, 28*(5), 930–930.

66. Nissle, A. (1959). Explanations of the significance of colonic dysbacteria & the mechanism of action of *E. coli* therapy (Mutaflor). *Medizinische, 4*, 1017–1022.

67. Ottman, N., Smidt, H., de Vos, W. M., & Belzer, C. (2012). The function of our microbiota: Who is out there and what do they do? *Frontiers in Cellular and Infection Microbiology, 2*, 104.

68. Ouwehand, A. C., Forssten, S., Nexmann Larsen, C., & Philipp, S. (2014). Probiotics and its effect on slow colonic transit. *Journal of Food & Nutritional Disorders, S1*, 1.

69. Rao, S. S., Kuo, B., McCallum, R. W., et al. (2009). Investigation of colonic and whole-gut transit with wireless motility capsule and radiopaque markers in constipation. *Clinical Gastroenterology and Hepatology, 7*(5), 537–544.

70. Riquelme, A. J., Calvo, M. A., Guzman, A. M., Depix, M. S., Garcia, P., Perez, C., et al. (2003). *Saccharomyces cerevisiae* fungemia after *Saccharomyces boulardii* treatment in immunocompromised patients. *Journal of Clinical Gastroenterology, 36*(1), 41–43.

71. Rolfe, V. E., Fortun, P. J., Hawkey, C. J., & Bath-Hextall, F. (2006). Probiotics for maintenance of remission in Crohn's disease. *Cochrane Database of Systematic Reviews, 4*, CD004826.

72. Sankar, S. A., Lagier, J. C., Pontarotti, P., Raoult, D., & Fournier, P. E. (2015). The human gut microbiome, a taxonomic conundrum. *Systematic and Applied Microbiology, 38*(4), 276–286.

73. Vanderpool, C., Yan, F., & Polk, D. B. (2008). Mechanisms of probiotic action: Implications for therapeutic applications in inflammatory bowel diseases. *Inflammatory Bowel Diseases, 14*(11), 1585–1596.

74. Vaughan, R. B. (1965). The romantic rationalist: A study of Elie Metchnikoff. *Medical History, 9*, 201–215.

75. Wagner, R. D., Pierson, C., Warner, T., Dohnalek, M., Farmer, J., Roberts, L., et al. (1997). Biotherapeutic effects of probiotic bacteria on candidiasis in immunodeficient mice. *Infection and Immunity, 65*(9), 4165–4172.

76. Waller, P. A., Gopal, P. K., Leyer, G. J., et al. (2011). Dose-response effect of *Bifidobacterium lactis* HN019 on whole gut transit time and functional gastrointestinal symptoms in adults. *Scandinavian Journal of Gastroenterology, 46*(9), 1057–1064.

77. Yan, F., Cao, H., Cover, T. L., Whitehead, R., Washington, M. K., & Polk, D. B. (2007). Soluble proteins produced by probiotic bacteria regulate intestinal epithelial cell survival and growth. *Gastroenterology, 132*(2), 562–575.

78. Zigra, P. I., Maipa, V. E., & Alamanos, Y. P. (2007). Probiotics and remission of ulcerative colitis: A systematic review. *Netherlands Journal of Medicine, 65*(8), 411–418.

79. Zocco, M. A., dal Verme, L. Z., Cremonini, F., Piscaglia, A. C., Nista, E. C., Candelli, M., et al. (2006). Efficacy of *Lactobacillus* GG in maintaining remission of ulcerative colitis. *Alimentary Pharmacology & Therapeutics, 23*(10), 1567–1574.

CHAPTER 4

GUT BIOME AND MENTAL HEALTH

Contents

INTRODUCTION

Disability-adjusted life years (DALYs), a measure of a person's life lost as a result of health-related conditions, show that mental disorders are the sixth leading cause of death worldwide. According to the Anxiety and Depression Association of America, depression affects almost 264 million people globally, with anxiety accounting for about half of all diagnoses. Depression is the most common psychiatric condition. According to a survey of 150,000 adults from 26 countries, the lifetime prevalence of anxiety disorders is 3.7%. The disparity between the number of psychiatrists per population in the developed and developing worlds is astounding. Just 32% of the 7,870 people with depression from 49 different countries surveyed recently had a formal diagnosis, and even fewer had ever received treatment. Roughly 75% of people in developing countries with mental health issues do not receive treatment, and nearly a million people commit suicide annually. The global prevalence of depression has increased approximately seven times since the COVID-19 outbreak. In 2017, the initial estimate was 3.44%.

When it comes to beneficial live bacteria (probiotics) or other substances (prebiotics, synbiotics) that may affect gut neurotransmitters, the term "psychobiotic" was first used. There have been conflicting results from studies looking into their impact on psychiatric disorders. They are commonly taken as OTC supplements, which are unregulated but heavily advertised goods that purport to be beneficial for mental health issues. Despite the lack of an objective cost-benefit analysis, many people who do not seek psychiatric assistance, do not have access to mental health care, or cannot afford it frequently use them. These puzzles prompted more methodical investigations concerning its applicability in underserved low-income households with limited access to mental health services and developing countries with lower psychiatrist-to-patient ratios.

The human digestive tract is home to tens of trillions of microorganisms, including different bacteria, viruses, phages, protozoa, fungi, and archaea. The theoretical model is based on decades of research on this environment. We call this the microbiome of the human gut. The number of these bacteria that live in the human gut usually exceeds the total number of human cells. This also applies to human DNA found in the body; there is roughly a 1:100 ratio between human and bacterial genes. According to recent research, the gut microbiota's great diversity affects a wide range of physiological functions, including mental and mood disorders as well as metabolic illnesses. To look into these preliminary results, the Human Microbiome Project was started in 2007. Later, the Human Microbiome Project Consortium and METAgenomics of Human Intestinal Tract (MetaHIT) were established. The autonomic nervous system (ANS), the neuroendocrine and immune systems, and the enteric nervous system (ENS)—

also referred to as the second brain—are some of the intricate bidirectional communication channels between the gut and the brain.

The gastrointestinal (GI) tract produces about half of dopamine and 95% of serotonin. A disruption in the gut microbiota is thought to trigger the well-known hypothalamic-pituitary-adrenal (HPA) axis. Early research has demonstrated that a diverse microbiome aids in the HPA axis's regulation and effective handling of external stressors during times of need. On the other hand, in situations of stress, a less diverse gut microbiome is more easily upset and loses equilibrium. After then, the gut becomes dysbiotic, which may cause the lining of the intestine to become permeable or leaky. Following incomplete digestion, food particles may enter the bloodstream. This upset of homeostasis can lead to inflammatory illnesses, as well as mental health conditions, according to recent research. More research has revealed that eating poorly and not exercising can exacerbate the symptoms of inflammatory gut disease and depression.

Moreover, major health issues like type 2 diabetes and cardiovascular diseases (CVDs) have been connected to the repeated activation of the HPA axis, which habituates the organism to the stressors. Numerous neurotransmitters, including histamine, gamma-aminobutyric acid (GABA), serotonin, and dopamine, have been linked to the pathophysiology of depression and anxiety. Since most contemporary antidepressants strengthen these neurotransmission networks, symptoms of depression, anxiety, and other mental health issues may be managed by keeping an eye on and regulating neurotransmitter levels.

Based on new findings, the gut and brain are closely related, suggesting that gut dysbiosis plays a part in a variety of mental illnesses. Peptic ulcer disease and irritable bowel syndrome (IBS) appear to be the most prevalent gastrointestinal issues that co-occur with mental health issues, according to studies. A different study discovered that the gut microbiome of the depressed group was undiversified and primarily colonized by Firmicutes, *Bacteroides*, and Actinobacteria (James et al., 2018).

4.1. GUT-BRAIN AXIS AND ITS RELATIONSHIP TO MENTAL HEALTH

The enteric and central nervous systems (CNSs) are connected by a bidirectional communication network called the gut-brain axis. This network encompasses not only anatomical pathways but also humoral, endocrine, metabolic, and immune communication pathways. The brain can affect intestinal activities, such as the activity of functional immune effector cells, through the autonomic nervous system (ANS), the HPA axis, and nerves within the GI tract. The gut can also affect mood, cognition, and mental health. Evidence from clinical, epidemiological, and immunological studies indicates that the gut-brain relationship—that is, mental state, emotional regulation, neuromuscular function, and HPA regulation—is greatly and extensively influenced by the enteric microbiota. The effects of microbiota on the brain's emotional and cognitive centers, both directly and indirectly, are still being studied. Research has shown that shifts in the microbiota are correlated with modifications in these communication systems.

For instance, there are now established connections between functional GI disruptions and a number of mood disorders, including anxiety, depression, and autism spectrum disorders (ASDs). In contrast, GI diseases, such as irritable bowel syndrome (IBS) and disease, frequently involve psychological comorbidities linked to changes in the gut microbiome. Furthermore, studies have shown that fetal and neonatal neurologic development may be influenced by the makeup of the gut microbiota. Not surprisingly, diet has also been demonstrated to affect the effect of gut microbiota on cognitive function.

4.1.1. Pathways of the Gut-Brain Axis

In 1998, it was discovered that giving rats subclinical doses of a single, distinct bacterium (*Campylobacter jejuni*) orally would cause anxiety-like behavior without triggering an immune reaction. Subsequent studies verified that feeding *C. jejuni* to mice resulted in anxiety-like behavior and the activation of brain regions that were dependent on information coming from the gut through the vagus nerve.

The animals used in the groundbreaking initial research that established the mechanisms of the gut-brain axis were housed in sterile environments. The question of whether postnatal microbial colonization could impact the development of brain plasticity and the ensuing physiological response was one that Sudo et al. attempted to address. They investigated the HPA axis response to stress by comparing germ-free (GF), specific pathogen free (SPF), and gnotobiotic mice in order to test the theory that gut microbes might influence the development of neural systems that regulate the endocrine response to stress. Similar to the brain-gut axis, they discovered that colonizing microbes changed the HPA response to restraint stress. This suggests that the relationship between gut bacteria and the brain is also bidirectional. This was the first report that demonstrated commensal microbes influencing the neural network in charge of regulating stress response. While both groups of mice did not exhibit any differences in their sensitivity to ether stress, the HPA response of the GF mice in this study was more sensitive to restraint stress than that of the SPF mice. Furthermore, the cortical expression levels of brain-derived neurotrophic factor (BDNF) were found to be lowered in GF mice. The authors concluded:

Because the HPA response to restraint stress is affected by the limbic system, including the prefrontal cortex, hippocampus and amygdala, and requires assembly and processing of signals from multiple sensory modalities before initiation of a stress response, whereas ether stress does not these results indicate that cognitive processing in the limbic system may be involved in an exaggerated HPA response by GF mice.

Their research demonstrated that the development of the HPA stress response is regulated by commensal microbiota, and that the sequence of events in the GI tract subsequent to postnatal microbial colonization can have a persistent impact on the neural processing of sensory data related to the endocrine stress axis.

These results have been expanded upon and confirmed in the 10 years following this ground-breaking study. For instance, it has been demonstrated that giving oral antibiotics to SPF mice temporarily changes the microbiota's composition and increases exploratory behavior and BDNF, a crucial protein involved in neuronal plasticity and cognition, in the hippocampus. Because these modifications were unaffected by vagal or sympathetic integrity, GI neurotransmitter levels, or inflammatory activity, the authors concluded that gut microbiota affects brain chemistry and behavior on its own. According to other related studies, giving *Lactobacillus rhamnosus* to zebrafish improved the expression of BDNF and genes related to serotonin signaling and metabolism in the brain. The four main gut-brain axis pathways—neurologic, endocrine, humoral/metabolic, and immune—have been further clarified by a substantial body of research that is well outside the purview of this review.

4.1.1.1. Neurological Pathway

The vagus nerve, the ENS, and the activity of neurotransmitters in the gastrointestinal tract (GIT) are all parts of the neurologic pathway. Neurologic modulation of afferent sensory nerves directly yields molecules such as acetylcholine, histamine, serotonin, melatonin, GABA, and other molecules that can function as local neurotransmitters; this pathway also produces biologically active forms of catecholamines in the gut lumen. In the gut, the ANS also affects immune system activation. For instance, it directly affects mast cell and macrophage responses to luminal bacteria. Furthermore, it seems that normal gut intrinsic primary afferent neuron excitability depends on the gut microbiota.

4.1.1.2. Endocrine Pathway

The gut-brain axis may be impacted by the gut microbiota's modification of nutrient availability, which in turn affects enteroendocrine cells' release of physiologically active peptides. For instance, it is believed that the neuropeptide galanin plays a role in a number of vital neurobiological processes, such as nociception, feeding, mood, blood pressure regulation, sleep/wake cycle regulation, and neurotrophic functions. Galanin increases glucocorticoid secretion from the adrenal cortex by activating the central branch of the HPA axis, which in turn influences the release of corticotropin-releasing factor and adrenocorticotropic hormone (ACTH). Furthermore, it has the ability to directly stimulate the release of norepinephrine from the adrenal medulla and the secretion of cortisol from the adrenal cortex, indicating its involvement in the HPA axis response to stress.

4.1.1.3. Humoral/Metabolic Pathway

The primary humoral determinants are bacterial metabolites, particularly short-chain fatty acids (SCFAs), which are generated by the fermentation of dietary carbohydrates by bacteria. They have a notable effect on enterocyte nutrition, exhibit strong hormone-like activity, possess immunomodulatory capabilities, and interact with nerve cells by inducing sympathetic branch stimulation in the ANS. Moreover, microbiota-derived SCFAs have been demonstrated to regulate microglia homoeostasis, which is necessary for normal brain development and brain tissue homoeostasis and is involved in behavior modulation. They can also cross the blood-brain barrier. Importantly, abnormalities in SCFA metabolism have been linked to autism development by impairing microglial function and communication. Additionally, SCFAs have been demonstrated to control the production of gut-derived serotonin from enterochromaffin cells as well as the release of gut peptides from enteroendocrine cells. These actions have an impact on the hormonal communication between the gut and the brain. Approximately 95% of the serotonin in the body is produced in the gut, and the majority of it is found in plasma.

Serotonin can activate afferent nerve endings that are directly connected to the CNS locally, despite its intrinsic roles in peripheral metabolism and the intestines. While most doctors understand that fluoxetine works by preventing gut serotonin from being transported into plasma, many are not aware that children with autism have been shown to have elevated plasma serotonin. Low serotonergic neurotransmission and high plasma serotonin have been shown to correlate negatively in young male adults with ASD. Lipopolysaccharide (LPS), which is primarily derived from the cell walls of Gram-negative enterobacteria, is another important bacterial metabolite. Leaky gut syndrome, or intestinal epithelial tight junction permeability defects, is the mechanism by which lipomatous polyposis enters the systemic circulation. It is known that patients with major depression have higher levels of antibodies produced by the human body against LPS compared to controls.

4.1.1.4. Immune Pathway

The gut microbiome affects the metabolism of inflammation in the GI tract, primarily through immune system release of cytokines (like interleukin [IL]-10 and IL-4) and other cellular communication mediators (like interferon-gamma) during dysbiosis. For instance, aberrant microbiota populations in IBS trigger mucosal innate immune responses, increasing gut epithelial permeability, triggering gut pain sensory pathways, and dysregulating the ENS; brain-gut and gut-brain dysfunctions are present, with the former being more prevalent. Intestinal motility and secretion are impacted by disruptions in the gut-brain axis, which also cause cellular changes in the immune and entero-endocrine systems and visceral hypersensitivity.

4.1.1.5. Epithelial Barrier Structure and Function

Stress has been shown to change the permeability of the intestinal epithelium, allowing LPS and bacterial antigens to enter the bloodstream and become humoral influencers with a variety of effects. According to *in vivo* research, acute stress affects the GI tract by altering colonocyte differentiation and lowering the expression of the mRNA that codes for tight junction proteins.

Many gut-related conditions, including necrotizing enterocolitis (NEC), irritable bowel syndrome (IBS), and the low-level inflammation frequently observed in metabolic syndrome, obesity, and diabetes, have been linked to defects in intestinal permeability. Leaky gut syndrome is observed in certain IBS subtypes, including those that typically have an infectious etiology, and in certain allergy types. Therefore, it should not be shocking that probiotic supplements may affect the integrity of the intestinal tight junction and aid in the treatment of illnesses brought on by or made worse by dysfunction of the gut barrier. Although there have been few probiotic clinical trials that specifically examine human epithelial barrier function, there is growing evidence that *lactobacilli* can be used in this capacity.

For instance, a multistrain combination of *lactobacilli*, lactococci, and bifidobacteria (Ecologic BARRIER, Winclove, Amsterdam, Netherlands) has been shown to reduce LPS load, inhibit mast-cell activation, stimulate the anti-inflammatory cytokine IL-10, and strengthen the epithelial barrier *in vitro* following pathogenic bacteria and inflammatory stressors. Later clinical evidence of the treatment's efficacy in preventing depression was derived from this preclinical work.

4.1.1.5. Inflammation, Depression, and the Microbiome

Independent research groups have confirmed the connection between gut microbiota and mood by studying gut dysbiosis in animal models and its association with monoamine disruptions observed in clinical depression. Furthermore, persistent low-grade inflammation linked to stress-related mental illnesses is believed to be caused by abnormalities in intestinal permeability. Proinflammatory cytokines like IL-1β, IL-6, tumor necrosis factor-α, interferon gamma, and C-reactive protein are often overexpressed in people who suffer from depression symptoms. The transcription of these same cytokines is influenced by gut microbiota; dysbiosis initiates the inflammasome pathway, while proinflammatory cytokines, like NF-κB, are lessened in production by beneficial metabolites, especially SCFAs.

It is commonly known that gut microbiota maintains the integrity of tight junctions between enterocytes. The fact that dysbiosis and related elevations in intestinal permeability are now known to occur in rheumatoid arthritis, Alzheimer's disease (AD), asthma, ASDs, and other systemic conditions—both inflammatory and non-inflammatory—should therefore not be surprising. A vast amount of research has been conducted in recent years to validate the mechanisms and the role of probiotics and the microbiome in managing inflammatory conditions, especially IBD.

The connection between inflammation and depression is becoming more widely acknowledged; in fact, anti-inflammatory medications like COX-2 inhibitors have been shown to be effective in treating major depression in the past. Studies have shown that an inflammatory phenotype modifies neurotransmitter metabolism by increasing the activation of the HPA axis and decreasing the availability of neurotransmitter precursors, both of which are involved in the etiology of clinical depression. Even though the gut-brain axis was first proposed in 1910, it has taken more than a century for converging research areas to confirm this important pathway's role in the effective prevention and treatment of clinical depression. The foundation for researching probiotics as a potential treatment for human mood disorders was established by preclinical studies.

4.2. EFFECTS OF PROBIOTICS ON MENTAL WELL-BEING AND MOOD

The wellness community frequently discusses the significance of gut health, and if you are familiar with probiotics—live microorganisms that, in sufficient dosages, can support the body and the microbiome—you have probably heard about their possible advantages for digestive health. But probiotics may have health benefits that go beyond the physical: research has connected a number of mental illnesses, such as depression and anxiety, to gut dysbiosis, or an imbalance of the microorganisms that make up the gut microflora. Probiotics used to treat mental health issues are referred to as "psychobiotics," a term that is relatively new. Continue reading to find out whether taking a psychobiotic could be a good idea for you, how doctors and researchers are utilizing them in healthcare, and any possible risks or benefits.

The term "psychobiotics" does not refer to a specific type of probiotic. Rather, it is used by experts when probiotics are being studied or used to improve mental health, says gastroenterologist Avanish Aggarwal, M.D. "Really, they are regular probiotics," he says.

Many people are already familiar with the potential benefits of probiotics, and about 4 million U.S. adults use probiotic supplements, according to the 2012 National Health Interview Study. Eating foods that contain scientifically tested probiotics, such as certain yogurts or fermented vegetable products, can also be beneficial to the gut flora.

Researchers refer to gut microbes that are beneficial to mental health as psychobiotics. Psychobiotics are still a relatively new concept, so it is unclear which particular bacterial strains are good for mental health. This is what scientists are currently attempting to ascertain, and understanding psychobiotics is complicated by a number of factors. He clarifies that probiotic supplements can differ substantially in composition and quality, and not all pertinent scientific research is of a high caliber or involved human subjects.

Psychobiotics are foods high in probiotics that are taken to enhance mental health. Probiotic foods have been linked by science to lower levels of anxiety and depression, but because they also contain a variety of other healthful nutrients, it can be challenging to isolate particular bacterial strains and determine how they affect the body. For instance, probiotic meals frequently contain fiber and antioxidants, both of which have been related to improved mental health. Meanwhile, without first comprehending the gut-brain connection, it is impossible to discuss the relationship between probiotics and mental health.

The role of psychobiotics in the gut-brain axis and mental health is still being investigated by experts. To ascertain the potential role of psychobiotics, further human research is required. According to a 2022 randomized, controlled study published in Translational Psychiatry, taking a high-dose probiotic containing eight different bacterial strains reduced depressive symptoms. Psychobiotics like the ones used in the experiment should be taken into consideration in addition to therapy, which the participants also underwent, the study's researchers stress, not in place of it. 10 clinical trials pertaining to probiotics and mental health were examined by a group of researchers in 2020. The subjects of every study they examined included those with anxiety or depression diagnoses or those who were under stress. Probiotic use was found to be positively correlated with a decrease in symptoms of depression; however, this correlation was not as strong when probiotics were used to treat anxiety.

Furthermore, those experiencing acute stress were unaffected by the probiotics. The inconsistent outcomes highlight the necessity for additional investigation concerning psychobiotics.

4.2.1. Ways Gut Bacteria and Mental Health, Probiotics and Depression are Linked

Gut microbes have a major role in your mental and emotional well-being. They may lessen stress, anxiety, and depressive symptoms, but they may also exacerbate them. The gut microbiome is an ecosystem made up of billions of bacterial cells that live in your colon. Their actions affect your brain in addition to permitting nutrients to enter the body and preventing opportunistic infections from entering. The body undergoes a number of adjustments in response to stress, focusing all of its energy and vital resources on the muscles and brain. The body releases cortisol in response to stress, and the gut microbiome may be impacted by all of these processes. Similarly, dysbiosis, or an imbalance in your gut microbiota, can impact your mood in general. That is because the activity of your gut bacteria affects stress and anxiety — a balanced microbiome can improve your stress resilience, but an imbalanced one can affect your mental health. Here's how gut bacteria, mental health, and depression are related to probiotics.

4.2.1.1. The Vagus Nerve Connects Your Gut and Your Brain

The vagus nerve, a vital part of the ANS that allows you to breathe, digest food, and swallow on autopilot, connects your gut to your brain. This nerve can communicate with your brain to control your colon and vice versa. Due to the relationship between the two organs, the gut-brain axis is starting to play a significant role in mental health, brain diseases, and even IBS. It explains why digestive issues can make you unhappy as well as how stress can negatively impact digestion.

The role of the vagus nerve in digestion:

- Motility — helps food move through the digestive tract
- Digestion — stimulates the release of digestive enzymes
- Appetite — communicates satiety to the brain

4.2.1.2. Gut Bacteria Talk to Your Brain Too

Food, especially dietary fiber, is broken down by gut bacteria and converted into metabolites like SCFAs. The nerve picks these up and relays the information to the brain, enabling the control of digestive functions. However, stress impairs the vagus nerve's ability to respond to inflammation, which is detrimental to the gut and its bacteria. Stress causes your body to focus its energy and attention on your muscles and brain instead of your gut. That is the reason the vagus nerve is so crucial.

4.2.1.3. Inflammation, Gut Bacteria, and Depression

Your gut microbiome needs to be varied in order to support your health, and diversity keeps it in balance. On the other hand, opportunistic microbes can take advantage of dysbiosis and multiply, causing inflammation, if it is not balanced. This is due to your body's reaction to opportunistic bacteria, which causes your immune system to become tense and cause inflammation. It is interesting to note that inflammation both causes and is caused by depression. However, a varied microbiome helps ward off inflammation. Thus, reducing inflammation can aid in elevating mood and reducing anxiety. One strategy to decrease inflammation and boost the diversity of microbes is through diet. Fiber is a vital source of energy for beneficial gut bacteria, which flourish on a natural, plant-based diet.

4.2.1.4. Gut Bacteria and Mental Health: The Butyrate Effect

A necessary SCFA called butyrate is created when you eat plants (fruit, vegetables, seeds, nuts, whole grains, legumes) and is produced by beneficial gut bacteria. Not only does it satisfy your stomach, but it also helps your brain. Atlas BioMed's microbiome test can tell you how much butyrate your gut bacteria are producing. Butyrate contributes to the strength and integrity of your intestinal lining because it is the primary fuel source for the cells in your lining. Additionally, it lessens inflammation, which has a negative impact on mood. According to a recent study, butyrate may even promote the growth of new brain cells. Your gut bacteria may produce butyrate and other less vital nutrients if you have dysbiosis. The good news is that you can actively increase butyrate production in your stomach by eating certain foods. Eating prebiotic foods—such as fruits, vegetables, whole grains, and pulses—directly feeds your gut bacteria is one method. These have fiber in them, which is converted into SCFAs such as butyrate. Thus, increasing your consumption will benefit your well-being!

4.2.1.5. Probiotics and Depression

The brain is one of the many health advantages of probiotic bacteria. In addition to being present in foods that have undergone fermentation, such as yoghurt and kefir, they are also naturally found in the gut. Probiotics, such as *Lactobacillus*, *Bifidobacterium*, and Lactococcus species, support your entire body and enhance mental health. The study of probiotics' impact on mental health is known as psychobiotics. Certain *Lactobacillus* species have been shown in some studies to reduce anxiety and enhance stress resilience. According to certain research, taking probiotics may even help reduce depressive symptoms. By maintaining the balance of the gut ecology and averting dysbiosis, probiotics contribute to human health. By doing this, you can support the production of butyrate and foster the growth of beneficial bacteria. Probiotics and depression appear to be positively correlated, and this finding is already quite promising.

4.2.1.6. Eat Prebiotics to Nourish Probiotic Bacteria

You have increased your consumption of probiotics, but in order to fully benefit from their health benefits, you must continue to feed them. Your gut bacteria need food to survive, be energized, and grow—just like you do. Prebiotics can help with that. Plant-based diets contain prebiotics, which are compounds that support healthy gut flora. Prebiotic fibers, resistant starches, and polyphenols all support gut flora, which then converts them into beneficial substances like SCFAs and vitamins (Zhou et al., 2015) (Table 4.1).

Table 4.1. Prebiotic Food List

Prebiotic	Component	Source	Function
Polyphenol	Blueberry polyphenol extract	Blueberry	Reduces weight and normalizes lipid metabolism
	Wine grape seed flour	Grape seed	Enhances intestinal permeability and alters adipocyte gene expression to inhibit high-fat-induced obesity and inflammation
Polyphenol	Orange albedo	Orange	Stimulates the growth, reproduction, and metabolism of *Lactobacillus acidophilus* and *Lactobacillus animalis*
	Catechin and punicalagin	Fermented pomegranate juice	Increases antioxidant capacity and improves survival of lactic acid bacteria
Polypeptide Polymers	Poly-gamma-glutamate (PGA)	Bacillus fermentation	Increases abundance of *Lactobacillus* and reduces abundance of *Clostridium*, helping to regulate the intestinal microbiota

Polysaccharides	Algae polysaccharides	Algae	Improves the activity of some beneficial flora and stimulates the production of functional metabolites in the intestinal microbiota
	Lotus seed resistant starch (LRS3-20%)	Lotus seed	Shows high probiotic activity against *Bifidobacterium* and *Lactobacillus acidophilus*
	Longan pulp polysaccharides	Logan	Promotes the growth of *Lactobacillus plantarum*, *Lactobacillus bulgaricus*, and *Lactobacillus fermentum*

Source: You S, Ma Y, Yan B, Pei W, Wu Q, Ding C and Huang C (2022) The promotion mechanism of prebiotics for probiotics: A review. Front. Nutr. 9:1000517. doi: 10.3389/fnut.2022.1000517

Additionally, studies have demonstrated a link between prebiotic consumption and a decrease in anxiety-related behavior. Therefore, it is critical to recognize the impact of nutrition on mental health. You can actually take a gut microbiome test like this one here to receive personalized food recommendations for your gut bacteria.

4.2.1.7. Gut Microbes Regulate Happy Hormones

You may be aware that the microbes in your stomach convert food into SCFAs. These SCFAs then interact with cells to produce serotonin, a neurotransmitter and hormone that controls mood, anxiety, and happiness. In essence, your body can produce more serotonin with the aid of your gut microbes. Similarly, GABA, a different neurotransmitter that helps to calm the nervous system and turn off stress reactions, controls and elevates mood. Surprisingly, your body can manufacture GABA on its own thanks to certain probiotic gut bacteria. Essentially, eating the right foods feeds happy bacteria, which is why your diet can help your bacteria protect your mental health. Furthermore, a microbiome rich in diverse healthy bacteria increases the production of chemicals that elevate mood, such as GABA and serotonin.

4.2.1.8. Your Microbiome Composition and Mental Health

It is evident that gut flora and depression are related. You can learn a great deal about what is going on inside your body by looking at the makeup of your gut microbiome. Although each person has a different gut microbiome, diversity has been shown to be important for maintaining the health of both your body and mind (Figure 4.1).

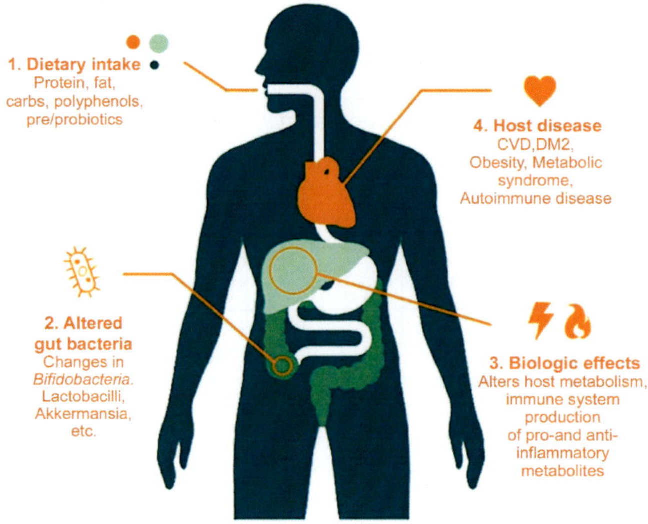

Figure 4.1. *Your gut bacteria crave plant goodness to boost your mood.*

Source: Singh, R.K., Chang, HW., Yan, D. et al. (2017). Influence of diet on the gut microbiome and implications for human health. J Transl Med., 15, 73. https://doi.org/10.1186/s12967-017-1175-y.

Thankfully, at-home testing has made it simple to determine the health status of your individual microbiome. You can see what foods you should eat to support a happy and healthy microbial ecosystem, as well as how diverse and productive your microbiome is in terms of butyrate production.

4.2.1.8. Balance Is the Key to Happy Gut Bacteria

It is common to consider the body's systems as distinct entities, and while this is true in certain cases, they are also intricately linked and have the ability to affect one another's functions. The brain and the gut are excellent examples of two organ systems that can change and impact the other. Dysbiosis, or an unbalanced gut microbiome, is linked to a number of illnesses, including mental health conditions like depression. Similarly, inflammation brought on by depression may impact the gut's natural ecology. Promising research, however, indicates that probiotics and prebiotics are beneficial for depression, anxiety, and stress tolerance.

4.2.2. Potential Benefits of Psychobiotics

Psychobiotics may be able to help with mood and digestion as well as mental and physical health issues.

Not only could psychobiotics potentially improve mental health, using psychobiotics in the form of food or a supplement may have additional health benefits, including:

- Improved digestion
- Decreased risk of cognitive disease, including Alzheimer's disease
- Decreased risk of heart disease
- Improved immune health

4.2.3. Ways to Incorporate Psychobiotics into Your Life

Probiotic supplements and including fermented foods in your diet are two strategies you can use to include psychobiotics in your life, suggests Dr. Naidoo. Some simple ways to up your daily probiotic intake include eating yogurt high in certain beneficial probiotics for breakfast or in a smoothie, using tempeh as a protein in savory dishes, adding miso to food, and topping sandwiches, salads, and burgers with kimchi or sauerkraut, according to her.

4.2.3.1. Probiotic Foods

There is ongoing research on probiotic foods as well as supplements. In a 2019 study involving over 26,000 participants, for example, it was discovered that those who regularly consumed probiotic-rich foods had lower rates of depression than those who did not eating foods high in probiotics to support mental health and other health benefits, such as improved digestion. Compared to probiotic foods, which you must eat anyhow, probiotic supplements are more costly. Some probiotic foods that contain potentially beneficial live organisms include:

- Kefir
- Yogurt
- Tempeh
- Miso
- Natto
- Kimchi
- Sauerkraut
- Pickled vegetables

4.2.3.2. Probiotic Supplements

The advantage of consuming probiotic foods, according to both experts, is that they contain other vital nutrients that support both the body and the brain, even though probiotic supplements might have higher concentrations of beneficial bacteria. For instance, fiber, which is present in high-fiber probiotic foods like kimchi and pickled vegetables, has heart and brain health benefits. Antioxidants, found in some probiotic foods like kefir, tempeh, and pickles, are comparable. According to studies, antioxidants protect brain cells from oxidative stress, which is beneficial to the brain. Due to the fact that probiotic supplements are not subject to U.S Food and Drug Administration (FDA), make sure any probiotic supplement you're thinking about is high quality by doing your homework. Online resources such as Consumer Lab, Pharmacopeia, and NSF International carry out independent testing on supplements, including probiotics. In the end, a study of one is crucial: you taking a supplement every day for a month to see if it has any effect. He advises switching to a different probiotic supplement or emphasizing probiotic foods if you do not see any improvement.

4.2.4. Potential Risks and Side Effects of Psychobiotics

Psychobiotics have few risks and few unfavorable side effects, with a wide range of possible benefits. Probiotic foods, like all foods, may put some people at risk for allergies. Furthermore, dangerous pathogens like *Salmonella* and *E. coli* could have a detrimental effect on health by causing foodborne illnesses if the fermentation process was not carried out safely. Regarding supplements, it is important to emphasize that since the FDA does not regulate them, research is essential. Diarrhea and atopic eczema are two possible risks and side effects of taking a probiotic supplement, according to the National Center for Complementary and Integrative Health. Probiotics may also conflict with certain medications, such as prednisone, so if you take any, consult your doctor before incorporating probiotics into your daily regimen.

4.3. PROBIOTICS AS POTENTIAL ADJUNCTIVE THERAPY FOR MENTAL HEALTH DISORDERS

Major depressive disorder (MDD) is a prevalent and severely incapacitating mental health condition that arises from the combination of susceptibility genes and unpleasant environmental experiences, specifically stressful ones. The pathophysiology of depression has been linked to many factors, including dysregulation of the hypothalamus-pituitary-adrenal (HPA) axis, an imbalance between pro- and anti-inflammatory cytokines, depletion of neurotransmitters (dopamine, norepinephrine, and/or serotonin) in the CNS, and altered glutamatergic and GABAergic transmission. Many different biological events that contribute to the pathophysiology of depression have led to the development of a large number of antidepressant medications, each with a unique set of pharmacological effects. However, clinical requirements remain unmet.

4.3.1. A Conceptual Framework of Probiotics Aligned to Precision Psychiatry in Clinical Depression

With the advent of increasingly reliable and validated prediction models that are specifically customized for treatment-response, prognostic, and diagnostic estimations, precision psychiatry has gained momentum in recent years. Nonetheless, the necessity of practical application in actual psychiatric practice serves as the primary impediment to knowledge expansion and research initiatives in this field. Creating unified frameworks for implementation purposes—which are essential to adopting this ground-breaking psychiatric care model—is one method to close the translational gap. Precision psychiatry's conceptual components pertain to the simultaneous exploration of neurobiological, genetic, environmental, and lifestyle bases in relation to depression. These bases may have implications for the potential health benefits of probiotics. These elements come from the precision medicine approach, which, fortunately, has been a cornerstone of clinical specialties other than psychiatry that have less translational gaps. In order to facilitate the incorporation of this concept into the clinical setting, potential future-oriented microbial-based clinical approaches have also been investigated.

MDD is the official term used to diagnose depression in a clinical setting. A wide range of diverse phenotypes are associated with MDD, a neuropsychiatric disorder. A licensed psychiatrist or clinical psychologist is required to diagnose a patient with clinical depression because it falls under the category of mood disorders. To diagnose clinical depression, a person must have at least two weeks of consistently present symptoms that significantly impair their ability to function, including the cardinal symptoms of depression or anhedonia, as well as other symptoms listed in the Diagnostic and Statistical Manual of Mental Disorders (DSM-5). Among the other depressive symptoms included in the DSM-5 are significant changes in weight or appetite, sleep disturbances, psychomotor changes, exhaustion, a decreased capacity for concentration, and suicidal or negative thoughts. When these psychophysiological changes are severe enough to impair an individual's functioning and persist for an extended period of time without a plausible external cause, clinical intervention is warranted. Integrating a patient's clinical history, physical examination, and laboratory results requires the collective judgment of a clinician. The fact that depression is such a diverse mood disorder makes it difficult to pinpoint its precise etiopathophysiology. Its neurobiological foundations, however, are mainly thought to be related to changed neurochemicals, compromised stress response systems involving neuroendocrine components, and neuroinflammation. The application of probiotics in depression may help to mitigate these neurobiological events. The strain-specific and epigenetic properties of probiotics mitigate any potential genetic components that may contribute to the onset of depression. Probiotics that focus on environmental and lifestyle variables that can be changed, such as stress and food, allow for a more comprehensive approach to treating clinical depression. All things considered, the conceptual application of precision psychiatry in the clinical management of depression is enhanced by this microbial-based strategy utilizing probiotics.

4.3.1.1. Neurobiological Bases

As time has gone on, our understanding of the pathophysiologic underpinnings of depression has expanded to include neurobiological bases of depression that go beyond the monoamine hypothesis. The long-standing monoamine theory is unable to fully account for the structural changes in various brain regions and other biological findings observed in patients with depression. The biological correlates of depression involving gut microbial-brain interplay that support the potential centric role of gut microbes in depression include dysregulated HPAaxis, altered neurochemicals, and neuroinflammation. The dysregulation of the HPA axis is largely attributable to chronic stress exposure that results in a persistent elevation of stress hormones, primarily cortisol. Long-term stress has been shown to have a significant impact on the HPA axis, which can lead to depressive symptoms. One of the pathways through which this effect occurs is the MGBA. A theoretical framework that explains this mechanism links the internalized disruptive impact of stress on the gut microbial ecosystem to an increase in intestinal barrier permeability and consequent decomposition.

On the other hand, the gut microbial ecosystem suffers when the HPA axis is overactive. An altered intestinal fatty acid metabolism has also been connected to stress-induced intestinal dysbiosis. These changes have an impact on adult hippocampal neurogenesis and neuroplasticity, which exacerbates hyperactivity of the HPA axis. The pathophysiologic events have a vicious effect on each other, ultimately resulting in the development of depression. Neurochemically speaking, low levels of a number of neurotransmitters, such as dopamine (DA), gamma-aminobutyric acid (GABA), norepinephrine (NE), 5-hydroxytryptamine (5-HT), and neurotrophin, BDNF, have been linked to the development of depressive disorders. By directly synthesizing neurotransmitters or indirectly by influencing their precursors and stimulating endocrine targets, gut microbes can alter the levels of these neurotransmitters. Gut-brain communication is mediated by neurotransmitters produced by the gut microbiota, which act through afferent and efferent vagus nerve fibers in signaling mechanisms. The neurotrophin

BDNF is essential to the neuroplasticity of the brain. BDNF preserves the health of neurons and the circuits that control mood and thought processes. It has been reported that neuroplasticity linked to changed BDNF levels is severely disrupted in MDD patients and stressed animal models. Hippocampal atrophy is also associated with impaired BDNF neuroregulatory effect and is frequently observed in MDD patients. Increased regional expression of BDNF has also been connected to the mechanism of antidepressant medications.

Pre-clinical research employing rat or mouse models established the majority of probiotics' anti-depressive potential. Environmental stress and genetic manipulation are used in the pre-clinical setting to mimic human depression-like phenotypes. Improved mobility in the forced swim test, an increase in sucrose preference, and a reduction in latency in the novelty-induced hypophagia test are among the behavioral outcomes that resemble the antidepressant effects. Serum corticosterone levels (human cortisol levels) to measure the involvement of the HPA axis, pro-inflammatory cytokines (TNF-alpha, IL-1, and IL-6), BDNF, and neurotransmitters (5-HT, DA, NE, and GABA) are among the frequently reported biomarkers of depression. Magnetic resonance imaging (MRI) is used to assess the structure and function of the hippocampus, amygdala, and critical frontolimbic regions.

Probiotics *L. helveticus* R0052 and *B. longum* R0175 are often studied together and are known to show strong anti-depressive effects in both animal and human models. It has been demonstrated that this combination reduces the activation of the HPA axis and the ANS in response to prolonged stress. In rodent depression models, this was indicated by a drop in plasma levels of the stress hormones corticosterone, adrenaline, and noradrenaline. Additionally, it has been demonstrated that this combination greatly reduces gut permeability and restores colonic epithelial integrity. In order to counteract stress-mediated insults involving the HPA axis, it increases adult hippocampal neurogenesis, prevents the degradation of tight junction proteins in the colonic mucosa, and restores synaptic plasticity. The same combination of probiotics significantly reduced cortisol levels and improved depressive scores in clinical studies, attenuating HPA axis hyperactivity. In a chronic unpredictable stress model of mice, the administration of *L. paracasei* CCF1229 and *L. rhamnosus* CFM1228 over a six-week period resulted in significant improvements in the behavioral and neurobiological outcomes of depression. These two strains increased hippocampal BDNF levels and prefrontal cortex 5-HT concentrations. *L. paracasei* CCFM1229 was found to have a similar effect on serum corticosterone levels in the same study, while *L. rhamnosus* CCFM1228 did not. These neurobiological effects were linked to the probiotic strains' advantageous modulatory effect on the host gut microbiome and to decreased xanthine oxidase activity in the cerebral cortex (Carabotti et al., 2015; Kali,

2016; Misra et al., 2019). The majority of pharmaceutical treatments for MDD aim to alter the brain's neurotransmitter activity. The most well-known probiotics linked to increased neurotransmitters are strains of *Lactobacillus* and *Bifidobacterium*, whose mechanisms resemble those of antidepressants. *L. paracasei* PS23, *L. helveticus* N8, *B. longum*, and *L. rhamnosus* were linked to higher hippocampal 5-HT levels in pre-clinical models of depression. It has been demonstrated that *L. plantarum* PS128 raises DA and 5-HT levels in the striatum. A correlation was found between *B. infantis* and lower NA levels. It has been determined that *Bifidobacterium* strains originating from the human gut are the primary contributors of GABA, with *Lactobacillus* strains coming in second. It has been determined that two of these strains, *L. plantarum* 90sk and *B. adolescentis* 150, are effective GABA producers with fluoxetine-like antidepressant properties. According to recent meta-analyses and systematic reviews, probiotics containing *Lactobacillus* and *Bifidobacterium* have the biggest impact on raising BDNF levels in depressed patients. Probiotic strains *L. helveticus* R0052 and *B. longum* R0175 are the most extensively researched.

Functional magnetic resonance imaging (fMRI) was used in a clinical trial of the probiotic *B. longum* NCC3001 in patients with IBS to record the pattern of brain activation in the hippocampus, amygdala, and front-limbic regions. Over the course of six weeks, the *B. longum* was administered, and depressive scores improved while limbic reactivity to unpleasant emotional stimuli decreased. Probiotics' potential as an antidepressant has been demonstrated in several other probiotic studies, mostly using behavioral outcome measures. In mouse models, *L. plantarum* 286—a probiotic originating from non-human sources—markedly ameliorated depressive-like behavior. When administered over eight weeks to patients with MDD, combined probiotics (*L. helveticus* and *B. longum*; *L. acidophilus*, *L. casei*, and *B. longum*) significantly improved depressive scores, as measured by the Beck depression inventory (BDI), when compared to placebo.

4.3.1.2. Genetic Bases

Since 2010, the reporting of probiotic designations with their strain type has been encouraged by advancements in genomic analysis. The distinct properties of each strain are responsible for the varying mechanisms of action on host pathogens, gut epithelial integrity, gut dysbiosis restoration, and immune response regulation. Probiotic use can be optimized by investigating the effectiveness of probiotics in relation to specific diseases, which is made possible by the identification of probiotic strains. In a pre-clinical investigation, separate 30-day administrations of strains of *L. plantarum* 286 and *L. plantarum* 81 of non-human origin produced distinct results, with the former exhibiting notable anti-depressive effects.

Probiotic effectiveness appears to be strain-specific, as the *L. plantarum* 81 group did not experience a similar effect. However, giving patients with MDD 8 weeks of a human-origin strain of *L. plantarum* significantly improved cognitive function. This was associated with lower kynurenine levels in these patients, but cortisol and pro-inflammatory cytokine plasma levels remained unchanged. A different study assessed the biochemical and behavioral effects of various *L. paracasei*, *L. rhamnosus*, *L. helveticus*, and *L. reuteri* probiotic strains. Merely *L. paracasei* CCF1229 and *L. rhamnosus* CCF1228 demonstrated noteworthy anti-depressive effects, exhibiting negligible variations in biochemical outcomes. These strains exhibited distinct neuroregulatory effects on genes related to the gut microbiota-inflammasome pathway and neuroplasticity.

No strains of *L. helveticus* or *L. reuteri* produced the best antidepressant effects. Understanding the potential genetic underpinnings of microbial agents in the amelioration of depression is made possible by the strain-specific probiotic mechanisms. This is accomplished by modulating the expression of genes related to inflammation and immunity as well as central gene regulation in the major brain regions in charge of mood regulation and cognitive processes. The hippocampus, amygdala, prefrontal cortex, and hypothalamus are some of these areas. In the end, this enhances the neurogenesis and neuroplasticity in these brain areas, which are pathophysiologically linked to depression disorders.

Adult neurogenesis in the dentate gyrus of the hippocampus has been strongly correlated with a causal factor of MDD and plays a role in the attenuation of the HPA axis. Hippocampal atrophy associated with reduced adult neurogenesis has been commonly reported in MDD patients. One of the causative mechanisms linked to reduced hippocampal neurogenesis is an impairment in the endocannabinoid (eCB) signaling system involving the cannabinoid receptor (CB1), which is instrumental in the regulation of hippocampal adult neurogenesis. A genetic study in MDD patients has also shown an increased frequency of a mutant allele for the CB1 receptor gene, CNR1. In essence, the host gut microbial dysbiosis affects the intestinal fatty acid metabolism, leading to a paucity of precursors, namely arachidonic acid (AA), thus downregulating hippocampal levels of 2-arachidonoylglycerol (2-AG), required in the activation of CB1. *L. plantarum*^{WJL} administration in a mouse model of depression associated with dysbiotic intestinal microbiota significantly ameliorates depressive-like behavior by normalizing reduced adult hippocampal neurogenesis. This was linked to the microbial restoration of fatty acid metabolism, which augmented hippocampal eCB precursors, 2-AG levels, and subsequent upregulation of CB1 receptors. *L. helveticus* R0052 and *B. longum* R0175 combination probiotics augmented adult hippocampal neurogenesis, leading to improved synaptic plasticity within the hippocampus and hypothalamic regions in animal models of depression. This genetic basis is correlated with the increased expression of several hypothalamic genes involved in neurotransmission and synaptic plasticity, thus enhancing the neuronal network within the hypothalamus to exert anti-depressive effects (Carabotti et al., 2015; Kali, 2016; Misra et al., 2019).

4.3.1.2.1. Epigenetics

It has been stated that depression is a pleomorphic illness, with genetic heritability making up only 40% of the cause, that results from gene-environment interactions. The way that environmental factors like stress and diet affect the neurobehavioral consequences of depression and how exogenous interventions like probiotics work is explained by the epigenetic mechanism. Nevertheless, given the novelty of the epigenetic subfield of genetic research, no investigation has solely examined and validated the epigenetic potential of probiotics in depression. However, the research that is currently available offers a strong foundation for comprehending the potential epigenetic underpinnings of probiotics' role in depression. Epigenetic mechanisms, which include histone modification, non-coding ribonucleic acid (RNA), and methylation of deoxyribonucleic acid (DNA), introduce heritable, long-lasting, but reversible phenotypic changes without involving genotypic alteration. Histone acetylation is linked to an induction effect on gene transcription, while DNA methylation is primarily associated with a suppressive effect. The gut-brain axis is altered by environmental factors, such as gut microbes and related metabolites, which modify the host's epigenetic signatures. This can result in different reactions and either positive or negative outcomes, including an inflammatory cascade. Probiotics and other gut-modulating substances function as epi-drivers to modify the host epigenome in a way that is advantageous. In order to prevent negative changes to the host's epigenetic signatures, probiotics have prophylactic and nullifying epigenetic effects.

A recent study demonstrated the strain-specific epigenetic potential of *lactobacilli* probiotics, *Limosilactobacillus fermentum* MTCC 5898 and *Lacticaseibacillus rhamnosus* MTCC 5897, on DNA and histone modifiers independently and following a challenge test using an inflammation-stimulating, opportunistic pathogenic commensal, *Escherichia coli* ATCC 14849. *E. coli* has also been shown to disrupt intestinal epithelial integrity. The mRNA gene expression of DNA and histone modifiers in the host IECs was assessed to determine the epigenetic ability of the *lactobacilli* probiotic strains using an *in vitro* model of intestinal inflammation and permeability, Caco-2 cells. *Limosilactobacillus fermentum* significantly increased the mRNA expression of both epigenetic modifiers in the Caco-2 cells independently at 12-h of incubation, while *E. coli* reduced the mRNA expression. The mRNA expression was even higher following a 12-h challenge test with *E. coli* after a 12-h pre-treatment with *Limosilactobacillus fermentum*, implying the epigenetic signatures of this probiotic strain. However, *Lacticillus rhamnosus* did not have any comparable effects. The same *lactobacilli* probiotic strains were found to significantly lower histone 3 and histone 4 acetylation at 6-hour incubation, both separately and together, when E coli was removed, outcompeted, or replaced by these probiotic strains in a different study. Conversely, after the *E. coli* challenge test, DNA methylation was markedly increased after pre-treating Caco-2 cells with *lactobacilli* strains in the competition and exclusion assays without these probiotics' pre-incubation. The capacity of these natural microorganisms to maximize the expression of tight junction genes linked to intestinal epithelial barrier functions determined the preset dose of the strains used for a 6-hour incubation period. Based on DNA-methylation and histone modifications, these studies support the ability of probiotic bacteria to modify host epigenetic patterns. This can be used to counteract pathogenic microbes that derange host epigenetic functions related to inflammation and gut epithelial cell barrier functions, either by mediating prophylactic effects or by nullifying them.

Other research has shown how probiotics can influence epigenetics by controlling BDNF and inhibiting histone deacetylase (HDAC) with its metabolites, which are primarily butyrate, propionate, and acetate. Histone acetylation, which is linked to the consolidation of learning and memory, is activated by HDAC inhibition, which operates at the levels of the hippocampus and amygdala. Strong HDAC inhibitors include butyrate, propionate, and acetate, whose actions have been likened to those of antidepressants. Through HDAC inhibition, these SCFAs can modulate hippocampal gene expression, neurochemicals (BDNF levels, GABA production, and 5-HT production), and neuroinflammation. They can also cross blood-brain barriers. This is in line with the widely documented decreased abundance of gut bacteria from the *Faecalibacterium* and *Coprococcus* genera that produce butyrate in MDD patients. Multi-strain probiotic supplementation for 28 days, consisting of *B. bifidum*, *B. lactis*, *L. acidophilus*, *L. casei*, *L. paracasei*, *L. plantarum*, *L. salivarius*, and *L. lactis*, significantly augmented butyrate-producing gut bacteria *Coproccocus* 3 and *Ruminoccocus grauvanii* and improved depressive outcome in depressed patients compared to the placebo arm (Carabotti et al., 2015; Kali, 2016; Misra et al., 2019).

4.3.1.3. Environment and Lifestyle Bases

4.3.1.3.1. Stress

Stress is frequently accompanied by triggers from the environment and life events, suggesting that stress is a lifestyle factor that can be changed as part of precision psychiatry. Numerous physiological functions in the human body are negatively impacted by stress. Depression is significantly more likely to develop in situations where stress is ongoing. Because chronic stress disrupts the integrity of the intestinal barrier, increases inflammatory responses, and causes a maladaptive stress-response system involving the HPA axis, it has devastating effects on the structure and functionality of neurons. It decreases glucocorticoid receptor expression in the hippocampus and hypothalamus and inhibits hippocampus neurogenesis, both of which are ultimately responsible for the pathogenic occurrence of depression. The frontolimbic structures and neurocircuits that are essential for controlling emotions and rewarding behavior suffer harmful changes when glucocorticoids, especially cortisol, are present at persistently high levels. Stress must therefore be taken into account as a crucial lifestyle factor in MDD prevention and management plans. Internal triggers have an impact on the degree of external.

Rodent stress models are frequently used in pre-clinical studies of probiotics to illustrate their anti-depressive effects. But when it comes to people, stress is a risk factor for MDD. In order to verify the effectiveness of probiotics in preventing MDD, psychological outcomes in stressed or healthy cohorts are used to assess the product's efficacy. In 66 healthy human volunteers, probiotic supplementation consisting of *L. helveticus* R0052 and *B. longum* R075 over 30 days significantly improved their mood and overall psychological well-being assessed through a few self-reported measures, including the hospital anxiety and depression scale (HADS). It reduced free urinary cortisol levels, indicating possible attenuation of the HPA axis. In students aged between 18 and 24 years old facing examination stress, a 28-day intervention with a multi-strain probiotic consisting of *Bacillus coagulans* Unique IS2, *L. rhamnosus* UBLR58, *B. lactis* UBBLa70, *L. plantarum* UBLP40, *B. breve* UBBr01, and *B. infantis* UBBI01 not only improved the stress outcome but also significantly reduced serum cortisol levels. In another study, a multi-strain probiotic consisting of *B. bifidum* W23, *B. lactis* W52, *L. acidophilus* W37, *L. brevis* W63, *L. casei* W56, *L. salivarius* W24, *L. lactis* W19, and *L. lactis* W58 significantly improved cognitive reactivity to either acute stress or sad mood in healthy individuals. Administration of *L. plantarum* P-8 over 12 weeks in stressed adults alleviated stress, modulated gut microbiota, increased abundance of *B. adolescentis*, *B. longum*, and *F. prausnitzii* and enhanced levels of microbial neuroactive metabolites SCFA, AA, and GABA. These studies demonstrate distinct stress-alleviating effects of probiotics, thus indicating their usefulness as a preventive intervention in MDD.

4.3.1.3.2. Diet

Microorganisms, both commensal and pathogenic, are present in the human gut and can have contradictory effects. The human host will continue to benefit physiologically from this interaction between gut microbes and the immune system as long as the immune system can distinguish between the pathogenic and beneficial activities of these microbes and mount a defense. However, homeostatic regulation will be severely impacted and immune responses will manifest if the balance of gut microbes is upset. Intestinal homeostasis and immunological regulation depend critically on diet, which is an environmental and lifestyle factor. A microbiota-targeting diet is included in the current concept of psychobiotics as a psychobiotic agent that modifies gut-brain communication to exert positive effects on mental health. Probiotics are "traditional" psychobiotics in this sense; they have strong antidepressant effects in a similar manner.

Probiotic-rich fruits and vegetables, fermented foods, whole grains, and legumes are components of a psychobiotic diet that have been shown to reduce symptoms of depression. Microbiological metabolites called SCFAs are created when dietary fibers ferment. They play important mediating and epigenetic roles in the MGBA, influencing neuronal transcription, inflammation, gut barrier functions, and neurotrophic factors and neurotransmitters. A recent groundbreaking study that examined how a psychobiotic diet affected the microbiota composition and perceived stress in healthy adults showed that there was a significant decrease in perceived stress with little to no change in the microbial composition. It was also deduced that the intervention's ability to reduce stress is dose-dependent. Further research on the effects of high-fiber, fermented foods and dietary interventions on microbial diversity and the immune system in healthy adults revealed significant alterations in microbial diversity along with increased *Lactobacillus* and *Bifidobacterium* abundance and decreased inflammatory markers. These investigations validate the impact of nutrition on gut microbiota (Carabotti et al., 2015; Kali, 2016; Misra et al., 2019).

4.3.2. Potential Microbial-Based Approach in the Clinical Management of Depression

While there are currently no clinical biomarkers for the diagnosis of MDD, the field of microbiome research adds a new layer of complexity by considering potential microbial biomarkers and providing corresponding neuroimaging evaluations. This supports an individualized therapeutic approach that employs probiotics as a stand-alone or adjunctive treatment modality. The pathomechanism of depression linked to microbial etiopathogenesis is commonly associated with biomarkers such as cortisol, kynurenine, BDNF, and pro-inflammatory cytokines. These are readily assessed by analyzing a patient's blood sample. While kynurenine levels are correlated with cognitive function,

BDNF evaluation is also correlated with suicide risk. The early 2000s saw significant progress in our understanding of the intricate, reciprocal relationship between the human brain and gut microbiota, which may signal the beginning of a new era of more precise diagnostic standards, eliminating some of the uncertainty surrounding the diagnosis of MDD.

One of the most widely recommended neuroimaging techniques for capturing structural and functional alterations in the various brain regions is magnetic resonance imaging. These results are correlated with neuropsychological scores and gut microbiota profiling. Numerous research works have validated the neuroimaging results in MDD patients, which typically entail anatomical and functional evaluations of the critical frontolimbic circuits and regions. One commonly reported hallmark of MDD patients is hippocampal atrophy. In depressed cohorts, hippocampal atrophy is linked to decreased adult hippocampal neurogenesis because of a disrupted HPA axis associated with microbial dysbiosis. It has also been proposed that the frontolimbic assessment is a useful predictor of suicide risk in depressed individuals. Thus, MRI-based functional and structural evaluations of the neuronal circuitry in the two main limbic regions (the hippocampal and amygdala) may serve as useful diagnostic neuro-biomarkers. This could accommodate an individualized probiotic selection that enhances traditional clinical intervention in MDD.

Through the integration of artificial intelligence, machine learning opens up new possibilities in the healthcare industry by providing a comprehensive, integrated, and futuristic approach to managing clinical depression. This approach complements clinical management of health conditions by enabling objective measurement of integrated large-scale datasets through computational software. By addressing a number of clinical issues, such as treatment selection, patient compliance, diagnosis prediction, and administrative duties, machine learning offers an integrated approach to patient care. In the end, this fits into the precision medicine and individualized patient care paradigm. Combining machine learning with the consortia of the characteristic, endophenotypic, neurobiological, and genetic censuses to optimize management and treatment options using probiotics, this application converges in the context of probiotics in clinical depression. A few studies have started to investigate the machine learning application in precision psychiatry, probiotics, and the microbiome, even though this is still a futuristic vision.

In the context of clinical depression, another microbial application to consider is metabiotics, which has been described as the evolutionary concept that builds upon the understanding of probiotics. In order to achieve desired health outcomes, it entails engineering the human gut microbiota through natural selection buffered with probiotics or synthetic manufacturing. Metabiotics are probiotic bacteria's metabolites, structural elements, and/or signaling molecules with a recognizable molecular structure that can enhance physiological functions in the host and regulate metabolic and/or behavioral outcomes related to the host microbiota's activity. By utilizing probiotics with established chemical formulas, dosages, safety profiles, and durability, this concept facilitates the selection of novel therapeutics and expands the scope of microbial-based applications beyond the psychiatric domain to encompass a wide range of human health (Carabotti et al., 2015; Kali, 2016; Misra et al., 2019).

4.3.3. Limitations and Translational Gaps

In medical research, the most important factor to take into account when creating and implementing a suggested clinical intervention is frequently the transferability of data from animal models to humans. Probiotics have shown promise in the treatment of depression, primarily in pre-clinical settings with rodent or mouse models. Rats and mice make excellent models for microbial manipulation with probiotics because their gut microbiomes are similar to those of humans. When compared to a human cohort, however, inducing depressive phenotypes in pre-clinical models is a one-dimensional exploration of depression due to the highly heterogeneous nature of both the human gut microbiota and depressive disorder, as well as the variable host and environmental factors in a human setting. In order to better understand how the gut microbiota affects clinical manifestations, it may be necessary to standardize the composition of the gut microbiota and the conditioning of research models, as indicated by the inconsistent response to probiotics in pre-clinical models.

But standardizing the human gut microbiome is not feasible. Notwithstanding these drawbacks, since the early 2000s, probiotic clinical trials have been emerging with encouraging outcomes in depression models, indicating the potential effectiveness of probiotics in treating depression. It has been shown that depressed human subjects have significantly more gut dysbiosis than healthy subjects do. This has opened the door for the use of probiotics as human gut microbiota-modulating agents to produce the intended anti-depressive effects. The ability of probiotics to exert anti-depressive effects primarily through gut microbiota modulation reflects the dynamic nature of probiotics' mechanism of action, notwithstanding the hosts' distinct indigenous microbial compositions, even though scientists have not yet established a definitive pattern of gut dysbiosis in depression. The varied age groups, health status, lifestyle, and receptivity of the study subjects, as well as the technical aspects of the studies, may be the cause of the observable discrepancies and inconsistencies among the findings of various studies on the gut microbial profile of depressed subjects, particularly in terms of identified bacterial genus and species.

Furthermore, a number of factors need to be taken into account before probiotics can be approved and validated as a clinical treatment modality. These include host factors, probiotic-related factors (such as strain type, dose, administration method, and clinical outcome), and the practical applicability and feasibility of probiotics in a clinical setting. It is difficult to choose probiotics specifically for depression because there are few studies, especially clinical trials, available. Depending on the target health condition and strain of the same bacterium, different probiotics have different beneficial qualities. The type of depression, the host's factors, and the probiotic strains are the categorical factors to take into account when choosing the right probiotics for depression. The ideal effectiveness of probiotics is determined by the kind and mix of strains as well as the origin of the strains (human versus non-human).

Compared to non-human origin strains, the human-origin strains are more adaptable to potentially hostile physiological environments because of the different pH levels along the digestive tract. The latest research has also shown that probiotic strains in combination produce better health results than probiotic strains alone. Together, probiotic strains of *Bifidobacterium* and *Lactobacillus* spp. continue to be a great choice due to their strong anti-depressant effects. While there have not been any documented side effects from probiotic use in MDD patients' clinical trials to date, it is still important to consider the possibility when administering probiotics to critically ill and immunocompromised patients. It is also important to pay attention to the possible risk that probiotics pose to elderly, pregnant, and cancer patients. Probiotic side effects have been documented to include gastrointestinal issues, skin issues, and, in the worst-case scenario, septicemia. These findings highlight the need for caution and additional research before administering probiotics to a patient population that is already at risk.

When it comes to MDD, medical professionals must assess the total clinical presentation as well as any potential biomarkers within the arrays of the microbiome in order to choose the right probiotics to address this extremely variable disorder. The host factors—which take into account the dynamic microbial changes that happen throughout an individual's lifespan due to aging, lifestyle, geographic influences, host immune status, and pre-existing comorbidities—would be the most difficult factor to take into account when choosing probiotics. This could have an impact on a person's level of probiotic responsiveness. Implementing the suggested microbial-based assessments may also be hampered by a lack of funding and resources, especially in remote clinical settings (Carabotti et al., 2015; Kali, 2016; Misra et al., 2019).

REFERENCES

1. Akkasheh, G., et al. (2016). Clinical and metabolic response to probiotic administration in patients with major depressive disorder: A randomized, double-blind, placebo-controlled trial. *Nutrition, 32*(4), 315–320. https://doi.org/10.1016/j.nut.2015.09.003.
2. Araya, R., Zitko, P., Markkula, N., Rai, D., & Jones, K. (2018). Determinants of access to health care for depression in 49 countries: A multilevel analysis. *Journal of Affective Disorders, 234*, 80–88.
3. Bäckhed, F., Ley, R. E., Sonnenburg, J. L., Peterson, D. A., & Gordon, J. I. (2005). Host-bacterial mutualism in the human intestine. *Science, 307*(5717), 1915–1920.
4. Bambury, A., et al. (2018). Finding the needle in the haystack: Systematic identification of probiotics. *British Journal of Pharmacology, 175*(12), 2491–2506. https://doi.org/10.1111/bph.14127.
5. Bruce-Keller, A. J., Salbaum, J. M., & Berthoud, H. R. (2018). Harnessing gut microbes for mental health: Getting from here to there. *Biological Psychiatry, 83*(3), 214–223.
6. Carabotti, M., et al. (2015). The gut-brain axis: Interactions between enteric microbiota, central and enteric nervous systems. *Frontiers in Neuroscience, 9*, 85. https://www.ncbi.nlm.nih.gov/pmc/articles/PMC4367209 (accessed on 9 December 2024).
7. Dinan, T. G., et al. (2013). Psychobiotics: A novel class of psychotropic. *Biological Psychiatry, 74*(10), 720–726. https://doi.org/10.1016/j.biopsych.2013.05.001.
8. Dinan, T. G., et al. (2017). The microbiome-gut-brain axis in health and disease. *Gut Microbes, 8*(2), 125–136. https://doi.org/10.1016/j.gtc.2016.09.007.
9. Evrensel, A., et al. (2015). The gut-brain axis: The missing link in depression. *Clinical Psychopharmacology and Neuroscience, 13*(3), 239–244. https://doi.org/10.9758/cpn.2015.13.3.239.
10. James, S. L., Abate, D., Abate, K. H., et al. (2018). Global, regional, and national incidence, prevalence, and years lived with disability for 354 diseases and injuries for 195 countries and territories, 1990–2017: A systematic analysis for the Global Burden of Disease Study 2017. *The Lancet, 392*(10159), 1789–1858.
11. Kali, A. (2016). Psychobiotics: An emerging probiotic in psychiatric practice. *British Journal of Psychiatry, 208*(6), 522–523. https://doi.org/10.1016/j.bj.2015.11.004.
12. Liu, C. S., et al. (2017). Evidence for inflammation-associated depression.

In *Inflammation-associated Depression: Evidence, Mechanisms and Implications* (pp. 85–106). Springer. https://doi.org/10.1007/7854_2016_2.

13. Misra, S., & Mohanty, D. (2019). Psychobiotics: A new approach for treating mental illness? *Critical Reviews in Food Science and Nutrition, 59*(8), 1230–1236.
14. Misra, S., et al. (2017). Psychobiotics: A new approach for treating mental illness? *Critical Reviews in Food Science and Nutrition, 57*(8), 1226–1235. https://doi.org/10.1080/10408398.2017.1399860.
15. National Center for Complementary and Integrative Health (2016). Probiotics: In depth. Retrieved from: https://nccih.nih.gov/health/probiotics/introduction.htm (accessed on 9 December 2024).
16. Pinto-Sanchez, M. I., et al. (2017). Probiotic *Bifidobacterium longum* NCC3001 reduces depression scores and alters brain activity: A pilot study in patients with irritable bowel syndrome. *Gastroenterology, 153*(2), 465–473. https://doi.org/10.1053/j.gastro.2017.05.003.
17. Rathod, S., Pinninti, N., Irfan, M., Gorczynski, P., Rathod, P., Gega, L., & Naeem, F. (2017). Mental health service provision in low- and middle-income countries. *Health Services Insights, 10*, 1178632917694350.
18. Ruscio, A. M., Hallion, L. S., Lim, C. C., et al. (2017). Cross-sectional comparison of the epidemiology of DSM-5 generalized anxiety disorder across the globe. *JAMA Psychiatry, 74*(5), 465–475.
19. Sagar, R., Dandona, R., Gururaj, G., et al. (2020). The burden of mental disorders across the states of India: The Global Burden of Disease Study 1990–2017. *Lancet Psychiatry, 7*(2), 148–161.
20. Sarkar, A., et al. (2016). Psychobiotics and the manipulation of bacteria-gut-brain signals. *Trends in Neurosciences, 39*(11), 721–723. https://doi.org/10.1016/j.tins.2016.09.002.
21. Wallace, C. J. K., et al. (2017). The effects of probiotics on depressive symptoms in humans: A systematic review. *BMC Psychiatry, 17*(1), 138. https://doi.org/10.1186/s12991-017-0138-2.
22. Wilkins, T., et al. (2017). Probiotics for gastrointestinal conditions: A summary of the evidence. *American Family Physician, 96*(3), 170–177. Retrieved from: https://www.aafp.org/afp/2017/0801/p170.html (accessed on 9 December 2024).
23. Zhou, L., et al. (2015). Psychobiotics and the gut-brain axis: In the pursuit of happiness. *Neuropsychiatric Disease and Treatment, 11*, 215–224. https://doi.org/10.2147/NDT.S61997.

CHAPTER 5

MATERNAL–FETAL GUT MICROBIOTA AXIS

Contents

INTRODUCTION

The entire population of microorganisms that live in a person's organs and tissues and primarily function in a commensal or symbiotic manner is referred to as their "microbiota." The microbiota is made up of bacteria, fungi, viruses, and archaea, of which the bacterial microbiota is the most well-studied and best-characterized.

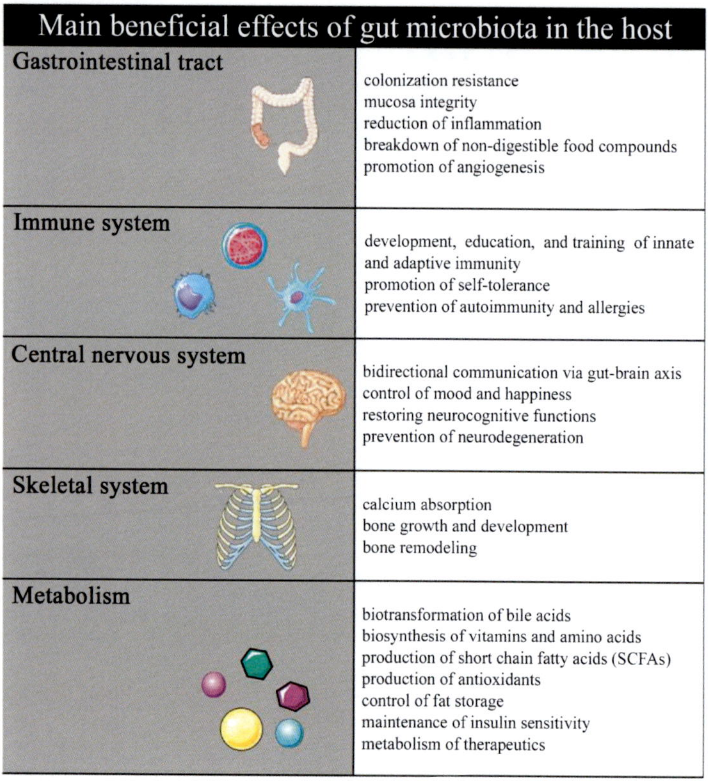

Figure 5.1. *Suggested functional effects of gut microbiota in host organism.*

Source: Miko E, Csaszar A, Bodis J, Kovacs K. (2022). The maternal-fetal gut microbiota axis: Physiological changes, dietary influence, and modulation possibilities. Life (Basel);12(3):424. doi: 10.3390/life12030424. PMID: 35330175; PMCID: PMC8955030.

The human body is home to trillions of bacteria, but the most densely colonized area is the gastrointestinal tract (GIT), where concentrations of bacteria range from 101–103 per gram of tissue in the upper intestine to 10^{11}–10^{12} per gram of tissue in the colon. Examining the proportion of intestinal bacterial taxa, the species Firmicutes (e.g., *Lactobacillus*, *Enterococcus*, Clostridial, and *Bacteroidetes* (e.g., *Bacteroides*) account for the majority, with Actinobacteria (Bifidobacteria), Proteobacteria (*Escherichia coli*), Fusobacteria, and Verrucomicrobiota having less dominance. The human microbiome, particularly the intestinal microbiome, has been shown in recent years to play a significant role in various physiological processes and to have a significant impact on human health. It is believed that the intestinal bacterial community takes part in immunological, metabolic, and biochemical processes (Figure 5.1).

It can be assumed that maternal and gestational factors as well as environmental exposures during pregnancy could affect the healthy development and composition of the fetal, neonatal, and infant gut microbiota and, consequently, the health issues of the offspring. This is because the human gut microbiota is thought to begin developing even before birth. Thus, the primary goal of this review is to provide an overview of the key traits of maternal gut microbiota and discuss potential implications for fetal development and the health of offspring. We will talk about the effects of maternal dietary factors and modulation opportunities on pregnancy microbiota because, in addition to genetic and environmental factors, nutrition is a major determinant factor affecting the composition and function of gut microbiota.

5.1. MATERNAL GUT MICROBIOTA

The state of pregnancy is difficult for the mother's body. It needs to go through a number of significant physiological changes in order to meet fetal requirements and maintain self-integrity. Changes in the immune system, metabolism, and endocrine system are the main aspects of maternal adaptation. It is commonly known that progesterone, estrogen, and thyroid hormone levels rise noticeably during pregnancy. The focus of metabolic changes is on the growing nutritional and energy requirements of neonates. This leads to marked increases in food intake, insulin secretion, and lipogenesis, which resemble the symptoms of the metabolic syndrome. Maternal-fetal immune tolerance refers to the extremely specific and tight regulation of tolerogenic and proinflammatory immune responses resulting from immunological changes. These immune mechanisms facilitate both adequate placentation and the successful implantation of the embryo, as well as the restoration of the mother's antimicrobial immunity. On the periphery, mild systemic inflammation is a characteristic of a healthy human pregnancy.

Over the past 10 years, it has become increasingly evident that pregnancy has an impact on the makeup of the mother's gut microbiota, suggesting yet another significant pregnancy-related maternal alteration that may have implications for the development of the fetus or neonate. The gut microbial community's makeup varies greatly from the first to the third trimester of pregnancy, although in the first trimester it is similar to that of healthy, nonpregnant women. Proteobacteria, Bifidobacteria, and lactic-acid-producing bacteria (some particular strains of *Lactobacillus*) start to grow in tandem with the decrease in butyrate-producing bacteria from the second trimester onward.

Overall, the microbiota of the gestational gut is low in terms of intraindividual bacterial diversity (alpha diversity index) and high in terms of interindividual bacterial diversity (beta diversity index), with the majority of the most noticeable changes taking place during the third trimester. Changes in the gut microbiota during pregnancy have functional implications, as demonstrated by the intestinal microbiota transfer from pregnant women to germ-free mice.

Mice with altered microbiota transplants gained weight and displayed insulin resistance-related impaired glucose tolerance. These findings imply that the well-known metabolic alterations seen in pregnant women may be influenced by the changes in the intestinal microbiota that have been documented during pregnancy. Furthermore, it is believed that a higher *Proteobacteria* ratio stimulates the immune system, resulting in stronger local inflammatory reactions. In turn, inflammation promotes bacterial translocation and increases the permeability of the gut mucosa.

This offers a plausible explanation, at least partially, for the mild systemic inflammation detected in the peripheral blood of healthy expectant mothers. It should be noted that alterations in the composition of the maternal microbiota may be caused by a variety of patient-related factors, including diet, BMI prior to conception, weight gain during pregnancy, and metabolic diseases, as well as population-level factors, such as geography, ethnicity, and environmental factors (Miko et al., 2022).

5.2. ESTABLISHMENT OF THE MATERNAL–FETAL GUT MICROBIOTA AXIS

One important area of research and theory is the effect of the mother's gut microbiota on the growth and development of the fetus. There are two main, distinct pathways that have been suggested for how the fetal side could be significantly impacted by the gestational intestinal microbiota (Figure 5.2).

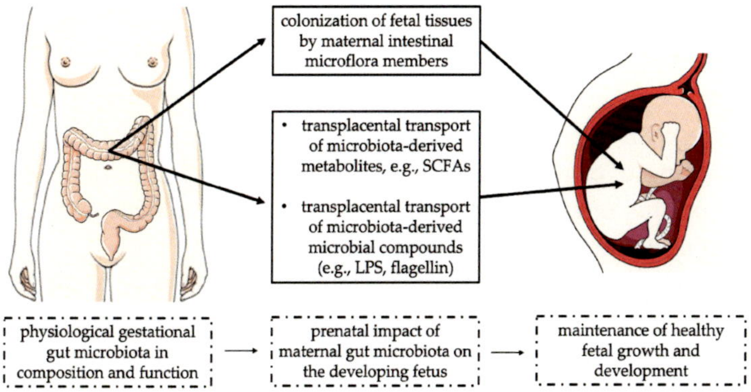

Figure 5.2. *Possible action of maternal gut microbiota on the developing embryo during pregnancy.*

Source: Miko E, Csaszar A, Bodis J, Kovacs K. (2022). The maternal-fetal gut microbiota axis: physiological changes, dietary influence, and modulation possibilities. Life (Basel); 12(3):424. doi: 10.3390/life12030424. PMID: 35330175; PMCID: PMC8955030.

5.2.1. Placental Microbiota

According to one theory, the presence of bacteria has direct and advantageous effects, assuming that maternal microbes colonize fetal tissues long before birth. The transfer of prenatal microbes from the mother's digestive system to the developing fetus is merely a theoretical idea that needs further empirical research. In fact, the prevailing assumptions state that microorganisms at the placenta are transferred in some way from the intestinal epithelium into the bloodstream. Pregnancy-related changes in the composition of the mother's gut microbiota may promote the bacterial transition. Pregnancy-related increases in the phylum Proteobacteria are thought to be linked to proinflammatory changes in the bacterial environment, including elevated levels of fecal calprotectin and serum endotoxin and zonulin. These changes also enhance mucosal permeability and microbiota translocation, which is the cellular uptake and occurrence of intestinal microbiota members in extraintestinal tissues and maternal circulation, likely leading to colonization of the fetal gut in utero.

Up until the last 20 years, there has been agreement on the postnatal establishment of the neonatal gut microbiota's origin. It was believed that the uteroplacental unit was sterile. It was previously believed that the presence of bacteria was linked to colonization and subsequent infection, usually by the ascending route, which resulted in chorioamnionitis. With the understanding that the sterile environment would shield the fetus from infections, the "sterile womb" doctrine was widely accepted.

New theories regarding the transmission of microbes from mother to fetus in utero have been supported by data derived from noncultivation, PCR, and DNA sequencing over the past 20 years. Strong evidence from animal studies has reinforced this mode of transmission. Mice that were nursing or pregnant showed increased bacterial translocation from the gut to extraintestinal tissues. It has been demonstrated that foreign bacteriophage DNA given orally to mice can withstand in the digestive tract and pierce the intestinal epithelium. It can also be found in fetuses and newborn animals via the transplacental pathway. Additionally, during pregnancy, genetically labeled bacteria were transferred to the fetal gut and amniotic fluid by oral inoculation of pregnant mice with an *E. faecium* strain. The murine fetus appears to be exposed to non-cultivable bacteria in late gestation and viable and cultivable bacteria in midgestation, which is an interesting observation.

It was discovered that the predominant phylum Proteobacteria dominated the human placental microbial community. The species of *Prevotella* and *Neisseria* suggested the hematogenic route of seeding from the oral cavity to the placenta. The composition was similar to the oral microbiota. It is noteworthy that the parodontopathogenic strains of *Porphyromonas gingivalis* and *Aggregatibacter actinomycetemcomitans* in the subgingival plaque rise concurrently with the number of viable oral microbiota during pregnancy. Inflamed placental and fetal complications were caused in pregnant mice by oral infections with *Campylobacter rectus*, *Porphyromonas gingivalis*, or *Fusobacterium nucleatum*. Pregnant patients with periodontal disease had a higher chance of experiencing pregnancy problems in humans. This observation may be explained by an enhanced bacterial transition to the uteroplacental unit from the inflamed oral mucosa with increased permeability.

Studies on the microbiota have not only expanded to include the placenta but have also revealed the presence of physiological bacteria in the meconium, fetal gut, amniotic fluid, and umbilical cord. They were all previously regarded as sterile. As more research is being done that suggests fetal exposure to mother microbiota while still in utero, questions have been raised about how to interpret the findings. One major problem with the potential for data distortion is the possibility of contamination of samples with low microbiota densities. Keeping oneself clean is extremely difficult because there are many sources. Samples may be contaminated at the time of or even prior to collection, and laboratory reagents (such as PCR reagents and nucleic acid extraction kits) may contain trace amounts of bacterial DNA. Another crucial matter is selecting the best analysis platform. Even among experts in the field, there is disagreement regarding the existence of in utero exposure of fetal tissues to members of the maternal microbiota. The reality that germ-free animals are typically created from non-germ-free pregnancies by embryo transfer after hysterectomy provides additional evidence against the in utero colonization theory. If members of the maternal microbiota had already interacted with mammalian fetuses during pregnancy, this could not be the case. Even for the harmless microbiota's entry, the intestinal epithelial barrier typically functions as a physiological barrier. Bacteria from the gut lumen can be taken up intracellularly by dendritic cells (DCs), which then localize the bacteria to lymph nodes. The bacterial spread may then proceed in a broad manner, e.g., into the bloodstream, which ultimately leads to the trafficking of pregnant women. Increased bacterial sequestering in mouse mesenteric lymph nodes (MLNs) supported the idea that the fetus is exposed to the hematogenic mother intestinal microbiota through translocation.

Bacteria most likely become noticeable once they reach the fetal site. Toll-like receptors (TLRs) are among the most viable means of identifying foreign structures (TLRs). Around 10 transmembrane proteins are represented by the family of human TLRs. In most human tissues, they are found in the intracellular vesicles or on the cell surface of the primary sentinel cells of innate immunity, which include mast cells, dendritic cells, and macrophages. Recognizing pathogen-associated molecular patterns (PAMPs), which are widely shared by bacteria but not by the host, TLRs serve as conserved innate immune receptors. Different TLR types can recognize different bacterial structures, which can trigger an inflammatory response and activate the classical immune system against the pathogen.

Although TLRs express themselves at similar levels on newborn monocytes, the TLR pathway is much less activated than in adults. Impaired phosphorylation activity of TLR-induced protein kinases, restricted expression of TLR-associated intracellular signaling proteins, and decreased levels of proinflammatory cytokines produced by neonatal monocytes all point to immature innate immunity and ongoing immune development during the perinatal period. Fetal recognition of members of the maternally derived microbiota exposed in utero would probably result in an insufficient immune response within this immune milieu, favoring immune tolerance of the translocated bacteria. Recent research has demonstrated that second-trimester fetal tissues contain effector memory T cells, which lends credence to this theory. It should be emphasized once more, though, that there is disagreement and controversy surrounding the idea that the prenatal microbiota exists before birth (Miko et al., 2022).

5.2.2. Effects of Microbiota-Derived Molecules

There exists an indirect pathway whereby the maternal intestinal microbiota regulates fetal growth and development. It is believed to be mediated by metabolites produced by the microbiota that pass through the placenta and into the developing fetus. These soluble factors are either metabolites of substances absorbed from the intestinal lumen or are produced endogenously by microbiota members.

Animal studies provide one of the strongest arguments in favor of this theory. A nonpathogenic *E. coli* strain was able to colonize germ-free pregnant mice in a reversible manner, which altered the development of the offspring's intestinal innate immune system. The observed proliferation of innate lymphoid cells type 3 (ILC3s) suggests the presence of aryl hydrocarbon receptor (AhR) ligands derived from the mother microbiota. ILC3s are an essential innate cell population for intestinal barrier functions and host defense. Furthermore, the offspring's gut epithelium's gene expression profile is changed by maternal colonization. Among offspring born to mothers who had undergone reversible colonization during pregnancy, there was modulation in the expression of genes encoding homeostasis, integrity, and differentiation (upregulated gene networks for cell division and differentiation, mucus and ion channels, metabolism of xenobiotics, bile acids, complex lipids, and sugars).

The primary soluble byproduct of bacterial metabolism, SCFAs are thought to have a significant influence on a person's health problems. The intestinal epithelium absorbs them, and the blood then carries them to the tissues. Pregnancy-related SCFA concentrations e.g., the cecum's concentrations of acetic, propionic, butyric, and caproic acids rise noticeably. Acetic acid is the predominant SCFA in both pregnant women and their fetuses. G-protein receptors (GPR), particularly GPR41 and GPR43, allow SCFAs to function as signaling molecules. GPR41 and GPR43 receptors in the uterus allow placental SCFAs from the mother's gut microbiota to be sensed. A number of studies conducted on mice showed that SCFAs are beneficial to embryo development. In muscle and brown adipose tissue, SCFAs are in charge of raising free fatty acid oxidation and mitochondrial activity. They also had a positive impact on metabolism in the fetus, primarily through regulating insulin levels. SCFAs have a significant effect on immune regulatory mechanisms in particular as the immune system develops. They might regulate and modulate immune reactions, which would stop both excessive inflammation and autoimmunity.

The influence of GPR41 signaling on the development of the nervous system is another major function of SCFAs in the fetus that has been suggested. The third trimester of pregnancy is marked by enhanced maternal gut microbiota and is also a critical time for the development of certain areas of the brain, including synaptogenesis and myelination. Enhanced levels of metabolites derived from the microbiota, like SCFAs, may be advantageous for the development of neurons. SCFAs have the ability to regulate the integrity of the intestinal barrier as well, primarily through the transcriptional regulation of proteins related to tight junctions.

Aside from living bacteria, many endogenous microbial compounds (e.g., lipopolysaccharide (LPS) or flagellin) can enter fetal tissues and bind to innate pattern recognition receptors (PRRs) (TLRs, for example, as previously discussed). Experiments on mice showed that the structural components of bacteria that are necessary for the maturation of the host immune system continuously penetrated various tissues. It is believed that antigen recognition from the mother's gut microbiota causes activated T cells, which are already present in the developing mouse fetal gut. Thus, it is possible that components of the mother's gut microbiota can enter fetal compartments and cause recognition even in the absence of bacterial trafficking. For the immune system to acquire the ability to differentiate pathogen types from microbiota in the future, it needs to be exposed to bacteria or molecules derived from them (Miko et al., 2022).

5.3. EFFECTS OF MATERNAL NUTRITIONAL FACTORS ON GUT MICROBIOTA AND OFFSPRING'S HEALTH

Adverse preconceptional, gestational, and postnatal factors, primarily of maternal origin, can be linked, at least partially, to the offspring's later health issues, according to epidemiological, clinical, and basic science studies. Dietary factors may have the greatest determining power with regard to the composition and function of the gut microbiota.

5.3.1. High-Fat Diet and Maternal Obesity

The type of diet, weight, and nutritional status of the mother have a significant impact on the developing embryo. Intestinal microbiota during pregnancy may be a conduit through which the mother influences the health of the fetus. This is believed to be regulated by dietary factors, at least in part. Excessive amounts of processed foods, dietary fat, and sugars characterize the typical Western diet. Such a diet is linked to poor mother and child health outcomes, excess weight gain, and dysbiosis of the gut. Fetal overgrowth, macrosomia, congenital defects, stillbirth, decreased neonatal Apgar score, preterm delivery, child morbidity, respiratory complications, and neonatal mortality are just a few of the many fetal developmental traits linked to maternal obesity.

It is commonly known that obese people have significantly changed gut microbiota. First, there is less microbial diversity and a greater abundance of the phylum Firmicutes than Brachyceroidetes. Rats that were fed an obesity-inducing diet showed similar results in terms of gut microbiota composition, with fecal *Bifidobacterium* species being less abundant overall and a greater relative abundance of Clusters XI and I of *Clostridium*. Furthermore, studies on mice showed that lean germ-free mice could acquire the obese phenotype through fecal microbiota transplantation (FMT). The gut microbiota is further altered by pregnancy itself. Pregnant women who were overweight had lower levels of *Bifidobacterium* and *Bacteroides* and higher levels of Staphylococcus, Enterobacteriaceae, and *Escherichia coli* when compared to normal-weight women. Reductions in butyrate, hydrogen, and methane production, as well as an increase in mucus degradation and local inflammation, are hypothesized to be linked to these microbiota changes. It is believed that during pregnancy, maternal adherence to the recommended dietary intake of fat and fiber is linked to advantageous changes in the composition of the gut microbiota, such as increased gut microbiota richness. Changes in the mother's microbiota may be passed on to the unborn child both during pregnancy and delivery. The bacterial microbiota pattern of infants born to obese mothers differs from that of infants born to lean mothers. These differences persist for a minimum of a year, indicating the long-term effects of obesity in mothers on the gut microbiota of their offspring. Similar results were seen in a primate model where the intestinal microbiome of the offspring in Japanese macaques was altered by a high-fat maternal diet that included 36% fat from lard, butter, animal fat, and safflower oil.

The primary alterations in humans are variations in the prevalence of *Bacteroides* species. Enterococci spp. Acinetobacter species. *Pseudomonas* spp. *Blautiana* spp. *Eubacterium* spp., Oscillibacter species along with *Faecalibacterium* spp. There are indications that *Lactobacillus* species may be more prevalent and a reduced prevalence of *Bacteroides* species. The gut microbiota during infancy may indicate a child's likelihood of becoming obese or overweight. These data collectively lend credence to the idea of a harmful generational cycle of microbiota patterns being passed down and linked to excessive weight gain and adverse metabolic development (Miko et al., 2022).

5.3.1.1. Gestational Diabetes Mellitus: A Special Case

Pregnancy is the primary cause of gestational diabetes mellitus (GDM), a disease characterized by abnormal glucose tolerance brought on by insulin resistance. The gold standard for diagnosing the condition is an oral glucose tolerance test (OGTT) conducted between 24 and 28 weeks. Metabolic disorder phenotypes are similar in obesity and GDM. A poor diet with high fat, high sugar, and low fiber intake is one of the main hypothesized mechanisms that could explain insulin resistance and the onset of GDM during pregnancy.

Changes in the diversity, composition, and imbalanced SCFA proportions of intestinal microorganisms are associated with GDM. There are noticeable variations in the microbiota in every trimester. A few studies concentrate on the dynamic changes in the mother's gut microbiota that occur during pregnancy and the development of GDM. The microbiota profile during pregnancy may serve as a biomarker to identify GDM early on and forecast the disease's course. The *Ruminococcaceae* family and glucose levels were found to positively correlate, with a higher odds ratio for the diagnosis of GDM.

The increased relative abundance of the *Ruminococcaceae*, *Lachnospiraceae*, and *Enterococcaecea* families; enrichment of *Bacteroides*, *Blautia*, *Collinsella*, and *Eggerthella* bacteria; reduction in *Faecalibacterium* levels; and decrease in alpha diversity in the GDM groups compared to healthy pregnancy were the main findings regarding microbiota changes. The ability of the GDM gut microbiota to function demonstrated a correlation with enhanced sugar membrane transport, oxidative stress responses, reduced butyrate biosynthesis, and branched-chain amino acid transport. Also investigated was the potential impact of GDM on the fetal and neonatal microbiota. The functional ability of the gut microbiota in GDM was found to be associated with enhanced sugar membrane transport, oxidative stress responses, reduced butyrate biosynthesis, and branched-chain amino acid transport. Lower concentrations of the Acinetobacter genus and *Pseudomonadales* order are found in the placental microbiota of women with GDM. Furthermore, a more detrimental metabolic and inflammatory phenotype was linked to a decrease in placental Acinetobacter. The meconium microbiota of children born to GDM mothers had higher abundances of *Lactobacillus* and *E. coli* and decreased alpha diversity. The intestinal microbiota of adult patients has been found to contain the same bacterial taxa that are enriched in the meconium microbiota of infants born to mothers with diabetes. The microbiota of newborns and mothers differed in the same way when analyzed from various body sites shortly after birth, indicating that the microbial variations seen in GDM are consistent across generations.

5.3.2. Vegetarian Diet

Vegetarian and plant-based diets are becoming increasingly popular. It is believed that following this diet will reduce the risk of obesity, diabetes mellitus, cerebrovascular disease, heart disease, and chronic kidney disease. The gut microbiota is believed to be altered by vegetarian diets, with positive effects on the host (e.g., a decrease in Proteobacteria and Firmicutes, an increase in Bifidobacteria, *Lactobacilli*, *Roseburia*, and *Ruminococcus*, and a higher synthesis of SCFAs). The gut microbiota of vegetarian pregnant women is a subject of limited information. In contrast to pregnant omnivorous women, one study indicated that pregnant vegetarians had lower beta diversity but no difference in alpha diversity of their intestinal microbiota. The relative frequency of a number of genera varied amongst vegetarians as well (*Roseburia* and *Lachnospiraceae* increased, *Holdemania* and *Collinsella* decreased). These modifications may lead to increased SCFA levels, which are linked to a healthier intestinal mucosa and reduced levels of inflammation. There is no information on how a vegetarian mother's gut microbiota affects her unborn child's health (Miko et al., 2022).

5.3.3. Artificial Sweeteners

In the United States, non-nutritive sweeteners (e.g., acesulfame-K, sucralose, and aspartame) are widely used in daily diets. This is presumably because people are becoming more conscious of the detrimental effects that sugar has on health, even though some artificial sweeteners appear to alter the makeup of the gut bacterial community while avoiding the colonic microbiota entirely. Artificial sweeteners have been demonstrated, mostly through animal experiments, to change the composition of the gut microbiota and impact adult and progeny bacterial taxa (depletion of *Akkermansia muciniphila* and increase in *Bacteroides*, *Lactobacillus*, and *Clostridiales*).

Moreover, they cause an increase in body weight concurrently with the activation of bacterial genes involved in energy metabolism, including those involved in glycolysis, sugar transport, and carbohydrate absorption. Genes for bacterial proinflammatory mediators have also been demonstrated to rise that were depleted in the microbiota following sweetener administration, *Akkermansia muciniphila* is a helpful bacterium linked to normal weight, balanced serum glucose, and anti-inflammatory effects in the intestines.

Human data, meanwhile, further lend credence to the idea that artificial sweeteners negatively impact the gut microbiota of their progeny. One-year-old infants whose mothers regularly consumed artificial sweeteners had higher BMIs, indicating that alterations in the infant gut microbiota may be partially responsible, as variations in certain metabolites linked to the microbiota may promote weight gain.

5.3.4. Alcohol

In addition to artificial sweeteners, alcohol consumption during pregnancy and its potential effects on the development of intestinal microbiota are also understudied. It is commonly known that frequent alcohol consumption during gestation has also been linked to a number of newborn disorders. Long-term and heavy alcohol use impacts the integrity of the gastrointestinal mucosa and, in turn, the composition of the gut microbiota. It was found to be linked to lower levels of butyrate-producing *Clostridiales*, *Bifidobacterium*, and *Lactobacillus*, increased gut permeability, and inflammation. It was also linked to lower levels of intestinal microbiota members like *Roseburia, Faecalibacterium, Blautia*, *Bacteroides*, and *Lachnospiraceae*. Local inflammation and enhanced alcohol metabolism are the outcomes of this altered microbiota pattern. After being exposed to ethanol, less *Bacillus* bacteria were seen in a study involving pregnant mice.

The effects of alcohol consumption during pregnancy on the pre- and neonatal microbiota are poorly understood. In a human study, the gut microbiota of newborns whose mothers acknowledged drinking alcohol during their pregnancies showed an elevated ratio of the Megamonas genus. It is interesting to note that Megamonas has been linked to major depressive disorders (MDDs), and it is commonly knowledge that alcohol consumption by mothers has an impact on the cognitive and behavioral development of their newborns, including autism and depression (Miko et al., 2022) (Table 5.1).

Table 5.1. Effects of Maternal Dietary Factors on Offspring's Microbiota Composition

Study Population	**Investigated Fetal Side Microbiota**	**Method**	**Main Findings**
Maternal High-Fat Diet, Obesity			
Infants of obese mothers (n = 16) vs. infants of normal-weight mothers (*n* = 26)	Infant fecal samples at 1 and 6 months of age	FISH qPCR	Higher weights of mothers were correlated with higher concentrations of *Bacteroides*, *Clostridium*, and *Staphylococcus*, and lower concentrations of the *Bifidobacterium* group prevalence of *Akkermansia muciniphila*, *Staphylococcus*, and *Clostridium difficile* groups were lower in infants of normal-weight mothers

Children of obese ($n = 26$) vs. nonobese mothers	Fecal samples from children 18–27 months of age	16S ribosomal RNA (rRNA) sequencing)	Effects of maternal obesity on offspring's gut microbiota were stronger among children of mothers of higher socioeconomic status Higher alpha and beta diversity in children of obese vs. nonobese mothers Children born to obese vs. nonobese mothers had greater abundances of *Parabacteroides* spp. and *Oscillibacter* spp., as well as lower *Blautia* spp. and *Eubacterium* spp.
Neonates ($n = 18$) born vaginally (5 to overweight mothers), neonates ($n = 56$) by elective C-section (26 to overweight mothers)	Second-day fecal samples from neonates	16S ribosomal RNA (rRNA) sequencing	Compared to neonates delivered vaginally to normal-weight mothers, microbiota of neonates born to overweight or obese mothers were enriched in *Bacteroides* and depleted in *Enterococcus*, *Acinetobacter*, *Pseudomonas*, and *Hydrogenophilus*
Gestational Diabetes Mellitus			
Newborns ($n = 23$): 5 from mothers with DM, 5 from mothers with GDM, 13 from mothers with no diabetes	Meconium samples	16S ribosomal RNA (rRNA) sequencing	The phylum *Bacteroidota* and the genus *Parabacteriodes* were enriched in the meconium in the DM group compared to the nondiabetes group

Placentas from women with GDM (n = 11) and from control women (n = 11)	Placenta	16S ribosomal RNA (rRNA) sequencing	*Pseudomonadales* and *Acinetobacter* showed lower relative abundance in women with GDM compared to control Increase in placental *Acinetobacter* ratio was associated with a more adverse metabolic and inflammatory phenotype
Pregnant women and their neonates with and without GDM	Oral, pharyngeal, meconium, and amniotic fluid samples	16S ribosomal RNA (rRNA) sequencing	In the amniotic fluid of the GDM group, a lower relative abundance of *Anoxybacillus* and a higher relative abundance of *Corynebacterium* were detected In the meconium of the GDM group, a lower relative abundance of *Corynebacterium* and a higher relative abundance of *Enterobacte*r were detected Microbes varied by the same trend between the maternal and neonatal microbiota
Vegetarian Diet			
Artificial Sweeteners			

Infants (*n* = 100) selected based on maternal sweetener consumption during pregnancy (50 nonconsumers and 50 daily consumers)	Infant fecal samples at 3 and 12 months of age	16S ribosomal RNA (rRNA) sequencing	Maternal sweetener consumption did not differ between clusters reflecting the maturation of gut microbiota but was associated with community-level shifts in infant's gut bacterial taxonomy structure and depletion of several *Bacteroides* sp. in a certain cluster Nine bacterial taxa from *Bacteroides* sp. were enriched or depleted at high levels of maternal sweetener consumption at 12 months of age. Daily maternal sweetener consumption is associated with higher infant weight and altered microbiota composition
Pregnant women and their neonates with (*n* = 10) and without (*n* = 19) alcohol consumption	Fecal samples of newborns within 48 h	16S ribosomal RNA (rRNA) sequencing	A positive relationship showed between *Megamonas* and newborns with maternal alcohol consumption

Source: Miko E, Csaszar A, Bodis J, Kovacs K. (2022). The maternal-fetal gut microbiota axis: Physiological changes, dietary influence, and modulation possibilities. Life (Basel);12(3):424. doi: 10.3390/life12030424. PMID: 35330175; PMCID: PMC8955030.

5.4. MODULATION OF MATERNAL GUT MICROBIOTA FOR OFFSPRING'S BENEFITS

Maternal microbiota has been shown to have an impact on fetal health and development both during and after pregnancy. Therefore, regulating gestational dysbiosis may offer protection against noncommunicable diseases like obesity, immunoinflammatory disorders, and neurocognitive issues. Because it is simple to modify the microbiota, oral probiotics and/or prebiotics given to pregnant mothers during pregnancy may be an inexpensive, safe, and effective interventional strategy to prevent disease in the offspring.

5.4.1. Probiotics

Live, helpful bacteria are called probiotics, and they can be found in some foods and supplements. It is believed that they aid in reestablishing the intestinal microbiota community's physiological equilibrium. The majority of research on probiotic interventions is limited to strains of Bifidobacteria and *Lactobacilli*. These strains offer a variety of advantageous effects. They reduce inflammation by strengthening mucosal integrity, promoting colonization resistance, limiting pathogen adherence to the mucosa, and boosting local immune response.

The majority of research on probiotics during pregnancy has concentrated on the clinical results in either the expectant mother or her unborn child. Regarding probiotics' beneficial and preventive effects on the onset of immune-mediated allergy disorders, there is disagreement. Although many clinical trials demonstrated the protective effect of probiotics for mothers in reducing the incidence of allergic conditions, other trials were unable to corroborate the benefits of probiotic therapy. In relation to obesity, perinatal probiotic therapy may alter the child's growth pattern by preventing excessive weight gain in the early years of life. Numerous studies have examined the impact of probiotics on the incidence and symptoms of GDM, with encouraging findings.

There is a dearth of information on probiotics' mode of action and how they affect the mother-fetal gut microbiota axis. There is strong evidence that probiotic bacteria derived from mothers can colonize an infant's gastrointestinal tract (GIT) and remain there for a year or two. Another study, however, found that when given to mothers in late pregnancy, the probiotic strain *Lactobacillus rhamnosus* GG increased the infant gut colonization by *Bifidobacterium* spp., but not by itself. This implies that probiotics might encourage the implantation of different bacteria in fetuses, most likely as a result of bacterial metabolites. Additionally, it is possible that different probiotic bacteria can transfer from the mother to the child in different ways. Additionally, the placenta and fetal intestinal tract expressed TLR-related genes differently as a result of the mother's dietary probiotic intake, which interfered with the development of the fetal immune system.

5.4.2. Prebiotics

Prebiotics are substances found in food that encourage the development and/or activity of healthy microorganisms. Oligosaccharides that are difficult for the small intestine to digest are the most prevalent example. Prebiotics may be a safer option than probiotic consumption for modifying the mother's gut microbiota during pregnancy, as evidenced by animal studies. Prenatal maternal supplementation with a fructo-oligosaccharide altered the intestinal microbiome of the offspring in a mouse model of atopic dermatitis and inhibited the rise in the offspring's scratching behavior and clinical skin severity score. It was discovered that prebiotic oligofructose treatment of diet-induced obesity in pregnant rats reduced gestational weight gain, prevented increased adiposity in dams and their offspring, and decreased maternal energy uptake. Mice fed a high-fiber diet showed a significant reduction in allergic respiratory disease in their progeny. This reduction may have been caused by changes in the mother's gut microbiota during pregnancy. There are very few studies describing the altered gut microbiota of the offspring and the health advantages of the mother's prebiotic consumption in mice (Miko et al., 2022).

5.5. THE ROLE OF MATERNAL MICROBIOMES IN OFFSPRING

A mother's microbiome is essential to her offspring's growth and well-being. The mother's microbiome changes significantly during pregnancy, and these changes are thought to be crucial for the fetal immune system and metabolism to develop.

5.5.1. Effects of Maternal Microbiomes on the Offspring's Immune System

Maternal microbiomes are essential to the health of the fetus in addition to being strongly linked to complications during pregnancy. The microbiomes that are already present in many areas of the mother's body have an impact on how babies' GMs develop. The microbiomes found in the mother's intestine, skin, breast milk, and vagina can be passed vertically to her offspring. Research indicates that the most critical times are during childbirth and the postpartum phase, particularly during which infants are exposed to the mother's skin, vagina, and feces. In recent years, there has been doubt about the notion that the uterine environment was sterile. The placenta contains distinct microbiomes that have never been documented before. Nonetheless, regardless of the presence or absence of placental microbiomes, current research indicates that the microbiomes of expectant mothers have a significant impact on the microbiomes of their offspring. Research has shown that compared to nonmaterially derived flora, the mother's GMs can survive longer and more effectively adapt to the baby's gut environment. The majority of the bacteria in the meconium sample matched the mother's oral bacterial sample (genetically labeled), according to a mouse experiment as well.

Moreover, research has demonstrated that the fetus's immune planning is influenced by brief alterations in the mother's microbiome throughout pregnancy. A perinatal study revealed that the small intestine of the offspring of pregnant mice treated with a high dose of antibiotics lacked both IL-17–producing cells and IL-17 transcripts. In a related study, the administration of three antibiotics to pregnant mice resulted in a decrease in the diversity of GMs in the offspring and a reduction in the number of CD4+ and CD8+ cells in the MLNs that produced IL-17-IFN-γ. These suggest that the enteric immune systems of offspring are significantly influenced by the maternal microbiomes.

Maternal microbiomes influence the offspring's peripheral immune system in addition to their enteric immune system. Gonzalez-Perez noticed that the distribution of dendritic cells and NK cell subsets altered in the pups whose moms had received antibiotic treatment during pregnancy and lactation, and the pups' CD8+ T cells did not produce IFN-γ. The skin responses of newborns to TNF-α and IL-1β were dependent on the maternal microbiomes in an autoimmunity-based mouse model based on the NLRP3 inflammasome mutation R258W. The impact of maternal microbiomes during pregnancy on the immune system of their offspring could potentially stem from the way microbiomes influence immune cell development. According to Joana Torres, pregnant women with IBD continue to have aberrant microbiomes, which also affect the diversity and abundance of bacteria in the baby's feces.

Consequently, the information indicates that mother microbiomes are essential for the development of the immune system in their children. The control of a pregnant woman's microbiome is critical to the health of her unborn child and could offer new perspectives on how to treat expectant patients in clinical settings.

5.5.2. Effects of Maternal Microbiomes on the Offspring's Neurodevelopment

Although the connection between the state of the brain and the GIT has been demonstrated previously, the microbial-gut-brain axis has only recently come to light. Numerous microbiomes coexist harmoniously with humans; these symbiotic microbiomes and the microbial communities within them are intimately linked to the development of the host's brain and nervous system. Research has shown that neurological disorders like anxiety, autism, depression, and stress are associated with structural alterations in the host's microbiome. The importance of the perinatal environment in heredity was recognized as early as 30 years ago, and this could have an impact on the adult health of the offspring. Comprehending the Barker hypothesis has led to the recent proposal that perinatal microbiomes are crucial in determining the health of the adult brain.

Furthermore, a growing body of research indicates that microbiomes play a crucial role in host neurodevelopment. A study on germ-free mice revealed that aberrant brain development in mice results from the absence of the regulatory effects of microbiomes. This includes aberrant microglia growth, high prefrontal cortex myelinization, and increased blood-brain barrier permeability. The expression of genes involved in neural development, such as those involved in neuronal plasticity and neurotransmission in the hippocampus, is impacted in mice lacking certain microbes. Changes in the structure of perinatal microbiomes regulate gene expression, function, and morphology of progeny microglia; these effects will manifest in the early stages of embryo development. These changes in neurophysiology will eventually translate into anxiety, increased stress response, cognitive deficits, visceral pain response, and changes in fear. These findings show that early-life microbiome colonization has a significant effect on the development and functionality of offspring's brains.

Synaptic complement proteins play an important role in the early development of neurons. For example, complement protein C_3 is localized to the axons of retinal geniculate cells and depends on upstream complement proteins C_1 and C_4 (Miko et al., 2022).

5.6. INFLUENCE OF MATERNAL MICROBIOTA DURING PREGNANCY ON INFANT IMMUNITY

Viruses, bacteria, fungi, archaea, and unicellular eukaryotes are among the many microorganisms that call the human body home. The microbiota, the group of bacteria that make up this community, is known to have a significant impact on human health, underscoring the symbiotic relationship that these microorganisms have with humans. The gut, vagina, skin, and breast milk of the developing infant are among the sites of maternal body sites from which microbes are vertically transmitted. Many of these start to affect the skin, feces, and vagina of the mother during labor and delivery or shortly after, but there is growing evidence that they may have an impact even before birth. The uterine environment has traditionally been thought to be sterile. With distinct placental microbiomes in some studies but not others, this dogma has since been refuted.

There is proof that the mother's gut microbiota during pregnancy has a significant impact on the microbiota of the fetus, regardless of the presence of a placental or fetal microbiome. Compared to gut strains from other sources, it has been demonstrated that maternal strains are more ecologically adapted and persistent in the developing baby's gut. Meconium samples from mice that matched those given to the mother orally contained genetically labeled bacteria. Furthermore, pups with altered innate lymphoid and mononuclear cells were born to germ-free pregnant dams that were temporarily colonized with a modified *Escherichia coli* strain and then returned to germ-free status before delivery. This suggests that fetal immune programming is driven by transient changes in maternal microbiota during pregnancy. Therefore, the time during pregnancy is the first instance of maternal microbes influencing fetal immunity, regardless of whether the fetus is colonized in-utero. It is still unclear how the mother's microbiota before delivery contributes to the development of the newborn's immunity.

5.6.1. Antibiotic-Induced Shifts in Microbiota

A substantial amount of research links prenatal exposures and events to the development of the propensity for disease in the offspring. The developmental origins of the health and disease (DOHaD) hypothesis is based on this idea. These factors, which can alter metabolism, epigenetics, and/or microbiota, include smoking, stress, toxins, and maternal diet. There is broad support for microbial-mediated processes. When comparing the MLNs of pups born to non-obese diabetic (NOD) dams treated with antibiotics during pregnancy to control pups, Tormo-Badia found significantly higher proportions of $CD8^+$ T cells. Similarly, compared to controls, pups born to mothers who received oral antibiotic treatment during pregnancy and after giving birth showed heightened vulnerability to vaccinia virus infection and decreased production of interferon (IFN)-γ by $CD8^+$ T cells.

Subsequent investigations demonstrated that the observed impairment of CD8+ T cells was caused by a change in the activation and expression of the T cell receptor (TCR), which is essential for the continuous production of cytokines. Using mouse models, we demonstrated how changing the mother's intestinal microbiota during pregnancy alone affected the pups' innate immunity 14 days after delivery. By giving oral vancomycin, which has a low oral bioavailability, during pregnancy, we hoped to modify the microbiota exclusively in the mother's gut and not at other sites of the body. In fact, two days following delivery, we found no vancomycin in the mother's plasma, suggesting that the antibiotic was not absorbed systemically. While the intervention was aimed at the mother's gut in our model, we noticed changes in the microbiota of breast milk and, to a lesser extent, the vaginal microbiota after giving birth. This implies that the microbiota of the mother's gut during pregnancy may also affect it at distal sites, which could be one way that maternal gut microbiota affects neonatal immunity (Arck and Hecher, 2013; Daniel, 2021).

5.6.2. Gut–Breast Milk Bacterial Axis

An enteromammary pathway allows certain gut bacteria from the mother's digestive tract to reach the mammary glands, according to a pioneering study by Martin. Some research has provided a scientific foundation for this kind of physiological translocation, despite its contentious nature. Dendritic cells (DCs) and $CD18^+$ cells, which remove non-pathogenic bacteria from the gut lumen and move them to other sites, such as nursing mammary glands, may be involved in the mechanisms. Mice that are late in their gestation have been shown to exhibit bacterial translocation from the gut to the mammary glands and milk. Two strains of lactic acid bacteria (LAB), *Lactococcus lactis* and *Lactococcus salivarius*, were given orally to expectant mothers and subsequently found in their milk and mammary tissue.

Although the postpartum immune development is influenced by the microbiota in breast milk, the prenatal gut microbiota plays a role in shaping the composition of these microbiota. Thus, through the gut–breast milk axis, the gut composition before delivery indirectly influences the immune development of the offspring at an early age. Nevertheless, more investigation is required to determine whether the bacterial enteromammary pathway actually exists. This new mode of mother-child communication may have an impact on how we currently perceive the development of the fetal immune system (Arck and Hecher, 2013; Daniel, 2021).

5.6.3. Vaginal Microbiota During Pregnancy and Infant Immunity

More than 170 species of bacteria have been found in the vagina; these communities are remarkably stable and less diverse throughout pregnancy. Microbes from the vagina may translocate across the choriodecidual plate to reach the placenta, amniotic fluid, and fetus. On the other hand, vaginal microbes are most likely essential in establishing newborn immunity. Numerous findings about immunological variations between infants delivered vaginally and via cesarean section point to the potential contribution of the vaginal microbiota to immune education in progeny. It is most likely during pregnancy that the mother's vaginal microbiota is established during labor and delivery. There is proof that the mother's vaginal microbiota during pregnancy affects the immunity of the unborn child even before it passes through the vaginal canal during delivery. *Lactobacilli* in the maternal vagina influence fetal immune development, as evidenced by the higher proportion of $CD45RO^+$ cells and lower level of IL-12 in the cord blood of newborns whose mothers were vaginally colonized with *Lactobacillus* during pregnancy. Benn discovered that a child's risk of wheezing at the age of 4–5 was correlated with the presence of specific maternal vaginal microbes. Although the exact mechanisms underlying this interaction are unknown, they may have to do with ascending organisms, bacterial metabolites, or the fact that women's vaginal microbiota is heavily influenced by their gut microbiota.

5.6.4. Maternal Microbiota During Pregnancy and Offspring Immune-Related Disorders

5.6.4.1. Allergy and Asthma

There is strong epidemiological evidence linking early life exposure to farming and pets to a lower prevalence of allergies and asthma. Furthermore, a sizable European study found that cord blood already contained the immunological tolerance to allergens and that mothers exposed to farm animals during pregnancy had a lower risk of their offspring developing allergies. Parallel to this, a recent study conducted on a Chinese cohort found that maternal exposure to farming during pregnancy affected the number and functionality of newborn regulatory T cells (Tregs), which may help to reduce the incidence of allergies and asthma in the offspring. Reduced immunoglobulin (Ig)E levels in cord blood and an increase in Treg cells, which counteract the effects of allergy-mediated T helper type 2 (Th2) cytokines, have been linked to maternal pet exposure. Douwes demonstrated how a mother's exposure to grain, hay, and/or animals during her pregnancy could lessen her child's symptoms of eczema, hay fever, and asthma. The low prevalence of these atopic diseases has been linked to prenatal exposure, and ongoing exposure has only occasionally provided further protection. The timing of these exposures is important because the effects are greatest when these exposures occur in utero and during the first year of life. In mouse models, the relationship between biological factors derived from farms and the host's immune system and disease susceptibility has also been examined. The protective effect of prenatal exposure to microorganisms derived from farms against asthma was examined in an intriguing study conducted by Conrad. Progeny protection from developing experimental asthma was contingent upon intact maternal TLR signaling, and it was conferred by intranasal exposure to the cowshed-derived bacterium *Acinetobacter lwoffii* F78. Although the precise process of allergy protection in humans is unknown, it is most likely due to altered gut microbial composition caused by both prenatal and postnatal pet exposure.

Probiotics have been demonstrated to influence the development of allergies in children by modulating their microbiome. The preventive effects of probiotics on infant eczema have been demonstrated to depend on combined pre- and postnatal probiotic supplementation; prenatal or postnatal supplementation alone was not effective. Some have discovered that the prenatal factor has greater bearing. While probiotic-induced modifications to a mother's gut microbiota during pregnancy and lactation may lower the frequency of childhood allergies, more study is necessary to establish the best time and dosage.

Antibiotic use during pregnancy can cause long-lasting changes to the microbiota and their metabolites, which provides more concrete evidence of the impact of the mother's microbiota during pregnancy on the atopic disease of the offspring. Prenatal antibiotic use was linked to a higher risk of atopic dermatitis at 18 months of age in infants whose mothers had atopy, according to a Danish birth cohort study. Furthermore, among children between the ages of 2 and 10 years, the use of antibiotics by the mother during her pregnancy was linked to a 1.3-fold rise in the likelihood that the offspring would develop asthma [95% confidence interval (CI) = 1.22–14.2]. The association between antibiotic use and atopy may be complicated by shared familial risk factors, according to a Swedish study by Ortqvist, which found that antibiotic exposure during fetal life was linked to an increased risk for asthma in a cohort analysis but not in sibling analyses. Additionally, a recent study found that antibiotic use by mothers prior to pregnancy and during the first nine months after giving birth is linked to a dose-dependent increase in the risk of asthma in the unborn child. Mulder discovered that prenatal antibiotic use was predictive of childhood asthma, even when compared to sibling controls, in a carefully planned sibling-control study. This study discovered that the influence is greater when antibiotic use occurs during the third trimester, despite some studies showing influence regardless of trimester of maternal antibiotic use. Several lines of evidence indicate that the mother microbiota during pregnancy plays a critical role in preventing an allergy-prone immune phenotype in infants, though more research is required to determine whether timing matters.

5.6.4.2. Type 1 Diabetes

In addition to allergies and asthma, type 1 diabetes and inflammatory bowel disease (IBD) are among the immune disorders for which the prenatal microbiome has been linked. Research on spontaneous type 1 diabetes (T1D) in rodent models has connected the gut microbiota to susceptibility to the disease. In comparison to untreated controls, Livanos's study showed that giving mice a subtherapeutic low-dose penicillin (STAT) during the late stages of pregnancy and the postpartum phase accelerated the incidence of Type 1 Diabetes in the male progeny. It is unclear, though, whether these effects were mediated in the antepartum phase. In any case, there were no variations in the frequencies of Th17 or Treg cells in the lamina propria between the STAT and control pups. Oral administration of a combination of neomycin, polymyxin B, and streptomycin to pregnant NOD dams delayed the development of diabetes in these susceptible offspring, according to a sophisticated study done on mice. Moreover, there was a notable decrease in the prevalence of diabetes in comparison to children born to untreated mothers.

The microbiota of both moms and pops showed changes that the authors noted, indicating that the microbiota directly influences the propensity of T1D in offspring. In addition, pups born to antibiotic-treated NOD dams had lower frequencies of IFN-γ production in the MLN from both $CD4^+$ and $CD8^+$ T cells than from pups born to untreated dams. It is interesting to note that pregnant women who received vancomycin had distinct changes in their gut microbiota and a higher risk of developing type 1 diabetes due to conflicting immune effects, which may have been caused by changes in antigen-presenting cells (APCs) (Arck and Hecher, 2013; Daniel, 2021).

5.6.4.3. Inflammatory Bowel Disease

In addition to Type 1 Diabetes, the microbiota has been linked to inflammatory bowel disorders (IBD). It has been demonstrated that the makeup of the gut during pregnancy affects the prognosis of IBD in the offspring. For instance, regardless of the presence of antecedent gastroenteritis, exposure to antibiotics during pregnancy but not during infancy was linked to an increased risk of very early onset IBD. In an investigation utilizing an IL-10 knock-out colitis model in mice, administering antibiotics to mothers from the third week of pregnancy until they were weaned resulted in heightened vulnerability to chemically induced dextran sulfate sodium (DSS) colitis and inflammation in their progeny, which persisted until adulthood. This implies that a key factor influencing the development of IBD during pregnancy is the mother's microbiota. When considered collectively, the T1D and IBD studies presented show that maternal dysbiosis during pregnancy affects the offspring's immune system and influences the likelihood that an infant will develop autoimmune disorders.

5.6.5. Potential Mechanism of Cross-Talk Between Maternal Microbiota and Offspring Immunity

Microbiota during pregnancy is thought to initiate offspring immune programming in various somewhat related ways that are not necessarily mutually exclusive, although the precise mechanisms are still unknown (Figure 5.3; Table 5.2).

Table 5.2. Correlation Between Meconium Microbiota and the Duration of Pregnancy

Time of Delivery	Phylum	Occurrence		
<33 weeks	*Firmicutes*	8.67% *Enterococcus* 0.82% *Lactobacillus*	Most of the neonatal meconium microbiota; the dominant ones: *Prevotella Bacteroides*	Detected
	Actinobacteria	*Bifidobacterium* 5.47%	–	Detected
	Proteobacteria	6.35% *Enterobacter* 0.98% *Photorhabdus*	–	Detected
	Tannerella	low bacterial count	–	–
	Bacteroidetes	–	Most of the neonatal meconium microbiota	Detected
>33 weeks	*Firmicutes*	0.41% *Enterococcus* 0.07% *Lactobacillus*	Most of the neonatal meconium microbiota	–
	Actinobacteria	0.35% *Bifidobacterium*	–	–
	Proteobacteria	0.06% *Enterobacter* 0.01% *Photorhabdus*	Lower compared to newborns delivered at lower gestational ages	–
	Tannerella	very low bacterial count	–	–
	Bacteroidetes	–	Higher compared to newborns delivered at lower gestational ages; the dominant ones: *Prevotella Bacteroides*	–

"–": non-occurrence.

Source: Dera N, Żeber-Lubecka N, Ciebiera M, Kosińska-Kaczyńska K, Szymusik I, Massalska D, Dera K, Bubień K. (2024). Intrauterine shaping of fetal microbiota. Journal of Clinical Medicine; 13(17):5331. https://doi.org/10.3390/jcm13175331.

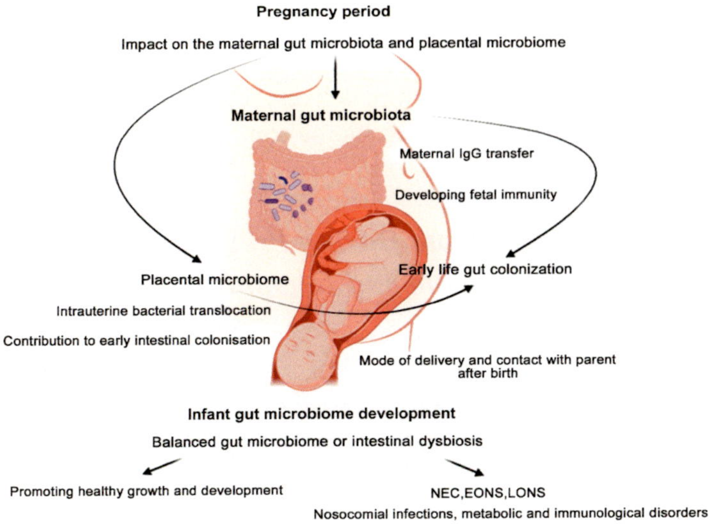

Figure 5.3. *Potential mechanisms of crosstalk between maternal microbiota and offspring immunity.*

Source: Dera N, Żeber-Lubecka N, Ciebiera M, Kosińska-Kaczyńska K, Szymusik I, Massalska D, Dera K, Bubień K. (2024). Intrauterine shaping of fetal microbiota. Journal of Clinical Medicine; 13(17):5331. https://doi.org/10.3390/jcm13175331.

5.6.5.1. Maternal Microbiota Influence Fetal Microbiota

There is evidence to suggest that the developing fetal immune system may be impacted by the fetal microbiota, which may enter the uterus through the placental barrier or through consumption of amniotic fluids. Certain bacteria from the mother's gut may relocate to extra digestive sites in healthy hosts, according to a number of studies. Studies on humans and mice have demonstrated that meconium becomes bacterially colonized during pregnancy. Jimenez found that after oral inoculation of the strain to pregnant dams, tagged *Enterococcus faecium* could be isolated from the meconium of offspring, suggesting that microbes from the mother's gut may potentially cross the placenta to the gut of the offspring.

Theoretically, amniotic fluid that is swallowed by the fetus exposes it to commensal bacteria and their byproducts, which could play a significant role in the early development of the immune system. For instance, in the human fetal gut, memory $CD4^+$ and $CD8^+$ T cells can be detected near the end of the first trimester. It has been demonstrated that memory CD4 T cells co-localize with APCs in the developing intestines of fetuses and secrete IFN-γ, IL-2, or tumor necrosis factor (TNF)-α, which stimulates intestinal development. All of these point to a potential influence on immunity from early fetal exposure to microbial antigens. It is possible that both of these microbiotas are essential in programming fetal immunity before delivery, even though it is unclear how much the mother's and the fetal microbiome contribute to the immunity of the offspring.

5.6.5.2. Maternal Microbiota During Pregnancy Influence Early-Life Infant Microbiota and Immunity

The mother's microbiota during pregnancy shapes the vaginal and breast milk microbiota, which will change the pioneer infant microbiota during a crucial period in immune development, even though it has no direct effect on the microbiota during pregnancy. As we have recently shown, changes to a mother's gut communities only during pregnancy have an indirect effect on the composition of her breast milk and, to a lesser extent, the vaginal microbiota. Furthermore, changes in the mother's gut microbiota brought on by vancomycin had a significant effect on the infant's gut microbiota 14 days after delivery. Compared to controls, pups born to dams treated with vancomycin during gestation had notably greater levels of $CD4^+$ T and B cells. All of the available data point to a complex relationship between the gut microbiota of expectant mothers, the microbiota in breast milk, the intestinal microbiota of babies, and the development of the postnatal immune system (Arck and Hecher, 2013; Daniel, 2021).

5.6.5.3. Bacterial Metabolites

Bacterial metabolites may affect an infant's immunity through the maternal microbiota. Numerous metabolites produced by gut bacteria play a crucial role as mediators in the production of energy, immunological modulation, and other host physiological processes. The immune system detects PAMPs and microbial products, including metabolites, and its ability to identify these molecules can affect host immunity. As byproducts of bacterial anaerobic fermentation, short-chain fatty acids (SCFAs) like acetate, propionate, and butyrate have been demonstrated to affect intestinal immunity, Treg development, DC biology, and epithelial integrity. The SCFAs that the mother's gut bacteria produce during pregnancy may have an indirect effect on the fetal immune system's development. Bacterial metabolites, for instance, have the ability to move from the mother's stomach to her mammary glands during pregnancy and may have an impact on the development of the baby's immune system during nursing. Additionally, it was demonstrated using a reversible mother colonization model that microbial components like aryl hydrocarbon ligands cause transcriptional alterations in the developing fetal gut, increasing the innate immune system's cellularity. Likewise, retinoic acid in mothers.

5.6.5.4. Epigenetics

Furthermore, through epigenetic modulation, metabolites or other molecules generated by the bacteria may have an effect on the developing fetal immune system. Thorburn noted that by increasing the quantity and functionality of Treg cells, exposing pregnant mice to acetate in their drinking water reduced the incidence of allergic airway disease in the offspring. Acetate likely worked by inhibiting histone deacetylases (HDAC) to increase acetylation at the promoter of forkhead box protein 3 (FoxP3). It has been demonstrated that maternal farm exposure affects the T_{reg}-specific demethylated region (TSDR) conserved element, which in turn affects the epigenetic regulation of neonatal FoxP3 expression in humans. The amount of demethylated TSDR was only significantly higher for mothers who consumed farm milk during their pregnancy, but it was still higher with any single exposure than it was with no exposure. Michel noticed alterations in the DNA methylation patterns of genes linked to allergies and asthma in kids. The DNA of children of farmers was found to have hypomethylated regions in ORMDL sphingolipid biosynthesis regulator 1 (ORMDL1) and signal transducer and activator of transcription (STAT)-6 when compared to non-farmers' children.

5.6.5.5. Maternal Immunoglobulins

Pregnancy-related changes in the mother's microbiota affect her immunity, which in turn affects the offspring's passive immunity. Researchers recently noticed changes in the amounts of IgG in breast milk from mothers whose intestinal microbiota was altered during pregnancy by vancomycin. Although the antibodies in milk are supposed to protect the newborn's immune system, it is possible that they are also transferring bacteria that are bound to IgG, which could harm the immune system's development. In fact, there has been prior evidence of IgA- or IgG-coated bacteria in a variety of bodily locations. Additionally, there is proof of IgG-mediated bacterial transfer occurring transplacentally. Gomez de Aguero, for instance, tested the role of IgG in mediating the effects by transferring serum from the transiently colonized dams to unexposed pregnant mice and saw a similar effect on offspring immunity in their previously mentioned maternal transient colonization experiments. Nevertheless, the effect on the immunity of the offspring was eliminated if the serum was IgG-depleted before transfer. This suggests that the transfer of bacterial components by IgG is a partial requirement for the programming of immunity in fetuses during pregnancy. In addition to IgG-mediated immunomodulation, cytokines, hormones, and bacterial components like LPSs can all mediate transplacental immune regulation (Arck and Hecher, 2013; Daniel, 2021).

REFERENCES

1. Aagaard, K., Ma, J., Antony, K. M., Ganu, R., Petrosino, J., & Versalovic, J. (2014). The placenta harbors a unique microbiome. *Science Translational Medicine, 6*(237), 237ra65.
2. Adak, A., & Khan, M. R. (2018). An insight into gut microbiota and its functionalities. *Cellular and Molecular Life Sciences, 76*, 473–493.
3. Arck, P. C., & Hecher, K. (2013). Fetomaternal immune cross-talk and its consequences for maternal and offspring's health. *Nature Medicine, 19*, 548–556.
4. Belo, L., Santos-Silva, A., Rocha, S., Caslake, M., Cooney, J., Pereira-Leite, L., Quintanilha, A., & Rebelo, I. (2005). Fluctuations in C-reactive protein concentration and neutrophil activation during normal human pregnancy. *European Journal of Obstetrics & Gynecology and Reproductive Biology, 123*, 46–51.
5. Bender, J. M., Li, F., Martelly, S., et al. (2016). Maternal HIV infection influences the microbiome of HIV-uninfected infants. *Science Translational Medicine, 8*(349), 349ra100.
6. Brown, J., de Vos, W. M., Distefano, P. S., Doré, J., Huttenhower, C., Knight, R., Lawley, T. D., Raes, J., & Turnbaugh, P. (2013). Translating the human microbiome. *Nature Biotechnology, 31*, 304–308.
7. Collado, M. C., Rautava, S., Aakko, J., Isolauri, E., & Salminen, S. (2016). Human gut colonisation may be initiated *in utero* by distinct microbial communities in the placenta and amniotic fluid. *Scientific Reports, 6*, 23129.
8. Coscia, A., Bardanzellu, F., Caboni, E., Fanos, V., & Peroni, D. G. (2021). When a neonate is born, so is a microbiota. *Life, 11*, 148.
9. Daniel, H. (2021). Diet and gut microbiome and the "chicken or egg" problem. *Frontiers in Nutrition, 8*, 828630.
10. Derrien, M., & van Hylckama Vlieg, J. E. T. (2015). Fate, activity, and impact of ingested bacteria within the human gut microbiota. *Trends in Microbiology, 23*, 354–366.
11. Dethlefsen, L., & Relman, D. A. (2011). Incomplete recovery and individualized responses of the human distal gut microbiota to repeated antibiotic perturbation. *Proceedings of the National Academy of Sciences USA, 108*(Suppl. 1), 4554–4561.
12. Dominguez-Bello, M. G., Godoy-Vitorino, F., Knight, R., & Blaser, M. J. (2019). Role of the microbiome in human development. *Gut, 68*, 1108–1114.

13. Eckburg, P. B., Bik, E. M., Bernstein, C. N., Purdom, E., Dethlefsen, L., Sargent, M., Gill, S. R., Nelson, K. E., & Relman, D. A. (2005). Diversity of the human intestinal microbial flora. *Science, 308*, 1635–1638.
14. Erlebacher, A. (2001). Why isn't the fetus rejected? *Current Opinion in Immunology, 13*, 590–593.
15. Ferretti, P., Pasolli, E., Tett, A., et al. (2018). Mother-to-infant microbial transmission from different body sites shapes the developing infant gut microbiome. *Cell Host & Microbe, 24*, 133–145.
16. Fink, N. R., Chawes, B., Bønnelykke, K., Thorsen, J., Stokholm, J., Rasmussen, M. A., Brix, S., & Bisgaard, H. (2019). Levels of systemic low-grade inflammation in pregnant mothers and their offspring are correlated. *Scientific Reports, 9*, 3043.
17. Freemark, M. (2006). Regulation of maternal metabolism by pituitary and placental hormones: Roles in fetal development and metabolic programming. *Hormone Research, 65*, 41–49.
18. Funkhouser, L. J., & Bordenstein, S. R. (2013). Mom knows best: The universality of maternal microbial transmission. *PLOS Biology, 11*, e1001631.
19. Gomaa, E. Z. (2020). Human gut microbiota/microbiome in health and diseases: A review. *Antonie van Leeuwenhoek, 113*, 2019–2040.
20. Kobyliak, N., Virchenko, O., & Falalyeyeva, T. (2016). Pathophysiological role of host microbiota in the development of obesity. *Nutritional Journal, 15*, 43.
21. Laukens, D., Brinkman, B. M., Raes, J., de Vos, M., & Vandenabeele, P. (2016). Heterogeneity of the gut microbiome in mice: Guidelines for optimizing experimental design. *FEMS Microbiology Reviews, 40*, 117.
22. Mackie, R. I., Sghir, A., & Gaskins, H. R. (1999). Developmental microbial ecology of the neonatal gastrointestinal tract. *American Journal of Clinical Nutrition, 69*, 1035S–1045S.
23. Miko, E., Csaszar, A., Bodis, J., & Kovacs, K. (2022). The maternal-fetal gut microbiota axis: Physiological changes, dietary influence, and modulation possibilities. *Life (Basel), 12*(3), 424. https://doi.org/10.3390/life12030424.
24. Nyangahu, D. D., Lennard, K. S., Brown, B. P., et al. (2018). Disruption of maternal gut microbiota during gestation alters offspring microbiota and immunity. *Microbiome, 6*, 1–10.
25. O'Hara, A. M., & Shanahan, F. (2006). The gut flora as a forgotten organ. *EMBO Reports, 7*, 688–693.

26. Qin, J., Li, R., Raes, J., Arumugam, M., Burgdorf, K. S., Manichanh, C., Nielsen, T., Pons, N., Levenez, F., Yamada, T., et al. (2010). A human gut microbial gene catalogue established by metagenomic sequencing. *Nature, 464*, 59–65.

27. Shanahan, F., Ghosh, T. S., & O'Toole, P. W. (2021). The healthy microbiome—What is the definition of a healthy gut microbiome? *Gastroenterology, 160*, 483–494.

28. Stinson, L. F., Boyce, M. C., Payne, M. S., & Keelan, J. A. (2019). The not-so-sterile womb: Evidence that the human fetus is exposed to bacteria prior to birth. *Frontiers in Microbiology, 10*, 1124.

29. Tormo-Badia, N., Håkansson, Å., Vasudevan, K., Molin, G., Ahrné, S., & Cilio, C. M. (2014). Antibiotic treatment of pregnant non-obese diabetic mice leads to altered gut microbiota and intestinal immunological changes in the offspring. *Scandinavian Journal of Immunology, 80*, 250–260.

30. Zhao, D., Su, H., Cheng, J., et al. (2015). Prenatal antibiotic use and risk of childhood wheeze/asthma: A meta-analysis. *Pediatric Allergy and Immunology, 26*, 756–764.

CHAPTER 6

PROBIOTICS AND IMMUNE HEALTH

Contents

INTRODUCTION

Innate immunity triggers the immune response when a person is exposed to foreign substances or sustains tissue damage. Part of how innate immunity maintains host homeostasis is by triggering inflammation and preparing adaptive immune responses to persistent insults. On the other hand, an unbalanced immune response results in extreme inflammation, uncontrollable tissue damage, and illness. The host mucosal immune system's perception of the gut microbiota is crucial for preserving intestinal homeostasis and triggering systemic defense reactions. Therefore, altering the gut microbiota may be a different strategy for preserving health and treating or preventing illness. "Live microorganisms which, when consumed in adequate amounts as part of food, confer a health benefit on the host" is the definition given to probiotics. In both humans and animals, *Lactobacillus*, *Bifidobacterium*, and *Saccharomyces* are three widely researched and frequently utilized probiotics.

Probiotics have been shown to have a number of advantageous effects on the intestinal mucosal defense system of the host. These include competing with pathogens and toxins for adherence to the intestinal epithelium and preventing the effects of pathogenic bacteria by generating bactericidal compounds. Probiotics support intestinal epithelial cell (IEC) survival, improve barrier performance, and elicit protective responses from IECs in order to maintain intestinal epithelial homeostasis. Most importantly, one of the most likely mechanisms for probiotics' positive effects on human health is immune system modulation. Probiotics have been shown to improve innate immunity and modify inflammation caused by pathogens through signaling pathways regulated by toll-like receptors (TLRs) latest research on the regulation of immune health by probiotics. Clinical applications and mechanisms of action are highlighted, including the roles and mechanisms of probiotics in vaccination, disease prevention, and treatment, as well as probiotic genes and probiotic-derived factors involved in the regulation of host immunity and the molecular targets of probiotic action responsible for the host immune responses.

6.1. OVERVIEW OF PROBIOTICS AND IMMUNE FUNCTION

Probiotics are live microorganisms that give the host health benefits when taken in sufficient amounts. Although they are mostly bacteria, they can also include some kinds of yeast. These good bacteria are found in the human gut naturally and are essential to preserving the proper balance of microbes in the digestive system. Probiotics' effects on digestive health are well known, but there has also been a growing body of research on their possible effects on immune function. To protect the body from dangerous pathogens like bacteria, viruses, and parasites, the immune system is a sophisticated network of cells, tissues, and organs. It is essential for preserving general health and delaying the onset of illnesses and infections. The immune system must be in good working order to promote the best possible health and wellbeing.

Probiotics and immunological function have a complex relationship that is still being researched. Probiotics have been shown to interact with the immune system through various mechanisms, including:

- **Modulation of Gut Microbiota**: Probiotics support the maintenance of a balanced population of gut bacteria, which is necessary for the best possible immune response. They keep harmful microorganisms from growing out of control in the gut by competing with them for resources and attachment sites. Probiotics assist in immune system development and function by fostering a varied and well-balanced gut microbiota.
- **Enhancement of Barrier Function**: Hazardous substances cannot enter the body because of the physical barrier that the gut lining provides. Through encouraging the synthesis of tight junction proteins, which support the preservation of the gut lining, probiotics have been shown to fortify the intestinal barrier. The likelihood that infections will enter the bloodstream and cause immunological reactions is decreased by a healthy gut barrier.
- **Regulation of Immune Cells**: Probiotics can interact with immune cells in other areas of the body as well as the gut-associated lymphoid tissue (GALT). They have the ability to affect the way immune cells respond to infections and lessen inflammation. Examples of these cells include dendritic cells, macrophages, and T cells. In addition to their antimicrobial qualities, probiotics can also promote the synthesis of antimicrobial peptides (AMPs), which strengthen the body's defenses against infections.
- **Production of Immunomodulatory Substances**: Bioactive substances with immunomodulatory properties, like SCFAs, can be produced by probiotics. These compounds have the ability to modulate immune responses, lower inflammatory levels, and enhance immune cell activity. Furthermore, probiotics have the ability to increase the synthesis of cytokines, which are immunological response-coordinating signaling molecules.

Probiotics may have a range of immune-related benefits, including:

- **Prevention and Management of Infections**: It has been demonstrated that specific probiotic strains can aid in the management and prevention of a variety of infections. Probiotics, for instance, can lessen the length and intensity of gastrointestinal, urinary, and respiratory tract infections. Additionally, they can strengthen the immune system's reaction to vaccinations, boosting their efficacy.

- **Allergy Prevention and Management**: Evidence suggests that early exposure to probiotics may lower a susceptible person's risk of developing allergic conditions like eczema and asthma. Probiotics have the potential to regulate immune responses linked to allergies and enhance immune tolerance.
- **Autoimmune Diseases**: Probiotics may help treat autoimmune conditions like multiple sclerosis, rheumatoid arthritis, and IBD, according to certain research. More studies are required to completely understand the effects of probiotics, but they may help control immune responses and lessen inflammation in certain conditions.
- **Gut-Brain Axis**: A connection between immune system performance, mental health, and gut microbiota may exist, according to new research. Probiotics may affect immunological reactions that affect mental health. Research has indicated encouraging results regarding the potential use of probiotics for conditions like stress, anxiety, and depression, even though the mechanisms are not fully understood.

It is significant to remember that probiotics' effects on immune function can differ based on the strain, dosage, and individual circumstances. Probiotics differ in the immune-related advantages they offer, and more study is required to fully comprehend their modes of action and therapeutic uses. Furthermore, it is always advised to speak with a medical professional prior to beginning any probiotic supplementation, particularly for people with weakened immune systems or underlying medical conditions. Probiotics have the ability to affect immune function in a number of ways. They can produce substances that modulate immunity, improve barrier function, and support a balanced gut microbiota. Probiotics have the potential to improve immune function and prevent certain infections and illnesses, but more study is required to completely grasp these advantages and determine the best ways to apply them in various demographics (Vieira et al., 2013).

6.1.1. Interactions Between Probiotics and the Immune System

Probiotics and the immune system interact through a variety of intricate and dynamic mechanisms that are vital to preserving immune homeostasis and promoting general health. Probiotics are good bacteria that work with the immune system in a number of ways. By affecting immune cell activity, encouraging the synthesis of anti-inflammatory cytokines, and controlling the ratio of pro- to anti-inflammatory signals, they can modify immune responses. Probiotics can also strengthen the integrity of the intestinal lining, improve the function of the gut barrier, and encourage the production of mucus and AMPs, all of which support immune defense. Probiotics can also affect the maturation, activation, and antigen presentation processes of immune cells in the GALT, including dendritic cells, macrophages, and T cells. These interactions support immune tolerance, help mold the immune response, and may aid in the treatment or avoidance of certain immune-related illnesses. However, a number of variables may affect the precise processes and outcomes of probiotic-immune interactions.

6.1.2. Probiotic Genes Involved in the Regulation of Host Immune Responses

Our understanding of the probiotic genes involved in the control of the host immune responses has increased thanks to metagenomic analysis. The ability of 42 *Lactobacillus plantarum* strains that were separated from various human and environmental sources to stimulate the production of interleukin 10 (IL-10) and IL-12 by peripheral blood mononuclear cells was assessed. Six potential genes with immunomodulatory properties were found by comparing the strain-specific cytokine responses with comparative genome hybridization profiles produced by *L. plantarum* WCFS1 DNA microarrays. These genes encode components of the transport pathway, the LamBDCA quorum-sensing system, the N-acetyl-glucosamine/galactosamine phosphotransferase system, and the biosynthesis of bacteriocin. These genes were deleted from L plantarum WCFS1, which eliminated the ability to induce the production of cytokines. Additionally, the same bacteria and techniques were used to investigate gene loci controlling dendritic cell production of IL-10 and IL-12. In peripheral blood mononuclear cells, several genes were found to be involved in the regulation of cytokine production. These genes included six that are involved in the production or secretion of bacteriocin, one that encodes a bile salt hydrolase, and another that encodes a transcription regulator. These findings thus imply that probiotic gene specificity also applies to the regulation of responses by distinct immune cells.

Three closely related strains of *Escherichia coli*, strains CFT073, a uropathogen, and strains 83972 and Nissle 1917, which are probiotic strains of fecal and urinary tract origin, respectively, were subjected to functional genomic analysis. These three strains' active genomic profiles are closely related, according to transcriptome profiling. The three genes yhaK, yhcN, and ybiJ were needed for *E. coli* Nissle 1917 to grow in urine and form biofilm, as demonstrated by this study. It appears from this data that pathogens and probiotics share functional gene profiles. It will be crucial to comprehend how these bacteria, despite having similar transcriptions, behave differently in different contexts, whether that be cellular or otherwise.

It has also been demonstrated that host factors influence how probiotic genes are transcribed. In an *in vitro* model of the gastrointestinal tract (GIT), genes related to adhesion and stress in *Lactobacillus acidophilus* NCFM were investigated. When *L. acidophilus L. acidophilus* NCFM was preincubated with acidified milk, the expression of the genes encoding the stress-related proteins, GroEL, DnaK, and ClpP, increased during gastric digestion and decreased during subsequent duodenal digestion. Genes encoding fibronectin- and mucin-binding proteins were not affected by gastric or salivary juice, but they increased dramatically when incubated in bile and duodenal juice. These findings offer sophisticated illustrations of the intricacy and usefulness of probiotics as they move through the digestive system (Vieira et al., 2013).

6.1.3. Probiotic Components Which Regulate Immune Responses in the Host

The active ingredients in probiotics have recently been investigated to identify effectors of probiotic action. *Bifidobacterium animalis* subsp. analysis using two-dimensional gel in conjunction with matrix-assisted laser desorption ionization time-of-flight mass spectrometry. 74 different proteins were found in the lactis BB-12 secreted proteins. 31 proteins, such as those that metabolize cell walls and bind oligosaccharides, amino acids, and manganese, are expected to perform their physiological function either on the outside of the cell or within it. Interaction with extracellular matrix proteins or human host epithelial cells is mediated by 18 proteins. Potential roles include plasminogen binding, fimbriae formation, collagen adhesion, mucin and intestinal cell attachment, and immunomodulatory response induction. These results imply a role for bacterial proteins in the immune system's immunomodulation, adherence to host tissues, and colonization of the GIT.

Probiotics exert their immunomodulatory effects through various components that interact with the host's immune system. These components include:

- **Surface Molecules**: Probiotics have unique surface molecules that can interact with immune cells and receptors, such as lipoteichoic acids (LTAs), peptidoglycans (PGNs), lipopolysaccharides (LPS), and surface proteins. By activating TLRs and other pattern recognition receptors (PRRs) on immune cells, these molecules can trigger immune responses by starting signaling pathways that control inflammation and immune activation.
- **Metabolites and Secretory Products**: Metabolites and secretory

products generated by probiotics have the potential to impact immune responses. Propionate, acetate, and butyrate are examples of SCFAs that are created when probiotics ferment dietary fibers in the stomach. Immunomodulatory effects of SCFAs include the promotion of regulatory T cells (Tregs) and control of T cell differentiation. Exopolysaccharides and bacteriocins are two more secretory products that have the ability to alter immune response.

- **DNA and RNA Fragments**: Microbial-associated molecular patterns (MAMPs), which are released by probiotics, are fragments of DNA and RNA that have the ability to trigger immune responses. Via receptors like TLR9, immune cells are able to identify these fragments, which causes the release of cytokines and changes in immune cell activity.
- **Extracellular Vesicles**: Extracellular vesicles (EVs) produced by probiotics can include a variety of bioactive substances, such as lipids, proteins, and nucleic acids. These EVs have the ability to interact with immune cells and impact their functions, including enhancing immune cell activation and antigen presentation or inducing anti-inflammatory responses.
- **Immunomodulatory Proteins**: Probiotics generate proteins that have the ability to modulate immune responses. For instance, certain probiotics generate particular proteins known as "moonlighting proteins," which have complementary roles in immunomodulation and metabolism. These proteins have the ability to influence immune signaling pathways and interact with immune cells.
- **Bioactive Peptides**: Bioactive peptides can be produced by probiotics during fermentation by means of enzymatic processes. It has been demonstrated that these peptides have immunomodulatory effects, such as the ability to reduce inflammation and control immune cell activity.

It is noteworthy that the effects of probiotic components on the immune system can differ based on host factors, strain, and dosage. These elements may be combined differently in different probiotic strains, resulting in distinct immunomodulatory profiles. Moreover, the host's unique immune status and genetic variables may also affect how the body reacts to probiotic ingredients. Further research is necessary to fully understand the specific roles that probiotic components play in regulating immune responses, as the precise mechanisms underlying these responses are still being worked out. However, taken as a whole, these elements support immune function and preserve immunological homeostasis in the host, contributing to the overall immunomodulatory effects of probiotics (Vieira et al., 2013).

6.1.4. Host Immune Responses Regulated by Probiotics

Probiotics help define and preserve the fine line that separates excessive defense mechanisms from those that are necessary, such as innate and adaptive immune responses. Probiotics can directly interact with IECs or, after being internalized by M cells, interact with dendritic cells and follicle-associated epithelial cells to trigger responses that are then carried out by macrophages, T and B lymphocytes, and other immune regulatory cells. Probiotic action that results in immunomodulation primarily stems from two mechanisms: host cell signaling pathways and regulation of gene expression.

6.1.4.1. Host Genes

The host's genetic diversity influences the range of responses to similar stimuli, and probiotic effects are no exception. Mucosal responses to *Lactobacillus casei* CRL-431, *L. rhamnosus* GG, and L acidophilus Lafti L10 were investigated in a double-blind, placebo-controlled study involving healthy volunteers. Transcriptomes grouped according to the individual rather than the intervention, indicating that the main factor influencing transcriptome differences was individual variation in gene expression. Furthermore, in the human proximal small intestinal mucosa, these three probiotic bacteria induced distinct gene-regulatory networks and pathways. L acidophilus controlled the expression of genes involved in immune response, ion homeostasis, and hormone regulation of tissue growth and development.

For instance, L acidophilus altered the transcriptional control of the IL-23 signaling pathway linked to mucosal inflammatory bowel disease (IBD). *L. rhamnosus* has been linked to ion homeostasis, wound healing, and the IFN response. Cellular growth, proliferation, and development were the main altered transcriptional networks and pathways regulated by *L. rhamnosus*, with major roles in JUN, JAK2 and STAT4, and IGF1. Mucosal responses to *L. casei* included blood pressure regulation by hormones, Th1–Th2 balance, and proliferation. *L. casei* induced an increase in IL-17D and IL-21, which stimulate the growth of natural killer cells, and a shift in the Th1/Th2 balance to a Th2 and/or Th17 type. These thorough analyses thus demonstrated that probiotic regulation of human mucosal immunity at the gene expression levels is strain- and host-specific as well as dependent on host genetics.

It has been demonstrated that the noncommensal, nonpathogenic *Saccharomyces cerevisiae* strain CNCM I-3856, which is used as a probiotic in gastrointestinal disorders, controls the expression of inflammatory genes in porcine intestinal epithelial IPEC-1 cells. Enterotoxigenic *E. coli* (ETEC)-induced expression of proinflammatory cytokines and chemokines, such as IL-6, IL-8, CCL20, CXCL2, and CXCL10, was inhibited by viable S cerevisiae at both transcriptional and protein expression levels. This inhibition was linked to an increase in the anti-inflammatory PPAR-γ nuclear receptor mRNA level, an agglutination of ETEC, and a decrease in the phosphorylation of MAPK and ERK1/2. *S. cerevisiae* did not, however, succeed in maintaining the integrity of the barrier in a monolayer exposed to ETEC, indicating that this yeast does not directly prevent the enterotoxin activity of ETEC.

6.1.4.2. Immune Cells

The immune system, which is in charge of protecting the body from infections, pathogens, and other foreign substances, is primarily composed of immune cells. The immune system consists of various cell types, each with a distinct role in identifying and getting rid of dangerous invaders. Here are some of the major types of immune cells:

- **Phagocytes:** Pathogens are engulfed and destroyed by phagocytes, which include neutrophils, monocytes, and macrophages. Through the use of specific receptors, they identify foreign substances and engulf them in a process known as phagocytosis. Initiating an immune response by presenting antigens to other immune cells is another function of phagocytes.
- **Lymphocytes:** B cells and T cells are the two primary subtypes of lymphocytes, which are white blood cells.
  - **B Cells:** Antibodies are proteins that attach to particular pathogens and identify them for elimination; B cells are involved in the synthesis of these proteins. Antigens can be recognized by B cells alone or in conjunction with other immune cells.
  - **T Cells:** T cells have several subtypes, including helper T cells, cytotoxic T cells, and regulatory T cells.
  - **Helper T cells:** An essential function of helper T cells is to orchestrate immune responses. When they identify antigens that other cells have presented, they release chemical cues known as cytokines, which stimulate additional immune cells.
  - **Cytotoxic T Cells:** Killer T cells, also referred to as CTLs, target and eliminate aberrant or infected cells within the body, including cancerous or virus-infected cells.
  - **Regulatory T Cells:** Regulatory T cells aid in preserving immunological tolerance and averting overreactions by the immune system. They help prevent autoimmune diseases by reducing the immune system's response to self-antigens.

- **Natural Killer (NK) Cells:** Another subset of lymphocytes that has the ability to eradicate cancerous or infected cells directly is the NK cell. By identifying alterations in cell surface markers, they identify aberrant cells and release poisonous chemicals to cause cell death.
- **Dendritic Cells:** Antigens are taken up by dendritic cells, which then process and deliver them to T cells and other immune cells. They are regarded as the immune system's sentinels and are essential in triggering immune reactions.
- **Mast Cells:** Both inflammatory and allergic reactions involve mast cells. Mast cells release chemicals such as histamine when they become activated by an allergen or infection. These substances cause inflammation and draw additional immune cells to the infection site.
- **Eosinophils and Basophils:** Both allergic reactions and parasite defense involve eosinophils and basophils. While basophils release chemicals like histamine during allergic reactions, eosinophils release noxious substances to combat parasites.

These are but a few illustrations of the various immune cells that cooperate to keep the body's defenses against illness intact. An effective immune response and an individual's general health depend on the interactions and coordination of various immune cells.

6.1.4.3. Dendritic Cells

Intestinal dendritic cells can be found diffusely throughout the intestinal lamina propria or within particular intestinal lymphoid tissues, which are collectively referred to as gut associated lymphoid tissues (GALT). The main cell type that acts as "sensors" of microbial ligands through the activation of innate immune receptors (such as c-type lectin receptors and Toll-like receptors (TLRs)) is dendritic cells. The integration of microbial and host metabolism with immune functions is facilitated by alterations in dendritic cell phenotypes and cytokine secretion, which are made possible by the signaling pathways triggered by molecules derived from bacteria. Retinoic acid production from vitamin A metabolism is a crucial immunomodulatory process linked to intestinal dendritic cells. Probiotic microbes can stimulate retinoic acid metabolism in human dendritic cells *in vitro* and in murine $CD103^{+}$ dendritic cells in the lamina propria of the small intestine, though not all of them can. Apart from the metabolism of vitamin A, it has been demonstrated that *Lactobacillus rhamnosus* induces the production of mucosal T regulatory (TREG) cells in the mesenteric lymph nodes (MLNs) through the induction of heme oxygenase 1 (HO-1), another metabolic enzyme of dendritic cells. The immunoregulatory effects on dendritic cells have been linked to metabolites and components of bacterial cell walls.

For instance, intestinal CD11c+ dendritic cells' presentation of segmented filamentous bacteria antigens in the major histocompatibility complex (MHC)-II-dependent manner is essential for the local induction of T_H17 lymphocytes. Furthermore, it has been demonstrated that *Bacteroides fragilis* capsular polysaccharide A (PSA) directly interacts with mouse plasmacytoid dendritic cells through TLR-2. Exposure to PSA induces the expression of molecules that protect against colitis and stimulates $CD4^+$ cells to secrete IL-10. *Bacillus* subtilis exopolysaccharide inhibits the TLR-4 and MyD88 signaling-dependent gut inflammation induced by Citrobacter rodential.

In the colon, dietary fibers like prebiotics ferment to produce short-chain fatty acids (SCFAs). Type 2 diabetes, obesity, IBD, colorectal cancer, and allergies may all have abnormalities in the metabolite production caused by dysbiosis and dietary factors. Of the SCFAs, butyrate appears to have a stronger immunomodulatory effect than either acetate or propionate. Because butyrate affects the activity of histone deacetylases (HDAC), dendritic cells are able to promote TREG cells while secreting less IL-12 and IL-6 cytokines. Propionate has the ability to stimulate dendritic cells to express Foxp3, while acetate lacks this ability, possibly due to the absence of HDAC activity. Moreover, butyrate inhibits HDAC in intestinal macrophages. The finding that butyrate stimulates dendritic cell regulatory activity, leading to the induction of T_{REG} cells and IL-10-secreting T cells, has been validated and expanded upon by a subsequent study. The Gpr109a G-protein-coupled receptor on colonic dendritic cells and macrophages mediated these effects. On the other hand, butyrate has also been demonstrated to increase the secretion of IL-23 by murine dendritic cells, which in some situations may increase TH17 responses.

Another significant mucosal metabolite secreted by gut microbes is histamine; patients suffering from irritable bowel syndrome (IBS) and IBD have elevated mucosal histamine levels. Histamine can increase the production of IL-10 while suppressing the release of proinflammatory and chemokine cytokines from dendritic cells stimulated by TLRs. Histamine produced this effect by binding to dendritic cells' histamine 2 receptor (H2R). Cyclic adenosine monophosphate (cAMP) and exchange protein directly activated by cAMP (EPAC) were needed for the signaling mechanism. Mice given a *Lactobacillus* strain that secretes histamine experienced rapid weight loss and increased secretion of cytokines from Peyer's patch.

6.1.4.4. Intestinal Epithelial Cells

Specialized cells called IECs border the inside surface of the intestine, creating a barrier that keeps the intestinal lumen and the surrounding tissue safe. They are essential for interacting with the immune system, preserving gut homeostasis, and absorbing nutrients. Here are some key features and functions of IECs:

- **Barrier Function**: Toxins and infections are kept out of the body by the IECs, which create a physical barrier. These cells' tight junctions work together to form a selective barrier that regulates the movement of ions and molecules through the intestinal lining.
- **Nutrient Absorption**: The finger-like projections known as microvilli on the surface of IECs increase their surface area and aid in the absorption of nutrients from the food that has been digested. These cells' surface is home to specialized transporters that facilitate the uptake of various nutrients, such as proteins, fats, carbohydrates, vitamins, and minerals.
- **Secretion of Mucus and Immunoglobulins**: Mucus is secreted by IECs, giving the intestinal lining lubrication and defense. Pathogens are trapped by the mucus layer, which keeps them from getting to the underlying tissue. These cells also generate immunoglobulins, like IgA antibodies, which are essential to the gut's immune defense system.
- **Interaction with Gut Microbiota**: The complex community of microorganisms living in the intestine, known as the gut microbiota, is closely interacted with IECs. They create a dynamic interface that controls how the host and microbiota communicate with each other. By identifying and reacting to microbial cues, IECs support the growth of beneficial bacteria in a symbiotic relationship and trigger the immune system to combat harmful pathogens.
- **Maintenance of Gut Barrier Integrity**: The integrity of the gut barrier is actively maintained by IECs. They generate a variety of proteins, including tight junction proteins that help control the permeability of the epithelial barrier, such as occluding and claudins. By keeping the barrier functional and unbroken, these proteins stop dangerous substances from moving from the gut lumen into the bloodstream.
- **Immune Surveillance and Regulation**: Sensing and reacting to pathogens and immune signals, IECs serve as immune system sentinels. They produce TLRs, which are PRRs that identify pathogen-associated molecular patterns (PAMPs) and trigger immune responses. Additionally, immune-modulating substances like chemokines and cytokines are secreted by IECs, and these helps control the activity of immune cells in the gut.
- **Repair and Regeneration**: To preserve the integrity of the intestinal barrier, IECs have a high turnover rate and the capacity to regenerate quickly. Intestinal stem cells within the epithelial lining differentiate

and multiply in response to injury or damage, replenishing damaged cells and maintaining the intestinal epithelium's ongoing renewal.

Because they are crucial for immune response, nutrient absorption, and gut barrier integrity, IECs are necessary for the healthy operation of the intestine. For gut homeostasis and general health, their interactions with the immune system and gut microbiota are essential. Researching gastrointestinal illnesses, immune system problems, and developing treatments that target the gut all depend on an understanding of the roles and regulation of IECs (Vieira et al., 2013).

6.2. ROLE OF GUT MUCOSAL BARRIER IN HOST DEFENSE

The digestive system's main job is to break down and absorb food in order to meet the needs of the body's metabolism and promote healthy human growth and development. Furthermore, the intestinal mucosa acts as a barrier for the host against the persistent antigens found in food and bacteria in the gut lumen. Saliva, stomach acid, peristalsis, mucus, intestinal proteolysis, intestinal flora, and epithelial cell membranes with intercellular junctional complexes are among the numerous components that guarantee protection against potentially harmful agents. The gut abruptly changes its function at birth from breaking down amniotic fluid to breaking down milk. The release of trophic hormones and the activation of secretion, motility, and absorption are triggered by food consumption. Further maturational and adaptive events in the gut defense barrier during postnatal development include the development of the intestinal flora and the appearance of mucosal proteins and digestive enzymes.

The early months of life are crucial for the development of the gastric mucosa's hydrochloric acid secretion, which is the stomach's primary defense against microbes. An essential physical barrier that prevents intestinal attachment of luminal antigens is the mucus called goblet cell mucus, which coats the GIT's epithelial surface. In the GIT, the establishment of normal bacterial populations can stop potential pathogens from growing out of control. The epithelial cell membranes, a significant mechanical interface between the host's interior environment and the luminal contents, are also impacted by maturational changes. Increased food-protein binding capacity has been linked to the postnatal maturation of small intestinal brush border membranes in experimental animals. It has been demonstrated that the ability of antigens to bind to epithelial cells affects the strength of mucosal immune responses and is connected to the rate and mode of antigen transfer.

6.2.1. Gut's Role in Immune Function

The role of the gut microbiota in immune system development has been well studied. According to one of these descriptions, the coordinated interaction between the immune system and the gut microbiota allows the host to withstand the high concentration of antigens in the gut.

We can now comprehend the mechanisms through which various microorganisms influence bodily functions and the consequences of these mechanisms thanks to advancements in our understanding of the biodiversity of the intestinal microbiota. Dysbiosis, or altered microflora, is typically linked to gastrointestinal disorders; however, common diseases that do not only affect the GIT are linked to microbial imbalances. Since germ-free (GF) mice have no microflora at all, studies on these animals provide a wealth of information regarding how microflora shapes the immune system. These mice exhibit severe immune abnormalities, including decreased B and T cell immune response, fewer and smaller Peyer plaques, and MLNs. Thus, the gut GF mice had lower serum levels of immunoglobulin (Ig) G and IgA. Furthermore, numerous investigations on humans and mice have demonstrated a correlation between altered microflora and specific inflammatory diseases.

Microbial dysbiosis is a common cause of inflammatory diseases. A prospective study involving children at high risk of developing asthma revealed that alterations in the microbiota take place prior to the onset of the illness. Certain species of gut commensal microbiota are necessary for the regulation of immune responses, and disruptions in the microbiota may result in the development of more pathogenic bacteria, the induction of inflammation, and a lack of immune regulation. The microorganism that a newborn is exposed to during their early years of life largely determines the microbial composition of the microbiota in the adult human gut. It has been demonstrated that techniques to control the microbiota during infancy can stop the emergence of allergic and atopic disorders in later adulthood. Therefore, it has been suggested that the purposeful modification of the microbiota composition can be achieved by using probiotics and prebiotics in the early postnatal period. Additionally, the metabolic and immunological profiles of the pregnant uterus as well as the likelihood that the offspring will contract the disease may be impacted by the mother's diet and exposure to microbes during pregnancy the use of prebiotics and probiotics when expecting (Vieira et al., 2013).

6.2.2. Probiotic Modulation of the Gastrointestinal Mucosal Immune System

The capacity of probiotic bacteria to both locally and systemically alter the host's GIT mucosal immune system is arguably one of their most significant qualities. The intricate and well-researched interaction between the probiotic microbe and the gut immune cells, gastrointestinal epithelia, and resident microbiota results in an immunomodulatory response.

Mitogen-activated protein kinase signaling cascades and activated nuclear factor kappa-B (NF-κB) mediated MAMPs and PRRS (including NOD-like receptor, TLRs, and C-type lectin receptors) were responsible for inducing the expression of cytokines and chemokine genes. Probiotic microorganisms primarily produce lipoteichoic acids (DMA), peptidoglycan, and S-layer proteins as MAMPs (Figure 6.1).

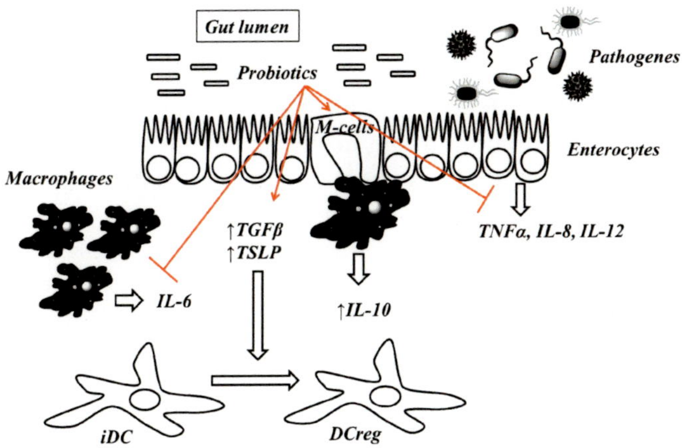

Figure 6.1. *Probiotic modulation of the gastrointestinal mucosal immune system.*

Source: Georgieva, M., Georgiev, K., & Dobromirov, P. (2015). Probiotics and Immunity. InTech. doi: 10.5772/61337

Probiotics inhibit the production of proinflammatory mediators like interleukin 8 (IL-8) and tumor necrosis factor a (TNF-a) in IECs exposed to pathogenic microbes or related stimuli. Instead, they induce the production of anti-inflammatory mediators like transforming growth factor b (TGF-b) and thymic stromal lymphopoietin (TSLP), which can help differentiate immature dendritic cells (iDCs) into regulatory dendritic cells (DCregs). Probiotics have the ability to reduce the amount of IL-6 that macrophages in the inflammatory mucosa produce while simultaneously increasing the production of IL-10.

Using functional genomic techniques, examine how these MAMPs affect immunity. Numerous studies have shown a marked decrease in proinflammatory cytokine productions along with a concurrent rise in anti-inflammatory IL-10 and downregulation of pro-inflammatory IL-12 and TJMF-H. The microbiota present in the intestine enhances the function of the mucosal barrier and fortifies the host's resistance against enteric infections. IL-1β, a cytokine that is normally produced during an active infection, is essential for neutrophil restoration and pathogen eradication. The production of pro-IL-1β at homeostatic levels by local intestinal macrophages is significantly influenced by microflora. Through mechanisms that are not dependent on MyD88 (Myeloid differentiation primary response gene 88), the gut microbiota can also improve host immunity. Notably, the development of Th-17 cells in the intestine is induced by commensal bacteria colonizing GF mice, and this is crucial for defense against *Citrobacter rodentium* infection.

6.2.3. Gut-Associated Lymphoid Tissue

The local adaptive immune system defends the mucosal membrane surface. The largest mass of lymphoid tissue in the human body is found in the gut-associated lymphoid tissue. As such, it is a crucial component of the host's overall immunologic potential.

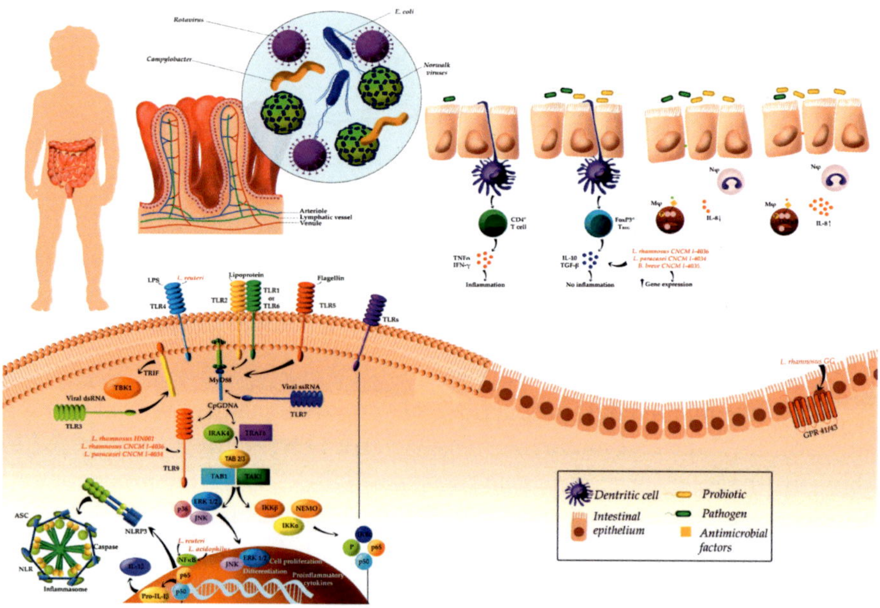

Figure 6.2. *General probiotic mechanism of action in pediatric gastrointestinal infections. Abbreviations: ASC, apoptosis-associated speck-like protein containing a CARD; ERK, extracellular regulated kinase; IKK, IκB kinase; IL, interleukin; IFN, interferon; IRAK4, IL-1 receptor-associated kinase 4; JNK, Jun N-terminal kinase; NF-κB, nuclear factor κ-B; NEMO, NF-κB essential modulator; TNF-α, tumor factor necrosis alpha; TLR, toll-like receptor, TAB1/2/3, TAK binding pro-*

teins; TAK1, ubiquitin-dependent kinase of MKK and IKK; TBK1, serine/threonine-protein kinase 1; TGF, transforming growth factor; TRAF6, Tumor necrosis factor receptor-associated factor 6.

Source: Plaza-Díaz J, Ruiz-Ojeda FJ, Gil-Campos M, Gil A. (2018). Immune-mediated mechanisms of action of probiotics and synbiotics in treating pediatric intestinal diseases. Nutrients; 10(1):42. https://doi.org/10.3390/nu10010042.

The intestinal immune response's regulatory processes take place in various physiological compartments, including the mucosa, intestinal epithelium, and secretory sites. They are also aggregated in follicles and Peyer's patches. While the lamina propria cells show a helper and inducer phenotype, the intraepithelial T lymphocytes primarily display a suppressor and cytotoxic phenotype. The cell lineage's lymphocytes are present in the lamina propria. Mucosal surfaces are sites of abundant production of immunoglobulin A (IgA) antibodies. Secretory IgA is different from serum IgA in that it is present in a polymeric or dimeric form.

Secretory IgA is perfect for shielding mucosal surfaces because it is immune to intraluminal proteolysis and does not trigger complement or inflammatory reactions. The isotype distribution of cells that produce immunoglobulins varies between the upper and lower regions of the human gut-associated immune system. While IgA2-producing cells are more common in the colon due to their increased resistance to bacterial proteases, IgA1 immunocytes are more common in the small intestine. The respiratory tract, lacrimal, salivary, and mammary glands are all included in the common mucosal immune system, which is comprised of the secretory IgA antibodies in the gut. Therefore, immune responses at other mucosal surfaces may be impacted by an immune response that starts in the gut-associated lymphoid tissue. During the lymphocyte maturation cycle, antigens are presented to T lymphocytes with helper and inducer phenotypes, which multiply and trigger a cell response. Antigens are also transported across Peyer's patches. The particular lymphocytes that secrete antibodies first show up in peripheral blood 2–4 days after being exposed to the antigen, peak in concentration after 6–8 days, and stay in the blood for 2–3 weeks. Study.

According to data, lymphocyte-intestinal epithelium interactions may be more significant than previously thought. The M cell-like cells found in Peyer's patches are a distinct type of epithelium made up of cuboidal epithelial cells, a rare number of goblet cells, and specialized antigen-sampling cells. Lymphocytes, especially those of the B cell lineage, can induce enterocytes into these cells. The gut lumen's microbes and particles are efficiently transferred into the follicles beneath by these cells. Probiotic-induced induction of gut-seeking B cells has the potential to impact mucosal immunity in ways other than IgA secretion.

Other yet-to-be-discovered mechanisms of mucosal immunity are provided by intraepithelial lymphocytes, which are mainly T cells with gamma-delta receptors. By destroying contaminated cells and drawing in additional immune cells to fight infection, these cells engage with the epithelial cells to protect the mucosa. These cells function as unique T cells in mice because they originate locally in cryptopatches, which are cell clusters beneath the gut's epithelial lining, rather than in the thymus or lymph nodes.

6.2.4. Control of Antigen Absorption in the Gut

Numerous antigens that enter the body through the enteric route present challenges to the small intestine. Furthermore, the composition of the antigen load is constantly changing, exposing the small intestine to rapid changes. In the gut, a functional mucosal barrier blocks the majority of antigens. The villous epithelium contains specialized antigen transport mechanisms in addition to immune exclusion, the first line of defense against pathogens in the gut. Transcytosis is the mechanism by which antigens are absorbed through the epithelial layer; here, the primary degradative pathway involves lysosomal processing of the antigen. Immune elimination, the second line of defense, aims to eradicate antigens that have made their way past the mucosa.

Unprocessed antigens can be transported via a minor pathway. The M cells cover Peyer's patches, which are important in determining the subsequent immune responses to the presence of the antigen. Antigen transport through this epithelium is typically characterized by quick absorption and minimal degradation. Subjacent T cells are exposed to antigens; these cells differentiate into a variety of effector cells that mediate active immune suppression and encourage the differentiation of B cells that secrete IgA. Dietary antigens undergo a transformation into a tolerogenic form during the absorption process across the intestinal mucosa (Figure 6.2). Consequently, one of the characteristics of the intestinal immune system is its hyporesponsiveness to antigens, such as food proteins and oral tolerance (Vieira et al., 2013).

6.3. ORAL TOLERANCE

The T-cell response's development is known to be significantly influenced by the type of antigen, the antigen's entry route, and its dosage. Additionally, the age of the host and the timing of the encounter influence the development of tolerance to oral antigen. In response to encountering an antigen, immune cells release a variety of cytokines, which in turn dictate the course of subsequent immune responses. The knowledge that T helper 1– and T helper 2–like cells generate cytokine patterns that are highly polarized has contributed to our understanding of the unique immune response to an antigen. These cytokines are produced more readily by early stages of immunologic activation. Atopic disease, eosinophilia, and increased IgE production can result from the T helper 2 phenotype, which requires interleukin 4. The immune response of T-helper-1 cells to intracellular pathogens is controlled by the cell-mediated system. A vital equilibrium is created and preserved in health between systemic hypo responsiveness particular to ubiquitous antigens, such as food, and protective mucosal immunity, which includes robust immune responses to pathogenic antigens.

The immunologic hypo responsiveness to enteric-route antigens is known as oral tolerance. Research conducted on laboratory animals demonstrates that the rate and amount of an antigen fed affects how long it takes an animal to develop tolerance. Feeding low doses of an antigen causes active suppression after the induction of regulatory T cells in Peyer's patches, while feeding high doses of an antigen causes clonal deletion or energy. The regulatory T lymphocytes function through the production of suppressive cytokines, including interleukin 4, interleukin 10, and transforming growth factor β. Clonal deletion or anergy is preceded by the local production of interleukin 12, interferon γ (with consequent suppression of interleukin 4 and transforming growth factor β generation), and involves the apoptosis of T helper 1 cells. It is therefore suggested that one of the major mechanisms by which the gut-associated lymphoid tissue maintains homeostasis is via local cytokine regulation, particularly transforming growth factor β–associated low-dose tolerance.

Oral tolerance is not induced by all intraluminal antigens. Particular reactions in the lymphoid tissue related to the gut are triggered by intraluminal bacterial antigens. The ability of intraluminal bacterial antigens to bind to epithelial cells explains this, as it permits antigen entry through enterocytes and circumvents the induction of tolerance in Peyer's patches. Such tonic immune responses in the lymphoid tissue associated with the gut may enable regulation of the metabolic process and the equilibrium of the gut microbiota. Probiotics that are currently on the market have been categorized based on the various capacities that antigens have been shown to have for adhering to epithelial cells. Strong antigen adhesion to epithelial cells is linked to an improved immune response in the gut. However, Duchmann's research demonstrated that while healthy people can tolerate their own microbiota, patients with IBD are unable to do so. Certain immunologic abnormalities linked to IBD were corrected when probiotic therapy changed the characteristics of the natural microbiota. These findings point to a paradoxical role for candidate probiotic bacteria in immune regulation: enhancement. Such paradoxical regulation of the immune response to enteral antigens seems to be a constant finding in the gut-associated lymphoid tissue, and oral tolerance is considered to be a concomitant effect of immune exclusion and suppression of systemic immune response, possibly attributed to the dual effect of the suppressor cytokine transforming growth factorβ.

6.3.1. Oral Tolerance or Allergic Sensitization

Early infancy is characterized by a large number of immunoregulatory aberrations that favor sensitization over tolerance induction. During a variable period after birth, the intestine's immune regulation, exclusion, and exclusion of antigens are incomplete, which predisposes it to aberrant antigen uptake. The underdeveloped immune system's defense shows up as a decreased ability to produce IgA-producing cells. Additionally, T-cell function is abnormal, and the generation of immunoregulatory cytokines varies significantly between the cells of infants and adults. The cytokine profile in newborns is biased toward humoral immunity rather than cell-mediated immunity, and the high concentration of cells that produce interleukin 4 at a crucial time may cause the immune system's T-cell memory to shift toward the T-helper-2 phenotype, which increases the production of IgE and may cause atopic sensitization. The vulnerability of oral tolerance breakdown at an early age can be explained by the immature gut barrier, which may cause aberrant antigen transfer and immune responses. It has been suggested that inadequate production of the anti-inflammatory cytokine transforming growth factor β by neonatal lymphocytes predisposes a person to sensitization by low doses of enteric antigen. At an early age, such antigens are frequently derived from food and allergic reactions to foods are common.

Oral tolerance may be abolished in the setting of inflammation due to the changed pace, path, and mode of antigen presentation. When bacteria, viruses, or food antigens cause inflammation in the intestinal mucosa, intestinal permeability may be secondarily increased. As a result, a significant number of antigens may pass through the mucosal barrier, changing the pathways of transport. The normal pattern of antigen handling is disrupted during the ensuing mucosal dysfunction brought on by immaturity, infection, or hypersensitivity reaction. This may cause aberrant immune responses and sensitization. These findings suggest that the inability of the gut-associated lymphoid tissue to develop or sustain oral tolerance to dietary antigens is what causes allergic reactions to these antigens (Vieira et al., 2013).

6.4. INTESTINAL FLORA: EFFECT ON THE GUT DEFENSE MECHANISMS

The process of developing the gut barrier and intestinal microflora is gradual, but microbial colonization starts right after birth. One source of bacteria that colonizes the newborn's intestine is the mother's intestinal flora. Additionally, contact with the environment influences colonization. The predominant strains at first are facultative anaerobic ones. Afterwards, variations in the species composition occur, primarily due to the kind of diet. The microflora of breastfed infants is more complex, consisting of bifidobacteria, *Bacteroides*, clostridia, and streptococci, while that of formula-fed infants is more simple. The adult flora's composition is similar to that of the microflora following weaning. Despite being found all over the intestine, the large intestine contains the majority of the microorganisms.

The human gut microbiota comprises both native and transitory species in adults. The oral cavity is home to a diverse range of facultative and strict anaerobes, including *lactobacilli*, yeasts, *Bacteroides*, and streptococci. There is very little microflora in the upper intestine (stomach, duodenum, and jejunum) with $\leq 1 \times 10^8$ colony-forming units/L. Bacterial concentrations progressively rise from the ileum through the rest of the digestive tract, arriving at 1×10^{11}–10^{12} colony forming units/g in the colon. There are potentially 500 different species of bacteria in the large intestine of an adult human. According to a number of reports, the majority of anaerobic bacteria that are still viable in the large intestine belong to five genera: Eubacterium, Fusobacterium, *Bifidobacterium*, *Peptostreptococcus*, and *Bacteroides*. There are a variety of facultative and aerobic organisms in the colon. According to estimates, bacteria collectively make up 35–50% of the volume of the contents of the human colon.

Increased antigen transport through the gut mucosa in the absence of an intestinal microflora demonstrates the critical role that the gut microflora plays in the intestine's defense barrier. This idea is further reinforced by evidence that the gut microbiota triggers particular immune responses both locally and systemically. Additionally, oral tolerance is induced and maintained by the gut microbiota in experimental animal models. The intestinal flora shortens the time it takes for the *Escherichia coli* toxin to abrogate systemic hypo-responsiveness and permits it to persist in response to an antigen.

Intestinal colonization contributes to tolerance induction and serves as a significant antigenic stimulus for the development of gut-associated lymphoid tissue maturation. The ability to produce IgA-producing cells gradually rises in response to intestinal antigenic stimulation, especially during the development of the gut microbiota. It has been demonstrated that after colonization, organisms move to the MLN; however, as the intestine's immune defense mechanisms mature, the number of translocating bacteria starts to decline with the onset of a particular IgA response.

In germ-free mice, the function of the intestinal microbiota in inducing the IgE response and oral tolerance was studied. When ovalbumin was given orally, germfree mice nevertheless produced IgE antibodies and other T-helper-

2-type immune responses, in contrast to control mice. The absence of intestinal flora caused the oral tolerance to be abolished. Restoring this flora during the neonatal stage could rectify the aberrant IgE response in germfree mice, but not when reconstitution is done later in life. These findings imply that the intestinal bacterial flora is a key factor in producing a T-helper-2 population that is amenable to oral tolerance induction, which in turn influences the development of gut-associated lymphoid tissue during the neonatal stage.

There are significant variations in the culturable microflora, according to our recent research examining the microflora development in infants delivered vaginally and via cesarean section whose mothers received prophylactic antibiotics. At six months of age, differences were still noticeable as a significant percentage of babies delivered via cesarean section did not have *Bacteroides fragilis* colonization. It seemed that colonization and humoral immune mechanism maturation went hand in hand. It is interesting to note that B. fragilis and, to a lesser extent, bifidobacteria were significant in this regard because the organisms they harbored increased the number of IgA- and IgM-secreting cells in the blood. These findings imply that human intestinal microbiota is significant and that variations in the microbiota's qualitative makeup may have an impact on immunologic homeostasis. Early infancy is the time when the mode of immune responsiveness to antigens is consolidated, marking the culmination of the effect of gut microflora on the maturation of the gut immune defense.

6.5. PROBIOTICS: IMMUNOMODULATION OF LOCAL AND SYSTEMIC IMMUNE RESPONSE

There is evidence that probiotic bacteria enhance the host's natural defense systems. Probiotic bacteria have also been demonstrated to improve humoral immune responses, supporting the intestine's immunologic barrier, in addition to their effects on nonimmunologic gut defense, which is typified by the stabilization of the gut microflora. Probiotic bacteria can also modulate the host's immune responses to potentially harmful antigens, potentially downregulating hypersensitivity reactions, and stimulating nonspecific host resistance to microbial pathogens, thereby aiding in immune elimination.

6.5.1. Nonspecific Immunomodulation

Lactobacilli can be introduced orally to increase nonspecific host resistance to microbial pathogens, which can help to exclude pathogens from the gut. Numerous live lactic acid bacteria (LAB) strains have been demonstrated to trigger the *in vitro* release of proinflammatory cytokines, such as tumor necrosis factor α (TNF-α) and interleukin 6, which indicate the stimulation of nonspecific immunity. In mice, the administration of *L. casei* and *Lactobacillus acidophilus* stimulates phagocytosis, while the oral introduction of *Lactobacillus casei* and *Lactobacillus bulgaricus* stimulates the generation of macrophages. Additionally, *L. acidophilus* Lal reported enhanced phagocytosis in humans. The inflammatory response is first triggered by phagocytosis, which occurs before the production of antibodies. When phagocytes undergo different types of inflammatory reactions, they release toxic agents such as lytic enzymes and reactive oxygen intermediates. Further immune-competent cell recruitment and the production of an inflammatory response are the outcomes of phagocytic activity. More recently, enhanced phagocytic activity was observed in atopic infants with food allergies compared with control infants, indicating that the capacity to generate and release functionally active.

Intestinal microbiota imbalance is typically present in conjunction with intestinal inflammation. Fecal urease, a proinflammatory mediator that predisposes gut mucosa to ammonia-induced destruction and consequently to the overgrowth of urease-producing bacteria, is found in higher concentrations in cases of rotavirus diarrhea. It has also been observed that the bacterial makeup of rheumatoid arthritis patients differs from that of healthy individuals, suggesting that the gut microflora is part of an ecosystem that reacts to inflammation outside of the gut. It has been demonstrated that patients with juvenile chronic arthritis have higher fecal urease concentrations. Fecal urease concentration is shown to return to normal in such inflammatory states of both infectious and noninfectious etiology when oral probiotic therapy is administered. Probiotic treatment may therefore aid in preserving the gut microbial environment and thereby inhibit the production of inflammatory mediators, which are produced continuously by the gut-associated lymphoid tissue in response to potentially hazardous intraluminal antigens that may compromise intestinal integrity. Proinflammatory cytokines, including interleukin 1, tumor necrosis factor α, and interferon γ, play a pivotal, yet paradoxical, role in inflammation. In the absence of interleukin 10 and interleukin 2, a benign immune response to commensal gut microbiota develops into a detrimental inflammatory state, according to experiments performed on cytokine transgenic knockout mice. This suggests that a dysregulated local or systemic cytokine milieu causes inflammation. Oral bacteriotherapy with *Lactobacillus rhamnosus* GG (ATCC 53103) was shown to reduce elevated fecal concentrations of tumor necrosis factor-α in patients with atopic dermatitis and cow milk allergy. Paradoxically, ingestion of *lactobacilli* in fermented milk products or as live attenuated bacteria was shown to potentiate the interferon-γ production by peripheral blood mononuclear cells. Interferon-γ can promote the uptake of antigens in Peyer's patches, where specific IgA-committed cells are generated.

An increase in systemic and mucosal IgA response to dietary antigens was shown after oral administration of *lactobacilli*. Therefore, ingestion of probiotic bacteria may stabilize the immunologic barrier of the gut mucosa by reducing the generation of local proinflammatory tumor necrosis factor γ and by reinforcing the systemic production of interferon-γ with physiologic protective effects in the gut. However, aberrant interferon-γ production was shown to interfere with the induction of oral tolerance and to disrupt epithelial barrier integrity in the gut. Therefore, it is interesting to observe that specific strains of probiotic bacteria can normalize aberrant antigen-induced production of interferon γ *in vitro*. These data indicate that the immunomodulating effects of probiotic bacteria may depend on the immunologic state of the host. They further suggest that differences between specific strains of probiotic bacteria may exist.

6.5.2. Specific Effects on Immune Response

The purpose of using probiotics specifically is to modify the host's immune response to potentially hazardous antigens. It has been demonstrated that *Bifidobacterium breve* stimulates the IgA response to cholera toxin in mice and that *Bifidobacterium bifidum* increases the antibody response to ovalbumin when given orally. Comparably, children with acute rotavirus diarrhea who received *L. rhamnosus* GG during the acute phase of diarrhea showed an enhanced humoral immune response, including an increase in rotavirus-specific antibody-secreting cells in the IgA class, in comparison to control studies. The convalescent stage mean serum rotavirus IgA antibody concentration was likewise higher in the *L. rhamnosus* GG-treated group. These findings are supported by the fact that oral *lactobacilli* administration increased the number of cells secreting antibodies to -lactoglobulin in suckling rats that had been sensitized to cow milk. In addition to immediate-type IgE-mediated hypersensitivity, cow milk allergy in humans is linked to delayed-type hypersensitivity to cow milk proteins and a malfunctioning production of local IgA responses. When atopic newborns are given challenge-proven cow milk allergens.

The gut's processing of food antigens is aided by the intestinal microbiota. Low-molecular-weight peptides released by specific bacterial species isolated from the gastrointestinal microbiota can elicit immune responses. Proteases derived from probiotic bacteria have the ability to break down cow milk casein, producing peptides that inhibit the proliferation of lymphocytes in healthy individuals. A study was designed to investigate whether caseins degraded by probiotic bacteria-derived enzymes could modulate the cytokine production with anti-CD3 antibody-induced peripheral blood mononuclear cells in atopic infants with cow milk allergy, in an effort to further characterize the immunomodulatory effect of probiotics. In cultures from patients with atopic dermatitis, casein without hydrolyzation increased the production of interleukin

4, while casein hydrolyzed by *L. rhamnosus* GG decreased the production of interleukin 4. These findings suggest that probiotics alter the immunogenicity of potentially harmful antigens by changing their structural makeup.

6.5.3. Anti-Inflammatory Properties of Probiotics: Comparison of Specific Strains of Probiotic Bacteria

Whole bacterial cells are shown to enhance proliferation of immune cells and induce production of proinflammatory cytokines, such as TNF-α and interleukin 6. Probiotic bacteria, on the other hand, mediate the suppression of T cell cytokine production and lymphocyte proliferation. We have endeavored to juxtapose the inhibitory impact of multiple probiotic bacterial strains in their nonviable configurations. After the probiotic strains were grown independently and subjected to sonication, the homogenates were filtered, and it was discovered that the enzymatic activity was negligible. Three 10-fold dilutions were prepared in accordance with the estimated total protein concentration of the homogenates. Tests for lymphocyte transformation were conducted on healthy adults to ascertain the mitogen-induced proliferative responses of peripheral blood mononuclear cells to these homogenates. In every experiment using probiotic homogenates, a dose-dependent suppressive effect on mitogen-induced lymphocyte proliferation was seen. A hierarchy of immunomodulation among probiotics was observed when the rate of proliferation was compared among cultures with the same protein concentration. The suppressive effect of 10 μmol dexamethasone/L was comparable with that of successful probiotics, indicating that specific probiotic bacteria possess significant anti-inflammatory properties comparable to a therapeutic pharmaceutical agent. These findings further implicate the potential use of probiotic bacteria as immunomodulatory agents.

6.6. PROBIOTICS AND PROBIOTIC FUNCTIONAL FOODS: LEGISLATION AND MODERN CHALLENGES

Probiotics are live microorganisms that, when given in sufficient quantities, offer health benefits to the host, according to the World Health Organization (WHO) and the Food and Agriculture Organization (FAO). Probiotic use in both humans and animals has increased dramatically in recent years. They can be used to make fermented foods (cheese, milk), as components of kid-friendly milk, as ingredients in baked goods or confections, or as raw materials for food additives. Probiotic use is another contemporary trend in agricultural science. It is being used in animal breeding and poultry farming to lower disease rates in young animals, maximize the nutritional value of fodder ingredients, promote rapid muscle growth without the need for hormonal preparations or antibiotics, and increase the amount of other products that animals and poultry provide, such as eggs, milk, and other products.

Microorganisms used as probiotics can be classified as members of the *Lactobacillus*, *Lactococcus*, *Bifidobacterium*, *Pediococcus*, or *Bacillus* families; the *Lactobacillus* family is the most commonly used. *Lactobacillus* family probiotics are beneficial components found in yogurt and other dairy products that help to maintain the health of the gastrointestinal system. Numerous studies support the benefits of probiotics for gastrointestinal health maintenance, immune system strengthening, prevention of colorectal cancer, lowering serum cholesterol, easing lactose intolerance, and other conditions. Probiotics also help prevent infections by competitively suppressing the development of pathogenic strains.

Before consuming any gram probiotic product, the minimum required concentration of live cells must be between 10^6 and 10^9 to produce any positive reaction. Probiotic products abound on the market, and the majority of them come in dry form. Dried probiotic products' expiration dates and the effects they have after consumption depend on a variety of factors. One of the ongoing challenges in the development of probiotic formulas is ensuring that the strains have longer shelf lives. Producers of probiotic products are always coming up with new ideas to keep probiotic strains alive during the drying process (lyophilization), rehydration, transit through the GIT, the biliary salts and stomach's sharply varying pH, and storage in retail locations. Probiotic strains' molecular properties and their impact on health have been the subject of several studies. Various methods to develop novel probiotic formulations with the strains' vitality preserved and beneficial effects.

Probiotic products into four generations:

- **First generation** – probiotic microorganisms without coating
- **Second generation** – probiotic microorganisms with single synthetic coating, which protects them when passing through the GIT
- **Third generation** – microcapsuled probiotic microorganisms with synthetic capsule
- **Fourth generation** – probiotic microorganisms with two synthetic coatings, protein and mucopolysaccharides, which protect them when passing through the GIT and against high pressure and temperature

6.6.1. Probiotic Functional Foods

Products with live microorganisms—usually yeast or bacteria—that provide health benefits when ingested in sufficient quantities are known as probiotic functional foods. Probiotics are these microorganisms that assist in keeping the proper balance of bacteria in the gut, which can improve immunity, digestion, and general health. Here are some examples of probiotic functional foods:

- **Yogurt:** One of the most popular probiotic foods is yogurt. It is produced by inoculating milk with certain bacterial strains, like *Bifidobacterium* and *Lactobacillus*. Yogurt can strengthen the immune system, promote gut health, and aid with digestion.
- **Kefir**: With a thinner consistency than yogurt, kefir is a fermented dairy product. It is prepared by fermenting milk using kefir grains, which are a blend of yeast and bacteria. Probiotic-rich kefir can improve gut health and aid in digestion.
- **Fermented Vegetables**: Fermentation is the process by which foods like pickles, kimchi, and sauerkraut grow good bacteria. These veggies that have undergone fermentation offer extra nutrients and fiber along with being a good source of probiotics.
- **Kombucha**: A culture of bacteria and yeast is used to ferment sweetened tea to create kombucha, a fermented tea beverage. It is frequently appreciated for its possible advantages to the immune system and digestive system and contains probiotics.
- **Tempeh**: A traditional Indonesian dish made from fermented soybeans is called tempeh. It is a good source of protein and has good bacteria in it. Tempeh is frequently incorporated into vegetarian and vegan diets in place of meat.
- **Miso**: Miso is a fermented paste made from soybeans, rice, or barley. It is commonly used in Japanese cuisine and is rich in probiotics. Miso soup is a popular way to consume miso, and it provides probiotics along with a range of nutrients.
- **Probiotic Supplements**: Probiotics can be obtained through natural food sources as well as supplements. Typically, probiotic supplements contain particular yeast or bacterial strains that have been shown to have positive effects on health. They can be taken as tablets, powders, or capsules.

It is critical to choose probiotic functional foods and supplements that list the strains and amounts of probiotics in each product. The effects of different strains may differ, and enough probiotics should be present to offer health benefits. Speaking with a healthcare provider is also advised, particularly if you have any dietary restrictions or specific medical issues (Borchers et al., 2009).

REFERENCES

1. Bolte, G., Knauss, M., Metzdorf, I., & Stern, M. (2008). Postnatal maturation of rat small intestinal brush border membranes correlates with increase in food protein binding capacity. *Digestive Diseases and Sciences, 43*, 148–155.
2. Borchers, A. T., Selmi, C., Meyers, F. J., Keen, C. L., & Gershwin, M. E. (2009). Probiotics and immunity. *Journal of Gastroenterology, 44*(1), 26–46.
3. Foster, J. A., & McVey Neufeld, K. A. (2013). Gut-brain axis: How the microbiome influences anxiety and depression. *Trends in Neurosciences, 36*(5), 305–312.
4. Frei, R., Akdis, M., & O'Mahony, L. (2015). Prebiotics, probiotics, synbiotics, and the immune system: Experimental data and clinical evidence. *Current Opinion in Gastroenterology, 31*(2), 153–158.
5. Hancock, V., Vejborg, R. M., & Klemm, P. (2010). Functional genomics of probiotic *Escherichia coli* Nissle 1917 and 83972, and UPEC strain CFT073: Comparison of transcriptomes, growth and biofilm formation. *Molecular Genetics and Genomics, 284*, 437–454.
6. Ianiro, G., Bibbò, S., Gasbarrini, A., & Cammarota, G. (2014). Therapeutic modulation of gut microbiota: Current clinical applications and future perspectives. *Current Drug Targets, 15*(8), 762–770.
7. Johnson, B. R., & Klaenhammer, T. R. (2014). Impact of genomics on the field of probiotic research: Historical perspectives to modern paradigms. *Antonie Van Leeuwenhoek, 106*(1), 141–156.
8. Kamada, N., Chen, G. Y., Inohara, N., & Núñez, G. (2013). Control of pathogens and pathobionts by the gut microbiota. *Nature Immunology, 14*(7), 685–690.
9. Kernéis, S., Bogdanova, A., Kraehenbuhl, J. P., & Pringault, E. (2000). Conversion by Peyer's patch lymphocytes of human enterocytes into M cells that transport bacteria. *Science, 277*, 949–952.
10. Kwon, H. K., Lee, C. G., So, J. S., et al. (2010). Generation of regulatory dendritic cells and CD4+Foxp3+ T cells by probiotics administration suppresses immune disorders. *Proceedings of the National Academy of Sciences of the United States of America, 107*, 2159–2164.
11. Li, J., Wang, W., Xu, S. X., et al. (2011). *Lactobacillus reuteri*-produced cyclic dipeptides quench agr-mediated expression of toxic shock syndrome toxin-1 in staphylococci. *Proceedings of the National Academy of Sciences of the United States of America, 108*, 3360–3365.
12. Masood, M. I., Qadir, M. I., Shirazi, J. H., & Khan, I. U. (2011). Beneficial

effects of lactic acid bacteria on human beings. *Critical Reviews in Microbiology, 37*(1), 91–98.

13. Palmer, S. (2011). Research suggests beneficial bacteria may support immune health. *Today's Dietician, 13*(1), 20.
14. Pfeiler, E. A., & Klaenhammer, T. R. (2009). Role of transporter proteins in bile tolerance of *Lactobacillus acidophilus*. *Applied and Environmental Microbiology, 75*(18), 6013–6016.
15. Qin, J., Li, R., Raes, J., Arumugam, M., Burgdorf, K. S., Manichanh, C., et al. (2010). A human gut microbial gene catalogue established by metagenomic sequencing. *Nature, 464*(7285), 59–65.
16. Saito, H., Kanamori, Y., Takemori, T., et al. (2001). Generation of intestinal T cells from progenitors residing in gut cryptopatches. *Science, 280*, 275–278.
17. Shroff, K. E., Meslin, K., & Cebra, J. J. (2005). Commensal enteric bacteria engender a self-limiting humoral mucosal immune response while permanently colonizing the gut. *Infection and Immunity, 63*, 3904–3913.
18. Tlaskalova-Hogenova, H., Stepankova, R., Hudcovic, T., Tuckova, L., Cukrowska, B., Lodinova-Zadnikova, R., et al. (2004). Commensal bacteria (normal microflora), mucosal immunity, and chronic inflammatory and autoimmune diseases. *Immunology Letters, 93*(2–3), 97–108.
19. Van Hemert, S., Meijerink, M., Molenaar, D., et al. (2010). Identification of *Lactobacillus plantarum* genes modulating the cytokine response of human peripheral blood mononuclear cells. *BMC Microbiology, 10*, 293.
20. Vanderpool, C., Yan, F., & Polk, D. B. (2008). Mechanisms of probiotic action: Implications for therapeutic applications in inflammatory bowel diseases. *Inflammatory Bowel Diseases, 14*, 1585–1596.
21. Vieira, A. T., Teixeira, M. M., & Martins, F. S. (2013). The role of probiotics and prebiotics in inducing gut immunity. *Frontiers in Immunology, 4*, 445.
22. Weiss, G., & Jespersen, L. (2010). Transcriptional analysis of genes associated with stress and adhesion in *Lactobacillus acidophilus* NCFM during the passage through an *in vitro* gastrointestinal tract model. *Journal of Molecular Microbiology and Biotechnology, 18*, 206–214.
23. Yan, F., Cao, H., Cover, T. L., et al. (2011). Colon-specific delivery of a probiotic-derived soluble protein ameliorates intestinal inflammation in mice through an EGFR-dependent mechanism. *Journal of Clinical Investigation, 121*, 2242–2253.

CHAPTER 7

GUT MICROBIOME COMPOSITION

Contents

INTRODUCTION

The bacteria, fungi, viruses, and archaea that inhabit the digestive tracts of animals are known as the gut microbiota, gut microbiome, or gut flora. All the genomes of the gut microbiota combined make up the gastrointestinal metagenome. Human microbiome is primarily found in the gut. The gut microbiome has a wide range of effects, influencing immunity, gut–brain axis regulation, colonization, pathogen resistance, intestinal epithelium maintenance, metabolizing food and medication compounds, and epilepsy. Depending on the area of the digestive tract, the gut microbiota's microbial makeup varies. Of all the human-associated microbial communities that have been studied to date, the colon has the highest microbial density, with between 300 and 1000 different species present. Around 30 to 40 species account for 99% of the gut flora, making bacteria the most numerous and well-studied component to date. In the dry mass of excrement, bacteria can make up to 60% of the mass. The gut contains more than 99% anaerobe bacteria, but the cecum is home to high densities of aerobic bacteria. There are approximately 100 times as many genes in the human gut microbiota as there are in humans.

7.1. GUT MICROBIOTA AND HUMAN MICROBIOME

The trillions of bacteria and other microorganisms that inhabit the human gut are collectively referred to as the gut microbiota. Their habitat is known as the microbiome. In the body, most microbes are beneficial, but when they get out of balance, they can become dangerous. Microbiome and microbiota are frequently used synonymously. This is untrue, though. The diverse range of bacteria, viruses, fungi, and other microorganisms found in a particular environment—such as the human digestive tract—make up the microbiota. The term "microbiome" describes the microorganisms, their genomes, and the surrounding environmental factors that make up the body's entire habitat.

When compared to other parts of the body, the human gut microbiota contains the greatest diversity and quantity of bacteria. One to two years after birth is when the gut flora in humans is established. By then, the intestinal epithelium and the intestinal mucosal barrier it secretes have co-developed in a way that is both a barrier to pathogenic organisms and tolerant of, and even supportive of, the gut flora. Certain gut microbiota and humans have a mutualistic relationship rather than just a commensal one (a non-harmful coexistence). Certain microorganisms found in the human gut assist the host by converting dietary fiber into short-chain fatty acids (SCFAs) like butyric and acetic acids, which the host can then absorb. The synthesis of vitamins B and K, as well as the metabolism of bile acids, sterols, and xenobiotics, are all aided by intestinal bacteria. The gut flora itself seems to function like an endocrine organ, and the systemic importance of the SCFAs and other compounds they produce is comparable to that of hormones. Numerous inflammatory and autoimmune diseases have been linked to dysregulation of the gut flora. The human gut microbiota's composition varies with time, diet, and state of general health. The probiotic bacteria strains that have been commercially available for use in preclinical and small-scale human trials have been the subject of a systematic review in 2016, which determined which strains had the greatest potential to be beneficial for specific disorders of the central nervous system (CNS).

7.1.1. Human Microbiome

In a symbiotic relationship, every human possesses between 10 trillion and 100 trillion microbial cells. According to some research, the number of microbial cells in the body is approximately 10 times greater than that of human cells, but other scientists argue the ratio is closer to 1:1. As long as the body is in good condition, both the microbes and their hosts gain from this symbiotic partnership. The number of different species of microorganisms that make up the human microbiota is estimated to be over 1,000. Individual differences can be significant in the human biome's composition. Additionally, distinct microbial communities will be found in various body parts.

7.1.2. Gut Microbiota

Human health is significantly impacted by the diverse and extensive collection of microorganisms known as the gut microbiota. The term "gut microbiota" was once used to describe the microflora of the gut.

The gut microbiota assists in a range of bodily functions, including:

- harvesting energy from digested food
- protecting against pathogens
- regulating immune function
- strengthening biochemical barriers of the gut and intestine

Modifications in the makeup of the microbiota may impact these roles. Although the gut contains helpful bacteria, the gastrointestinal tract (GIT) can become infected due to harmful bacteria. Food poisoning and other gastrointestinal illnesses that cause vomiting and diarrhea are examples of these infections.

7.1.3. Importance of Human Microbiota

Microorganisms perform a variety of essential roles that are essential to human life. In actuality, the GIT contains about 10 times as many bacterial as human cells.

Research suggests that bacterial population makeup and potential disturbances have links to:

- asthma
- autistic spectrum disorder
- cancer
- celiac disease
- diabetes
- eczema
- heart disease
- malnutrition
- multiple sclerosis
- obesity

7.1.3.1. Nutrition

Gut microbes are critical to human nutrition absorption as well as energy absorption from food. For example, the body uses gut bacteria to aid in the breakdown of complex molecules found in meats and vegetables. Plant cellulose cannot be digested by the body without the assistance of gut bacteria. Additionally, the metabolic activities of gut microbes may affect feelings of fullness and food cravings. The diversity of a person's diet affects the diversity of their gut.

7.1.3.2. Immunity

Some research suggests that the body's first exposure to microbes may occur before birth.

Adaptive immunity would not exist without these initial microbial visitors. This essential defense system picks up on microbe responses after coming into contact with them. This makes it possible to respond to organisms that cause disease more quickly and effectively. The first microbial exposure causes the gut microbiota to develop, and it usually takes three to five years for it to reach its full composition. Disturbances to these initial exposures can impede the microbiota's development.

7.1.3.3. Behavior

The brain and gut microbiota of an individual are in constant communication with one another. Intestinal function is the primary effect of the gut-brain axis. Yet, correlations between the gut microbiota and psychological conditions like ASD and depression have also been noted by researchers.

7.1.3.4. Disease

According to research, bacterial populations in the GIT may be involved in the development of inflammatory bowel diseases (IBD), which include ulcerative colitis (UC) and Crohn's disease. Type 2 diabetes and obesity have also been linked to low gut microbial diversity. There are connections between metabolic syndrome and the state of the gut microbiota. Prebiotics, probiotics, and other dietary supplements can help lower these risk factors by altering the diet. Antibiotic-induced disruption of the microbiota can also result in disease, including antibiotic-resistant infections. Additionally, the microbiota is crucial in preventing the overgrowth of foreign populations in the gut that would otherwise lead to illness because the "good" bacteria outcompete the "bad," and some of them even release compounds that reduce inflammation.

7.2. COMPOSITION OF THE GUT MICROBIOME

Thousands of different microorganisms, including bacteria, viruses, eukaryotes, and archaea, inhabit the human body. These microorganisms can be found in the skin, genitalia, gut, and mouth, among other parts of the body. Different microbial environments with various functional roles are present at each body site. The human gut microbiome, which is primarily made up of bacteria and bacteriophages, is the most studied ecosystem in humans to date.

More than 1,000 distinct bacterial species make up the human gut microbiome, with a few prominent phyla—*Bacteroidetes*, Firmicutes, Actinobacteria, and Proteobacteria—being the most prevalent. In a typical, healthy gut, *Bacteroidetes* and Firmicutes are the most prevalent of these. At the lower taxonomic levels, a small number of taxa are typically dominant; however, the composition of these microbes varies greatly, resulting in significant inter-individual variation. Apart from the handful of prominent genera, numerous less frequent taxa are also prevalent, such as *Escherichia*, *Bifidobacterium*, and *Akkermansia*. Studies have demonstrated that the functional significance of a bacterium's abundance is not always reflected in its abundance, since host health and function are increasingly linked to variations in both common and rare taxa.

7.2.1. Good or Bad Microbes

The great intra-species functional variation makes it challenging, if not impossible, to categorize certain bacteria as beneficial or detrimental to the host above the strain level. Variation at the strain level can reveal whether a taxon is harmful or benign. One example of this is *Escherichia*, which is frequently found in healthy human guts where it helps with metabolism. However, strains can be pathogenic due to gene-based variation if they encode toxins. Because of their capacity for horizontal gene transfer, which is the transfer of a gene from one living bacterial cell to another, bacteria are able to introduce intra-strain variation that is missed by taxonomy based on marker genes. The ability of bacteria to transfer genes for antibiotic resistance indicates that it is a significant source of functional variation, though.

7.2.2. Pro- and Prebiotics

One of the gut microbiome's many vital roles during the fermentation of carbohydrates is the production of acetate and lactate, which is carried out by strains of the *Bifidobacterium* clade. One of the most important roles the gut microbiota plays in preserving gut health is its ability to ferment carbohydrates and produce short-chain fatty acids as a result. Thus, Bifidobacteria and other lactic acid bacteria (LAB) are regarded as probiotics, or beneficial bacteria.

Numerous pre- and probiotic products are available on the market; they are typically taken as dietary supplements or added to yogurts. To be referred to as probiotics, products containing live microorganisms, such as *Saccharomyces boulardii* and *Lactobacillus acidophilus*, must improve human health. Probiotics have a lot of potential to alter the human microbiome in a positive way, but more research is needed before we can fully enjoy their advantages. Many placebo-controlled trials have been conducted for a variety of indications, and while some indicate a beneficial effect, others—such as a preventive measure for travel-related diarrhea—have not produced results that are compelling. Prebiotics are a substrate that host microbes use to impart a health benefit; they are not live microbes. Contrary to popular belief, all substances regarded as prebiotics are fermentable dietary fiber or carbohydrates that are accessible to the microbiota. There is currently discussion surrounding the prebiotic idea.

7.2.3. Factors Shaping the Gut Microbiome

The host immune system, dietary practices, and the intricate interactions among its microbial constituents all influence the gut microbiome. Shaping factors encompass both exogenous (diet, exercise, medication) and endogenous (immune system, bile acids, hormones) components to the host. These taken together make up the environment that the gut microbiota lives in and needs to function in. In an intriguing study led by Professor Kevin Foster, the role of competition versus cooperation in forming the microbiome is one of the many variables influencing the microbiome that are still being actively researched.

7.2.4. The Role of the Gut Microbiome

Both humans and animals are hosts to the gut microbiome, which is vital to their health. Its primary roles include immune system instruction and regulation, metabolism, and defense against pathogen invasion. The human body would be deficient in several nutrients, such as vitamins and amino acids, if it weren't for the gut microbiota. Fermentation of non-digestible fibers is one of the gut microbiome's vital functions. A group that produces short chain fatty acids (SCFAs) benefits from these fibers. The three most significant SCFAs that are produced are butyrate, propionate, and acetate, with acetate being the most prevalent. Colonocytes, the cells that make up the inner lining of the human colon, depend on butyrate as a major energy source. It is thought that a deficiency in SCFA is a major factor in many microbiome-related conditions, including leaky gut and local inflammation. Butyrate has been connected to a number of critical processes, including the capacity to cause colon cancer cells to undergo apoptosis and the stimulation of intestinal gluconeogenesis, which is pertinent to diabetes and energy balance. Propionate is also crucial for maintaining glucose homeostasis because it controls the liver's gluconeogenesis.

The human gut's microbiota takes over every niche that becomes available, scavenging space from potentially harmful microbes to protect the host. One of the main theories explaining the effectiveness of fecal transplantations in treating *Clostridium difficile* infections is that a balanced community is necessary to keep pathogens at bay. In addition to its space-occupying defensive function, the microbiome aids in host defense by educating the immune system. Numerous immune cells can be found in the intestine, and they actively engage with microorganisms to enhance immune tolerance. It is believed that dysbiosis, or the absence of good bacteria, causes changes in the makeup of microbiological components that enter the body and damages the intestinal epithelium, leading to leaky gut syndrome. Lipopolysaccharide (LPS), a component of bacteria, can exit the intestine and alter the kinds or amounts of LPS that the immune system comes into contact with, which can lead to both local and systemic inflammation.

The microbiome plays numerous roles in normal human function, but it has also been closely linked to the onset and course of numerous diseases. Recent research has revealed a direct connection between mental health and the gut microbiome. The normal development and function of the brain can be impacted by the gut microbiota, according to research. Patients with ADHD and autism have a different gut bacterial composition than those without these conditions, according to a comparison of children with and without these conditions. However, there is also evidence linking the gut microbiota to cancer, autoimmune and inflammatory diseases, and metabolic disorders like diabetes. The capacity of the gut microbiota to influence drug metabolism and, consequently, the quantity of active drugs made available to the host, has also attracted growing attention.

7.3. GUT MICROBIOTA VARIATIONS

The gut microbiota is made up of a variety of microorganism species, such as viruses, yeast, and bacteria. Bacteria are categorized taxonomically based on their phyla, classes, orders, families, genera, and species. Out of over 160 species, only a few phyla are represented.

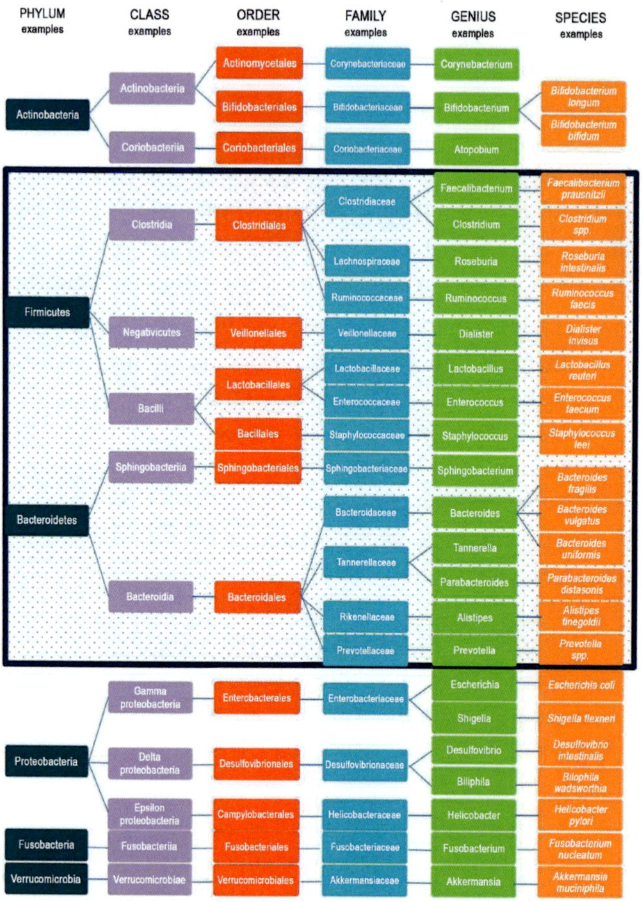

Figure 7.1. *Examples of taxonomic gut microbiota composition. In the box are cited examples of bacteria belonging to Phyla firmicutes and Bacteroidetes, representing 90% of gut microbiota.*

Source: Rinninella E, Raoul P, Cintoni M, Franceschi F, Miggiano GAD, Gasbarrini A, Mele MC. (2019). What is the healthy gut microbiota composition? A changing ecosystem across age, environment, diet, and diseases. Microorganisms; 7(1):14. doi: 10.3390/microorganisms7010014. PMID: 30634578; PMCID: PMC6351938.

Around 90% of the gut microbiota is made up of the two phyla Firmicutes and *Bacteroidetes*, which are the predominant gut microbial phyla along with *Actinobacteria*, *Proteobacteria*, *Fusobacteria*, and *Verrucomicrobia*. More than 200 genera, including *Lactobacillus*, *Bacillus*, *Clostridium*, *Enterococcus*, and *Ruminicoccus*, make up the Firmicutes phylum. 95% of the phyla of Firmicutes are made up of the *Clostridium* genera. *Prevotella* and *Bacteroides* are two of the most common genera in the *Bacteroidetes* group. The *Bifidobacterium* genus is the primary representative of the Actinobacteria phylum, which is comparatively less abundant. Figure 7.1 shows examples of the composition of the taxonomic gut microbiota (Rinninella et al., 2019).

7.3.1. Variations in the Same Individual

The human gut microbiota exhibits taxonomic and functional variability across the GIT, and these variations can occur within an individual due to factors such as age, environmental factors like antibiotic use, and infant transitions. Table 7.1 summarizes the differences in microbiota between individuals.

Table 7.1. Microbiota Variations Within Individuals

		Gut Microbiota Abundance							**Bacteria Diversity**
		Actinobacteria	***Bacteroidetes***	**Firmicutes**	**Proteobacteria**	**Fusobacteria**	**Verrucomicrobia**	**Euryarchaeota**	
Anatomical Part of Gut Tract	**Small intestine**			*Lactobacillus*	*Enterobacteriaceae*				
	Colon		*Bacteroidaceae Prevotellaceae Rikenellaceae*	*Lachnospiraceae Ruminococcaceae*					

Gestational Age	**Preterm birth (<37 weeks of gestation)**	*Bifidobacterium* spp.↓ *Atopobium* spp.↓	*Bacteroides* ↓ (non-secretor mothers)	Firmicutes ↓ (non-secretor mothers) *Lactobacillus*↑ *Ruminococcus* spp. *Lachnospiraceae Peptostreptococcaceae Clostridiaceae*	*Enterobacteriaceae* ↑ *Enterococcus* spp.↑				↓
	Full-term birth	*Bifidobacterium* spp.↑	*Bacteroidetes* ↑	*Ruminococcus* spp. *Lachnospiraceae Peptostreptococcaceae Clostridiaceae*	*Enterobacteriaceae*				↑
Type of delivery	**Vaginal delivery**	*Bifidobacterium* spp.↑ *Bifidobacterium longum*↑ *Bifidobacterium catenulatum*↑	*Prevotella*↑ *Bacteroides fragilis*↑	*Lactobacillus*↑ *Staphylococcus*↑ *Streptococcus* ↑	*Escherichia*↑	*Sneathia*↑			↑

	C-section	*Corynebacterium*↑ *Propionibacterium*↑	*Bacteroides* ↓	*Staphylococcus*↑	*Escherichia*↓ *Shigella*↓				↓
Methods of (milk) feedings	**Breast milk**	*Bifidobacterium*↑↑		*Lactobacillus*↑ *Staphylococcus* ↑	*Enterococcus*↑				↑
	Artificial milk	*Bifidobacterium*↑	*Bacteroides*↑	*Clostridium*↑ *Clostridium difficile*↑ *Lactobacillus*↑	*Escherichia*↑				↓
	Introduction of solid food	*Bifidobacterium*↑	*Bacteroidetes* ↑ *Bacteroides*↑	Firmicutes ↑ *Lactobacilli*↑ *Clostridium coccoides*↑					↑
Human age	**Childhood (first year of life)**	*Bifidobacterium*	*Bacteroides*	*Veillonella C. coccoides C. botulinum*			*Akkermansia muciniphila*		↑
	2–3 years old to adult	*Bifidobacteriaceae Coriobacteriaceae*	*Bacteroidaceae Prevotellaceae Rikenellaceae*	*Lachnospiraceae Ruminococcaceae*	Proteobacteria	*Fusobacteria*	*Akkermansia muciniphila*	*Methanobrevibacter smithii*	↑

	Over 70	*Bifidobacteriaceae*↓		*Clostridium* ↓	Proteobacteria ↑				↓
Antibiotic treatments	**Macrolide**	Actinobacteria ↓	*Bacteroides*↑	Firmicutes ↓	Proteobacteria ↑				↓
	Clarithromycin	Actinobacteria ↓	*Bacteroides*↑	Firmicutes ↓	Proteobacteria ↑				↓
	Vancomycin			*Lactobacillus*↓ *Clostridium*↓					↓
	Ciprofloxacin	*Bifidobacterium*↓	*Alistipes*↓ *Bacteroides*↑	*Faecalibacterium*↓ *Oscillospira*↓ *Ruminococcus*↓ *Dialister*↓					↓
	Clindamycin	*Bifidobacteriaceae*↓ *Lactobacillus*↓							↓

Source: Rinninella E, Raoul P, Cintoni M, Franceschi F, Miggiano GAD, Gasbarrini A, Mele MC. (2019). What is the healthy gut microbiota composition? A changing ecosystem across age, environment, diet, and diseases. Microorganisms; 7(1):14. doi: 10.3390/microorganisms7010014. PMID: 30634578; PMCID: PMC6351938.

7.3.1.1. Intestine Anatomical Regions

The gut microbiota differs depending on the anatomical regions of the intestine, which differ in terms of substrate availability, host secretions, digesta flow rates (rapid in the mouth to the caecum, slower afterwards), pH and O_2 tension, and physiology. The relatively short transit times (3–5 h) and high concentrations of bile make the small intestine a more difficult environment for microbial colonizers. The largest microbial community is found in the large intestine, which is distinguished by slow flow rates and a pH that is neutral to slightly acidic (dominated by obligate anaerobic bacteria). Table 7.1 shows that there is a progressive decrease in aerobic bacteria to the benefit of only anaerobic bacteria, accompanied by a qualitative and quantitative microbiota gradient.

7.3.1.2. Infant Transitions

7.3.1.2.1. Birth Gestational Age

One of the main factors influencing gut microbiota colonization is birth gestational age. Preterm infants (less than 37 weeks gestation) have a different microbiota composition than their term counterparts (Table 7.1). After birth, the microbiota colonization in preterm infants faces challenges due to organ immaturity and environmental factors like hospital stays, enteral feeding, and antibiotic use. For these reasons, preterm birth may have a significant impact on the postnatal development of the gut and systemic immunity.

The gut microbiota composition of term and preterm pigs in a newborn pig model was analyzed, and the results indicated that *Ruminococcus* spp. and several Enterobacterium species. The most common genera in both term and preterm pigs were *Lachnospiraceae*, *Peptostreptococcaceae*, and *Clostridiaceae*. Yet, a greater level of *Enterococcus* spp. Preterm pigs were found to be more abundant than term pigs.

Low diversity is seen in preterm infants, who have lower concentrations of strict anaerobes like *Bacteroides*, *Atopobium*, and *Bifidobacterium* and higher colonization of potentially pathogenic bacteria from the Proteobacteria phylum (Enterobacteriaceae family).

The composition of human milk is determined by genetic factors, the mother's secretory protein, and Lewis blood groups, which define four phenotypes with varying oligosaccharide levels. Premature infants born to non-secretor mothers have been found to have higher levels of Proteobacteria and lower levels of Firmicutes. Just recently, Praticò et al., Analyzing the composition of human breast milk revealed that the gut microbiota composition of infants is modulated by human milk oligosaccharides (HMOs) associated with various mother phenotypes. In particular, by reducing pathogens linked to sepsis and necrotizing enterocolitis (NEC), HMOs associated with secretor mothers may have a protective prebiotic effect. This suggests that in preterm infants, HMOs may affect the intestinal microbiota and prevent gut dysfunction and NEC.

Another well-known ingredient in human milk is lactoferrin (LF), which promotes the colonization of the gut by beneficial bacteria in infants and, as a result, plays a critical role in the microbiota of preterm infants. It has been shown that there are relationships between LF and the good microbiota in breast milk and baby feces. High fecal LF levels, especially in preterm infants, may play a significant role in the genesis, growth, and/or makeup of the newborn gut microbiota, which in turn influences the newborn's immunologic maturation and general wellbeing.

Thus, a mother's milk contributes to the shaping of the developing core gut microbiota, which can enhance neurodevelopment and growth while lowering the risk of late-onset sepsis and NEC. Premature infants should consume human milk as their main enteral diet due to the following reasons (Rinninella et al., 2019).

7.3.1.2.2. Type of Delivery

The intestine is bacteria-free and sterile at birth. Following delivery, a diverse and ever-changing ecosystem emerges from the mother's skin, feces and vaginal microbiota, and interactions with the surrounding microbiota. The type of delivery affects the colonization of microbiota (Table 7.1).

Newborns that are delivered vaginally develop a microbiota composition that is similar to that of their mother. In fact, a study of newborn infants' meconium showed a significant relationship between the *Lactobacillus*, *Prevotella*, and *Sneathia* microbial communities in the mother's vagina and the microbiota of the newborn digestive tract. Biasucci et al. showed that prominent groups like *Bifidobacterium longum* and *Bifidobacterium catenulatum* are also present in the microbiota of infants delivered vaginally. The infant gut is colonized by other facultative anaerobic species, including *Bacteroides fragilis*, *Streptococcus*, *Staphylococcus*, and *Escherichia coli*.

Conversely, newborns delivered via cesarean section (C-section) pick up bacteria from the mother's skin and the hospital environment, including Staphylococcus, Corynebacterium, and Propionibacterium species. In babies delivered via cesarean section, the species *Escherichia*, *Shigella*, and *Bacteroides* are underrepresented. Compared to infants delivered vaginally, the intestinal microbiota of neonates delivered by cesarean section has a lower diversity of bacterial species. Analyses conducted at the age of seven have also revealed differences in the microbiota of infants born vaginally and those born by C-section. Chronic immune disorders like asthma, systemic connective tissue disorders, juvenile arthritis, IBD, and obesity have been linked to a higher risk of cesarean delivery.

7.3.1.2.3. Methods of Milk Feeding

Research has indicated that compared to breastfed infants, formula-fed infants have higher colonization rates of *Bacteroides*, *Clostridium difficile*, and *Escherichia coli*. About the abundance of Actinobacteria, *Bifidobacterium* spp. has been connected to formula milk and breastfeeding. Nonetheless, compared to formula-fed infants, breastfed babies typically have a more complex and varied *Bifidobacterium* microbiota. Compared to formula-fed infants, breastfed infants have a microbiota that contains more than twice as many *Bifidobacterium* cells. Breastfed babies have a more diverse and richer population of *Bifidobacterium* spp. in their gut microbiota, which is beneficial and fewer instances of *Escherichia coli* and *Clostridium difficile* than infants fed formula (Table 7.1).

Different species of *Bifidobacterium* are in charge of fermenting galactooligosaccharide (GOS), which is one of the primary ingredients in breast milk, to create SCFAs. Yes, *Bifidobacterium* species possess the lacto-N-biosidase enzyme, which makes GOS absorption easier. A decrease in *Bifidobacterium* quantity results in a decrease in HMOs in feces, which raises acetate and lactate concentrations and lowers pH. The consumption of HMOs by Bifidobacteria in an infant's gut may indicate a prebiotic effect of HMOs on the gut microbiota of infants by specifically stimulating *Bifidobacterium* spp. β-palmitate, an innate fatty acid found in human milk and a constituent of certain baby formulas, has been shown to have a prebiotic effect on the gut microbiota of infants by favorably affecting *Bifidobacterium* spp. and *Lactobacillus* species.

Furthermore, the maintenance of a healthy and balanced mother's gut microbiota during pregnancy is also considered as an important factor in positively influencing the milk microbiota composition. Additionally, oral probiotic supplementation in mothers with vaginal delivery may increase breast milk *Bifidobacterium* spp. and *Lactobacilli* spp. abundance.

According to these studies, the native core microbiota are shaped in the early stages of life. A full-term, vaginally delivered, breastfed infant with a balanced and healthy mother's milk microbiota is thought to have a healthy intestinal microbiota profile. Studies should aim to show how the composition of this native core microbiota affects gut health and ensuing illnesses (Rinninella et al., 2019).

7.3.1.2.4. Weaning Period

The transition from breast milk to solid foods and the end of weaning occurs at the same time as significant alterations in gut microbiota. After weaning, the genera *Bacteroides*, *Clostridium coccoides*, and *Bifidobacterium* are more common. Children who have been breastfed exhibit a more similar microbiota to those who have previously been fed formula milk (Table 7.1).

Gut microbiota variations are significantly influenced by dietary habits, infant weaning techniques, and feeding practices, which act as determinants. Traditional foods high in carbohydrates and fiber lead to an increase in *Firmicutes* and *Prevotella*, while foods high in animal protein and fiber lead to an increase in *Bacteroidetes*. Additionally, Fallani et al. showed that while southern European countries are more likely to have higher levels of *Bacteroides* and *Lactobacilli* in their infant gut microbiota, northern European countries are linked to a higher proportion of Bifidobacteria.

7.3.1.3. Age

A child's microbiota composition at one year old exhibits a characteristic abundance of *Bacteroides*, *Veillonella*, *Clostridium coccoides* spp., and *Akkermansia muciniphila* and species of *Clostridium* botulinum. Age-related increases in microbiota diversity lead to a stable adult composition of microbiota that is dominated by three bacterial phyla (Table 7.1): Actinobacteria (*Bifidobacteriaceae* and Coriobacteriaceae), *Bacteroidetes* (*Bacteroidaceae*, *Prevotellaceae*, and *Rikenellaceae*), and Firmicutes (*Lachnospiraceae* and *Ruminococcaceae*). These bacteria are the product of maturation brought about by genetics, environment, diet, lifestyle, and gut physiology. A child's gut microbiota composition and diversity are most similar to those of an adult at around three years of age.

The composition of the gut microbiota in older adults over 70 years of age may be influenced by changes in nutrient absorption and digestion as well as a decrease in immune function. More monotonous dietary changes may also reduce the diversity of the gut microbiota. Anaerobic bacteria, like *Bifidobacterium* spp., are lower and there has been a noticeable rise in Proteobacteria and *Clostridium*. Considering how important *Bifidobacterium* spp. is, a decline in Bifidobacteria may contribute to the explanation of malnourishment and low systemic inflammatory status in older adults by stimulating the immune system and metabolic processes.

7.3.1.4. Antibiotics

The composition of the gut microbiota can be influenced by the use of antibiotics.

A follow-up investigation into the impact of antibiotics with varying modes of action on the composition of the human gut microbiota revealed that antibiotic treatments alter the composition of the gut microbiota, causing some species to become more abundant or visible while others decrease in number or become less noticeable (Table 7.1). Antibiotics with a broad spectrum of action cause an imbalance between *Bacteroidetes* and Firmicutes. Throughout the treatments, both the diversity and abundance of these bacteria decline. The antibiotic class, dosage, exposure duration, pharmacological action, and target bacteria all affect how the microbiome is composed (Table 7.1). The intestinal microbiota is selectively influenced by certain antibiotic properties, such as their mode of action or antimicrobial effects, which also play a role in the changes in bacterial composition that occur during antibiotic therapy. Each class of antibiotics has different properties and excretion systems, resulting in different patterns of alteration in the microbiome composition.

Three people were the subjects of a cultivation-independent survey to determine the composition of their gut microbiota prior to, during, and following two exposures to the same antibiotic (ciprofloxacin). This study demonstrated a significant resistance to the administration of antibiotics, but it also raised the possibility that the system occasionally remembers previous disruptions and that repeated disruptions always result in a persistent regime shift. Therefore, the effect of antibiotic disruption on the microbiota's resistance to subsequent antibiotic treatments can also differ significantly among individuals (Rinninella et al., 2019).

7.3.2. Gut Microbiota Variations Between Individuals

We have shown that the composition of the gut microbiota varies within an individual and will be observed to vary between individuals as well. Enterotypes, body mass index (BMI), and external factors like lifestyle, frequency of exercise, ethnicity, and dietary and cultural practices are the main causes of these inter-individual variations. Table 7.2 shows the variations in microbiota abundance between individuals.

Table 7.2. *Microbiota Variations Between Individuals*

		Gut Microbiota Abundance							**Bacteria Diversity**
		Actinobacteria	***Bacteroidetes***	**Firmicutes**	**Proteobacteria**	**Verrucomicrobia**	**Euryarchaeota**	**Spirochaetes**	
Enterotype	**I**	*Slackia*	*Bacteroides Parabacteroides*	*Clostridiales Alkaliphilus Lactobacillus Catenibacterium*	*Geobacter*		*Methanobrevibacter smithii*		
	II	*Eggerthella*	*Prevotella*	*Veillonella Ruminococcaceae Holdemania Peptostreptococcaceae Staphylococcus Leuconostoc*	*Desulfovibrionaceae Rhodospirillum Helicobacter Escherichia Shigella*	*Akkermansia muciniphila*			

	III	*Gordonibacter*	*Sphingobacterium*	*Ruminococcus Staphylococcus Marvinbryantia Symbiobacterium Ruminococcaceae Dialister*		*Akkermansia muciniphila*			
BMI level (among children and adolescents)	**Over-weight to obese level**	*Bifidobacterium*↑ *B. vulgatus*↓	*Bacteroidetes* ↓	*Lactobacillus*↑ Firmicutes ↑	Proteobacteria ↑				↓
	Under-weight level			*Roseburia* spp.↓ *Ruminococcus*↓ *Clostridium* spp.↓	*Enterobacteriaceae* ↑		*Methanobrevibacter smithii*↑		↓
Exercise frequency	**High performance sport**		*Bacteroides*↓	*Ruminococcaceae* ↑ *Lactobacillus*↓ *Lactobacillaceae* ↓	*Succinivibrionaceae* ↑	*Akkermansiaceae*↓			↑

	Children and adolescents daily exercise			*Clostridiales*↑ *Roseburia*↑ *Lachnospiraceae*↑ *Erysipelotrichaceae*↑					↑
Type of diet, communities, climate, and geographical sites	**High-fiber diet (African diet)**	*Bifidobacterium*↓	*Prevotella*↑ *Bacteroides*↓	*Eubacterium*↑ *Oscillibacter*↑ *Butyricicoccus*↑ *Sporobacter*↑ *Blautia*↓ *Dorea*↓ *Lachnospiraceae* ↓ *Roseburia*↓ *Faecalibacterium*↓ *Ruminococcus*↓ *Erysipelotrichaceae* ↓	*Succinivibrio*↑			*Treponema*↑	↑

	High-fat and animal protein diet (Western diet)	Actinobacteria ↑	*Bacteroides*↑ *Alistipes*↑ *Barnesiella*↑	*Roseburia*↓ *Eubacterium rectale*↓ *Ruminococcus bromii*↓	*Bilophila*↑				↓

Source: Rinninella E, Raoul P, Cintoni M, Franceschi F, Miggiano GAD, Gasbarrini A, Mele MC. (2019). What is the healthy gut microbiota composition? A changing ecosystem across age, environment, diet, and diseases. Microorganisms; 7(1):14. doi: 10.3390/microorganisms7010014. PMID: 30634578; PMCID: PMC6351938.

7.3.2.1. Enterotypes

Enterotypes are specific bacterial clusters that characterize an individual's gut microbiota. Table 7.2 lists the three predominant bacterial clusters that define the three enterotypes: *Ruminococcus* (enterotype III), *Bacteroides* (enterotype I), and *Prevotella* (enterotype II). Different bacterial genera are present in each enterotype. These three enterotypes are precisely regrouped based on functions, rather than just being lists of bacteria. In actuality, an enterotype is not so much a systematic addition of bacteria species as it is a functionally harmonious association of several bacteria species. Unlike blood groups, which are definitive identities, enterotypes are characteristics that do not change throughout adulthood and can be restored through modification. Each enterotype defines a unique method of producing energy from fermentable substrates present in the colon, thanks to its unique bacterial clusters and corresponding functional traits. In fact, enterotype I bacteria clusters mainly use the pentose phosphate and glycolysis pathways to obtain energy from carbohydrates, whereas enterotype II and III bacteria clusters can degrade.

7.3.2.2. Body Mass Index (BMI) Classes

Numerous investigations into the relationship between childhood BMI and gut microbiota composition have shown that children with normal or overweight BMI are exposed to a greater variety of microorganisms than children who are underweight. Based on BMI class, the diversity of the gut microbiota decreases stepwise. Bervoets along with others showed that, in comparison to the microbiota of lean children, the microbiota of obese children has an increased Firmicutes-to-*Bacteroidetes* ratio. Also, there were high concentrations of *Lactobacillus* species and low relative proportions of *Bifidobacterium vulgatus* are found in the microbiota of obese people. Riva et al. verified that obesity is linked to decreased levels of *Bacteroidetes*, such as *Bacteroidaceae* and *Bacteroides*, and increased levels of Firmicutes, such as *Ruminococcaceae*. Children who are obese have higher SCFA levels, which may indicate higher substrate utilization. These results imply that the pathophysiology of obesity may be influenced by the dysbiosis of the gut microbiota, and that the higher ratio of Firmicutes to *Bacteroidetes* is linked to the increased generation of SCFAs and the extraction of energy from colonic fermentation.

Significant differences in gut microbiota were also found in the underweight BMI level. Borgo along with others carried out a thorough data analysis comparing the anthropometric characteristics and gut microbiota of 15 anorexic women with AN and healthy controls. In contrast to healthy controls, the intestinal microbiota of individuals with AN exhibited a notable increase in Enterobacteriaceae and *Methanobrevibacter smithii*. Conversely, the depletion of the genera *Clostridium*, Ruminococcin, and *Roseburia* was consistent with the observed decrease in AN of the total SCFAs.

A reliable predictor of gut microbiota dysbiosis is BMI levels. Research has indicated a correlation between variations in gut microbiota and either elevated or decreased production of SCFAs, which could potentially play a role in the pathophysiology of obesity or anorexia. Prebiotics and probiotics are examples of interventions that might be used to treat patients with AN as well as childhood obesity.

7.3.2.3. Ethnicity, Dietary Habits, and Cultural Habits

A healthy person's microbiome is generally stable, but dietary and lifestyle choices can undoubtedly affect the dynamics of gut microbiota. African children's microbiota has an astonishing abundance of *Prevotella* and *Xylanibacter*, according to a study done on children from Burkina Faso who were fed a diet rich in millet/sorghum + local vegetables containing very few lipids and animal proteins, and children from Europe who were fed a Western diet.

Additionally, the prevalence of *Escherichia* and Shigella is significantly low. The gut microbiota of Italians and Hadza hunter-gatherers was compared in another study. Proteobacteria and Spirochaetes, which are exceedingly rare in the Italian gut microbiota, are heavily concentrated in the Hadza gut microbiota at the phylum level, while Actinobacteria, a significant subdominant component of the Italian gut microbiota, is nearly absent. At the genus level, the gut microbiota of Hadza is relatively depleted in *Bifidobacterium*, *Bacteroides*, *Blautia*, *Dorea*, unclassified *Lachnospiraceae*, *Roseburia*, *Faecalibacterium*, *Ruminococcus*, and unclassified *Erysipelotrichaceae* and comparatively enriched in *Prevotella*, *Eubacterium*, *Oscillibacter*, *Butyricicoccus*, *Sporobacter*, *Succinivibrio*, and *Treponema*.

These two studies demonstrated that the enterotype *Prevotella* (enterotype II) is clearly present in the gut microbiota of Africans. The African diet, which is high in millet/sorghum and local vegetables with low lipid content and animal proteins, promotes mucin degradation in concert with *Desulfovibrionaceae*. The predominant enterotype in the gut microbiota of people who follow a Western diet is *Bacteroides* (enterotype I). This is because the high fat and animal protein content of the European diet enables the fermentation of carbohydrates and proteins to produce energy. David et al. compared the microbiomes of diets based on plants and animals, showing that the microbiomes of the former had higher levels of fiber and lower levels of fats and proteins, while the latter had higher levels of fats and proteins. The study also showed that the microbiomes of the two diets changed with time, with a shift toward an animal-based diet leading to an increase in bile-tolerant bacteria (Alistipes, Bilophila, and *Bacteroides*) and a decrease in firmicutes—*Roseburia*, *Eubacterium rectale*, and *Ruminococcus bromii*—and lower levels of microorganisms that metabolize plant polysaccharides in food. Moreover, when the type of diet varies, microbial activity alternates between the fermentation of proteins and carbohydrates. This alteration happened just one day following the death. David et al. showed that diet alters human microbiota rapidly and reproducibly.

7.3.2.4. Exercise Frequency

Bai et al. suggested correlations between young children's and adolescents' gut microbiota composition and how frequently they exercise. Frequent exercise produces more SCFAs, which may increase the expression of tight junction proteins in colon epithelia to heighten the resistance of the intestinal barrier, reduce mucosal permeability, and inhibit inflammatory cytokines. This results in an enrichment microbiota of *Firmicutes*, *Clostridiales*, *Roseburia*, *Lachnospiraceae*, and *Erysipelotrichaceae* (Table 7.2).

Clarke et al. examined how professional athletes from an international rugby union squad's gut microbiota was affected by their diet and level of exercise. Athletes have been shown to have a higher alpha diversity of gut microorganisms than both high and low BMI controls, with these microorganisms belonging to over 20 different phyla. In fact, this study demonstrated that protein consumption positively correlates with microbial diversity and that exercise increases gut microbial diversity (Table 7.2). These findings show that exercise has a positive effect on gut microbiota diversity, but they also suggest that there may be a connection between exercise and microbiota diversity and concomitant dietary extremes (Rinninella et al., 2019).

7.3.3. Gut Microbiota Variations in Health and Disease

The composition of the gut microbiota varies greatly. In the framework of a healthy gut microbiota, the variation itself is regarded as physiological, taking into account factors such as age, ethnicity, lifestyle, and dietary habits. These physiological differences in gut microbiota, however, have a significant impact on intestinal and extraintestinal disorders. In fact, dysbiosis is frequently described as a cause or an effect of disorders as well as a change in the composition of the gut microbiota. Determining whether a change is beneficial or detrimental is often a challenging task. Regarding human conditions, shifts can strengthen resilience by providing a challenge for improved context adaptation. On the other hand, an unstructured ecosystem may experience stress and disruption from persistent external stimulation. We'll talk about the connections between a number of illnesses and changes in the composition of the gut microbiota.

7.3.3.1. Intestinal Disorders

7.3.3.1.1. Irritable Bowel Syndrome

The bacterial component of gastrointestinal (GI) disorders, such as irritable bowel syndrome (IBS), has been extensively studied. The investigation of colonic and fecal mucosal biopsy samples from IBS patients and healthy controls showed that the concentration of aerobic bacteria, such as the *Lactobacillus* species, was significantly lower in the fecal samples from IBS patients than in the healthy controls. An additional investigation revealed that, in contrast to control subjects, the microbial communities of individuals with IBS exhibit an enrichment of *Proteobacteria* and *Firmicutes*, while *Actinobacteria* and *Bacteroidetes* are reduced. Specifically, there are more 16S rDNA sequences from the phylum Firmicutes in the IBS clone library. These sequences belong to the family *Lachnospiraceae.*

Additional research supported these findings, pointing to the following phyla and genera variations between IBS patients and healthy controls: a rise in the Firmicutes-to-*Bacteroidetes* ratio, a fall in the populations of Actinobacteria (*Bifidobacteria, Collinsella*), a decrease in some Firmicutes families (*Lactobacilli*, *Faecalibacterium*), and an increase in some Firmicutes families (*Veillonella, Streptococci*, and *Ruminococcus* spp.) and in Enterobacteriaceae species of Proteobacteria). Thus, these results revealed a reduction in microbial diversity that could be related to the production of amino acids, the integrity of cellular junctions, and the inflammatory response, implying a partial explanation of IBS symptoms by a dysfunction of the epithelial barrier.

7.3.3.1.2. Inflammatory Bowel Disease (IBD)

Idiopathic, persistent, and recurrent inflammatory GIT conditions, such as UC and Crohn's disease (CRD), are collectively referred to as IBD. The results of Frank et al. showed that the gut microbiota of IBD patients was different from that of control patients in that the *Lachnospiraceae* and *Bacteroidetes* microbial populations were smaller and the Proteobacteria families were larger.

There seem to be differences in the gut microbiota between UC and CRD. *Faecalibacterium prausnitzii* and *Roseburia hominis*, two bacteria that produce butyrate, have been found to be less abundant in UC patients when compared to controls. In contrast, CRD patients have been found to have higher levels of *Faecalibacterium prausnitzii* along with lower overall diversity. An unidentified species of *Clostridium* spp. called *Dialister invisus* was found to be less prevalent in CRD patients, according to another study as well as an increase in *Ruminococcus gnavus*, *Bifidobacterium adolescentis*, and *Faecalibacterium prausnitzii*.

The report by Sokol et al. verified that *F. prausnitzii* had been reduced in CRD patients, as well as showing that giving live *Faecalibacterium prausnitzii* orally tended to improve the dysbiosis and lessen the severity of colitis. These findings point to a potential probiotic approach for treating CRD: utilizing *Faecalibacterium prausnitzii* to counterbalance dysbiosis. These findings don't prove that microbial dysbiosis and IBD pathophysiology are related; rather, they point to the possibility that gut microbial imbalances may be a factor in the severity of the illness.

7.3.3.1.3. Celiac Disease

A genetically predisposed person's aberrant immune response to dietary gluten proteins results in celiac disease (CD), a chronic intestinal inflammatory disorder. In fact, dysbiosis is associated with an inflammatory environment in celiac disease patients, suggesting that variations in the gut microbiota may be important in the pathophysiology of CD. The interaction of genetic factors, environmental factors like gut microbiota, and gluten results in celiac disease.

Research has shown that *Bifidobacterium* spp, *Clostridium histolyticum*, *Bifidobacterium longum*, compared to healthy controls, the proportions of the *Faecalibacterium prausnitzii* and *C. lituseburense* groups are lower in untreated CD patients. In fact, compared to healthy subjects, patients with CD exhibit a decrease in beneficial species (*Lactobacillus* and *Bifidobacterium*) and an increase in potentially pathogenic species (*Bacteroides* and *E. coli*). Patients with CD have a mucosal layer that is unable to both keep the gut microbiota stable and shield the host from pathogens and harmful antigens. This dysbiosis is lessened after a gluten-free diet, though it might still persist.

7.3.3.1.4. Colorectal Cancer (CRC)

The third most common cause of cancer-related death worldwide is CRC. It has been shown that there are structural fecal bacterial segregations between CRC patients and healthy volunteers. The CRC patients' gut microbiota was deficient in *Roseburia* and other butyrate-producing bacteria of the *Lachnospiraceae* family and rich in *Bacteroides fragilis*, *Enterococcus*, *Escherichia/Shigella*, *Klebsiella*, *Streptococcus*, and *Peptostreptococcus*. The gut microbiome of the healthy participants exhibited an increased presence of *Bacteroides uniformis* and *B. vulgatus*. Consequently, a significant structural imbalance of the gut microbiota in CRC patients is caused by variations in the gut microbiota, such as a decrease in butyrate producers and an increase in opportunistic pathogens.

Shen et al. showed that when comparing CRC cases to controls, there was a greater abundance of Proteobacteria and a lower abundance of *Bacteroidetes*. *Dorea* spp. were more prevalent in CRC cases at the genus level and species of

Faecalibacterium and reduced percentages of *Bacteroides* species and species of Coprococcus compared to controls. *Fusobacterium* spp. were found to be associated through genomic analysis with CRC; *Fusobacterium* spp. were more prevalent in the microbiota of CRC patients. While there was a reduction in the Firmicutes and *Bacteroidetes* phyla. In addition, Fusobacterium spp. may use an inflammatory-mediated mechanism to contribute to tumorigenesis, but more research is needed to determine the exact role Fusobacteria play in the pathogenesis of colorectal carcinoma. These results show changes in the colon cancer microbiota that could be related to the development of CRC. The application of these findings could result in methods for identifying and preventing CRC by manipulating microbiota (Rinninella et al., 2019).

7.3.3.2. Extra-Intestinal Disorders

We emphasized that many intestinal disorders are related to the critical relationship between the gut microbiota and the host. Research has recently shown that this mutualistic relationship between the gut microbiota and the host appears to be important for many extra-intestinal diseases, including neurological disorders and metabolic diseases with interactions to the "gut-brain axis," as well as complex multifactorial diseases like age-related macular degeneration with interactions to the "gut-retina axis."

7.3.3.2.1. Metabolic Disorders

1. **Obesity:** The bulk of the causal evidence connecting alterations in microbial composition to obesity comes from studies conducted on animals; human data, however, is more inconsistent. Furthermore, the microbiomes of mice and humans are similar at the division level, with Firmicutes and *Bacteroidetes* predominating, even though the majority of mouse gut species are unique.

Genetically obese ob/ob mice and lean mice were given the same polysaccharide-rich diet. The gut microbiota of the obese mice showed a lower relative abundance (50%) of *Bacteroidetes*, while the Firmicutes were higher. In fact, compared to their lean littermates, leptin-resistant obese (leptin-promoting satiety) and diabetic mice (db/db) have higher numbers of *Ruminococcaceae* and Rikenellaceae. Furthermore, the increased prevalence of Rikenellaceae and *Ruminococcaceae* in mice given a high-fat diet is associated with type 2 diabetes and obesity in addition to being dependent on the food consumed. Moreover, it has been shown that there is a correlation between the abundance of *Desulfovibrionaceae* and type 2 diabetes and obesity. Additional research supported the notion that proteobacteria contribute to obesity by generating pro-inflammatory compounds like LPSs, harvesting energy, and increasing the amount of fat that is stored by the host.

Conversely, it has recently been shown by researchers that obese mice induced by diet or genetics have lower levels of *Akkermansia muciniphila*, a mucin-degrading bacteria that lives in the mucus layer. The results of these studies all point to the association between obesity and modifications in the species diversity and abundance of genes involved in metabolism within the gut microbiota. These changes in the gut microbiota are transmissible and impact the microbiome, which can absorb more energy from the diet. The pathophysiology of obesity may be influenced by gut microbiota, which may also have therapeutic implications.

2. **Type 2 Diabetes (T2D):** It has been shown in multiple studies that patients with type 2 diabetes have altered gut microbiota compositions; however, it is unclear if these changes are the actual cause of the disorder or just one of its effects. Larsen along with others revealed that, in comparison to the healthy group, the proportions of phyla *Firmicutes* and *Clostridia* are significantly lower in T2D patients. Additionally, compared to healthy patients, T2D patients have significantly higher ratios of *Bacteroidetes* to *Firmicutes*, *Bacteroides-Prevotella*, and *Betaproteobacteria*, all of which have a positive correlation with plasma glucose.

A metagenome-wide association study of the gut microbiota of Chinese patients with type 2 diabetes revealed that the presence of butyrate-producing bacteria, such as *Roseburia* spp., was decreased in these patients and a rise in many opportunistic infections, including *Clostridium* spp, *Bacteroides caccae*) in addition to an enrichment of additional microbial functions that provide reduction of sulfate (*Desulfovibrionaceae* spp). The gut microbiota has the ability to interact with multiple host sensing and signaling pathways, which can alter immune responses, the endocrine system, nervous system activity, and ultimately the susceptibility to metabolic diseases. Dysbiosis of the gut microbiota is the cause of diabetes and other metabolic diseases and suggests new treatment approaches.

7.3.3.2.2. Central Nervous System (CNS)-Related Disorders

The gut–brain axis connects the brain and the gut, allowing the CNS, enteric nervous system (ENS), and GIT to interact in both directions. The brain can affect both immunological and gastrointestinal processes thanks to these reciprocal interactions.

1. **Alzheimer's and Parkinson's Diseases:** Globally, more than 46 million people suffer from dementia today. In older adults, Alzheimer's disease (AD) and Parkinson's disease (PD) are thought to be the most prevalent types of dementia. AD is a neurodegenerative condition that develops slowly over time and

is characterized by amyloid-beta peptide buildup in the brain and cognitive impairment. Cattaneo et al. showed that in patients with brain amyloidosis and cognitive impairment, a decrease in *E. rectale* abundance and an increase in *Escherichia/Shigella* abundance may be related to a peripheral inflammatory state. Vogt et al. found that the gut microbiota composition of patients with and without an AD diagnosis showed a lower microbial diversity that was compositionally different from control subjects who were age- and sex-matched. The microbiome of AD participants also showed an increase in *Bacteroidetes* abundance and a decrease in Firmicutes and *Bifidobacterium* abundance.

The pathophysiology of PD also involves variations in the gut microbiota. Hopfner et al. indicated that compared to healthy controls, Parkinson's patients had a higher abundance of *Lactobacillaceae*, which is consistent with the majority of other studies. Furthermore, no differences in beta diversity have been found for any of the three bacterial families (*Lactobacillaceae, Barnesiellaceae*, and *Enterococcaceae*) while no differences in alpha diversity have been found between PD cases and controls. Gut microbiota dysbiosis may be the mechanism of neuroinflammation that results in PD pathology, as PD is linked to changes in gut microbiota.

2. **Hepatic Encephalopathy:** Hepatic encephalopathy (HE) is a range of neuropsychiatric abnormalities in patients with liver dysfunctions and is considered a complication of cirrhosis. Bajaj et al. connected the composition of the gut microbiota in HE with inflammation and cognition. When compared to controls, the fecal microbiota of cirrhotic had significantly higher levels of *Enterobacteriaceae*, *Alcaligenaceae*, and *Fusobacteriaceae* and lower levels of *Ruminococcaceae* and *Lachnospiraceae*. Additionally, HE patients had altered higher levels of *Veillonellaceae* compared to cirrhotics without HE. *Alcaligenaceae* and *Porphyromonadaceae* showed a positive correlation with cognitive impairment in the cirrhosis group. *Ruminococcaceae* had a negative correlation with inflammation, while *Enterobacteriaceae*, *Fusobacteriaceae*, and *Veillonellaceae* had positive correlations. Hence, cirrhosis is linked to notable changes in the gut microbiota when compared to healthy people, particularly when it is complicated by HE. In HE, there is a strong correlation between cognition and inflammation and specific bacterial families, namely *Alcaligenaceae*, *Porphyromonadaceae*, and *Enterobacteriaceae*.
3. **Autism Spectrum Disorders:** The gut microbiota of children with ASD is less diverse and shows higher levels of *Lactobacillus*,

Clostridium, *Bacteroidetes*, *Desulfovibrio*, *Caloramator*, and *Sarcina*, and lower levels of *Bifidobacterium* and Firmicutes. This is in contrast to the gut microbiota of children without ASD. An investigation by Wang and colleagues revealed a drop in *Bifidobacterium* species and mucolytic bacteria *Akkermansia muciniphila* in the gut microbiome of individuals with autism in comparison to those without the condition.

Williams et al., found an increase in the *Firmicutes*-to-*Bacteroidetes* ratio and a significant increase in *Sutterella* compared to their abundance in controls. These findings demonstrated that gut microbiota dysbiosis has close relationships with gastrointestinal and behavioral manifestations of autism.

4. **Stress:** The gut microbiota's composition can be changed by the brain due to the two-way interactions between the gut and the brain. An experiment involving adult mice that caused social disruption revealed that stress causes significant alterations in the composition of the gut microbiota, specifically a reduction in *Bacteroides* species and a rise in the abundance of *Clostridium* spp. Compared to controls, abundance has been seen in mice under stress. Stress can alter the microbial habitat by causing changes in intestinal mobility and mucin secretion. The exact mechanisms by which stress affects the composition of the gut microbiota are unknown.

Additional research showed that the microbiota's composition reflects long-term psychological stress. In fact, depression brought on by ongoing stressful life events is linked to a rise in Enterobacteriaceae, while psychological stress is linked to a decrease in *Lactobacilli* species together with a rise in Pseudomonas species and *Escherichia coli*. These neurological disorders are closely correlated with changes in the gut microbiota, and therapeutic interventions may target the gut bacterial communities (Rinninella et al., 2019).

7.4. FACTORS THAT INFLUENCE GUT BACTERIAL COMPOSITION

A complex community of bacteria living in the intestine is called the gut microbiota. The gut microbiota's composition can be greatly influenced by a number of factors, as animal models have shown. These include genetics, the way an infant is fed, the mode of delivery at birth, the use of medications, particularly antibiotics, and diet. Numerous extrinsic and intrinsic factors have the potential to impact gut flora composition and, in turn, overall health.

7.4.1. Method of Delivery at Birth

The CDC estimates that as of 2014, 32.2% of deliveries in the US were made by cesarean section. When an infant is delivered via cesarean section, their gut bacterial community is not the same as when it is delivered vaginally. Vaginal birth exposes the newborn to the mother's bacteria, which changes the infant's gut flora and activates white blood cells and other immune system components. According to studies, babies delivered by cesarean section have a higher chance than vaginally delivered babies of becoming obese and/or diabetic. According to a recent cross-sectional study of 8,900 preschoolers, children born via cesarean section had odds of 1.35 being overweight and 1.25 being obese. In a related study, 672 preschoolers who were delivered via cesarean section had prevalence rates of overweight and obesity of 15% and 12%, respectively, in a small cohort. Contrary results have also been documented, though. Numerous factors, such as the study population and the offspring's sex, can affect the results of these studies, even though more research seems to support the association.

7.4.2. Infant Feeding

Since mother's milk is not sterile, breastfeeding is another crucial element in the establishment of the bacterial community in the gut. It has been established that human breast milk contains commensal and possibly probiotic bacteria that affect how an infant's gut flora develops. There are over 700 different species of bacteria in human breast milk. The median bacterial load in human milk is approximately 106 bacterial cells/mL over time, despite the fact that bacterial communities are typically complex and vary individually. According to current estimates, a lactating infant who consumes 800 mL of breast milk per day may ingest up to 8×10^8 bacterial cells per day—a daily intake that is approximately 100 times higher than previous estimates—and the composition of these cells varies throughout the lactation period. Both Streptococci and Staphylococci seem to be the most common bacterial genera in human milk, and they are also most common in the skin microbiota.

As a result, some skin bacteria may also be present in human milk. However, in baby colostrum samples, *Weissella*, *Leuconostoc*, *Staphylococcus*, *Streptococcus*, and *Lactococcus predominate*, while *Veillonella*, *Leptotrichia*, and *Prevotella* significantly increased in milk taken at one and six months. Research indicates that a mother's microbiota can have an impact on her infant's development and growth. It has been discovered that the bacteria in milk from obese moms differ and are less varied from those in milk from mothers of normal weight. Additionally, milk from obese moms exhibited increased proinflammatory qualities. Furthermore, compared to milk samples from mothers who delivered their babies vaginally, the breast milk of mothers who had cesarean sections contained different bacteria. One of the first microbes to enter an infant's body are those found in breast milk and on the mother's skin. These bacteria may have a significant impact on the infant's health. Additionally, a rich source of IgA antibodies against various pathogens is breast milk. Nonetheless, research has demonstrated that certain baby formulas with probiotics have a positive impact on the gut microbiota and immune systems of young children.

7.4.3. Genetics

The genetic composition of the host has an impact on the quantity of particular bacteria present in the gut microbiota, which in turn influences host metabolism and may have an impact on health. It has been discovered that the gut microbiota of family members is more similar than that of unrelated people, and that monozygotic twins have a more similar gut microbiota than dizygotic twins. Though certain immune system genes are linked to IBD, there are presently no genome-wide studies that have identified particular genes and pathways that control the composition of the gut microbiome.

7.4.4. Infections

Infections caused by bacteria and viruses are influenced by the gut microbiota, but the opposite is also true. According to one study, which examined the impact of an enteropathogenic *Citrobacter rodentium* infection on the microbiota of mice, the infection causes changes in the relative abundance of *Lactobacillus* and other bacterial groups in the gut. By using 16S ribosomal RNA (rRNA) gene pyrosequencing in a human study, researchers discovered that both asymptomatic carriers and patients with *Clostridium difficile* had lower levels of microbial richness and diversity when compared to healthy subjects. Severe dysbiosis in the gut microbiota is often associated with *C. difficile* infection. Remarkably, transplanting the gut microbiome from healthy donors to patients with infection enhanced the diversity and richness of microorganisms, and this technique is being used in clinical settings. According to these studies, bacterial infections change the composition and diversity of the gut microbiota. Chou and colleagues employed a mouse model of hepatitis B virus infection demonstrated that gut microbiota establishment is necessary for the eradication of hepatitis B virus infection. That shift is clearly identifiable.

7.4.5. Medications

A growing body of research indicates that many nonantibiotic medications, including those used to treat Type 2 diabetes, may have an effect on gut microbiota. Similarly, the effectiveness of drugs is also influenced by the gut microbiota. Millions of lives have been spared from infections thanks to the widespread prescription of antibiotics, which also have a significant impact on the normal gut microbiota. The result is immediate and occasionally long-lasting. While broad-spectrum antibiotics decrease the number of helpful bacteria and increase the abundance of some bacteria that opportunistic pathogens can use, they also decrease bacterial diversity. The use of broad-spectrum antibiotics, like clindamycin, in young children and infants has been shown to have the longest-lasting effects on the gut microbiota's composition. Early exposure to antibiotics in newborns can cause microbial dysbiosis, which may put them at risk for developing IBD. There seems to be a connection between diet and the use of antibiotics. Research conducted on mice and humans has revealed that early antibiotic use may influence obesity later in life through changes to the gut microbiota.

Nevertheless, those studies have certain shortcomings. The inherent limitations of human studies are their retrospective nature, the diversity of diets, and the heterogeneous (or not heterogeneous) population. Because male mice gain more weight than female mice, most mouse studies on obesity caused by a high-fat diet with or without antibiotic treatment only used male mice; in contrast, human obesity does not appear to be skewed towards any particular gender. According to a recent study, antibiotics changed the gut microbiota of the hosts without affecting their metabolism. Antibiotics have been shown in other studies to decrease body weight and improve insulin sensitivity. The primary ingredient in a Chinese herb extract used to treat bacterial diarrhea, berberine, also has antidiabetic properties through regulating gut microbiota and reducing insulin resistance and glucose levels. Therefore, it is unclear if the sharp increase in obesity, particularly in childhood, is the result of antibiotic use. Nonetheless, it is evident that the gut microbiota's alteration is significantly impacted by the use of antibiotics.

In people with type 2 diabetes, metformin is frequently used to help control hyperglycemia. The medication makes body cells—particularly hepatocytes, muscle cells, and fat cells—more sensitive to insulin. Metformin also stops hepatocytes from producing too much glucose. Moreover, metformin postpones the absorption of glucose during postprandial digestion. It's interesting to note that metformin administration has been shown to change the microbiota's makeup. In a study where mice were fed a high-fat diet to induce obesity, the authors discovered that metformin caused the mucin-degrading bacterium *Akkermansia* to be more abundant in the obese mice than in the obese mice who did not receive metformin. A human study conducted recently validated the impact of metformin on the gut microbiota. The common side effect of the medication in the digestive system may be caused by an altered gut microbiota, which may also affect the medication's ability to effectively treat diabetes.

7.4.6. Diet

It is possible that methodological constraints that have been resolved recently led to the underestimation of the role that food-ingested bacteria play in the gut microbiome. Studies on humans and research mice have demonstrated the role that high-calorie diets play in the development of T2D and obesity. But more and more data points to the gut microbiota as the mediator between diet and obesity. Diet is the most logical target to manipulate because it is known to play a significant role in the composition of the gut microbiome. Studies using interventional methods demonstrate that alterations in diet lead to significant and quick changes in the composition of the gut microbiome. Research on mice has indicated that a diet high in fat (60% fat) reduces the diversity (α) of bacterial species in the gut microbiome. Additionally, there is a significant difference in the composition of the gut microbiome between mice fed the high-fat (unpurified) diet and those fed a regular unpurified diet (β diversity). According to one study, prebiotic feeding of *A. muciniphila* restored its abundance and enhanced metabolism. However, the abundance of *A. muciniphila* declined in obese and type 2 diabetic mice.

Human gut microbiota has been studied using 16S rRNA sequencing, which has classified it into several enterotypes based on the kinds of bacteria that are present. Enterotypes and long-term diets, especially those high in protein and animal fat, were significantly correlated. Wu et al. revealed that *Bacteroides* was linked to protein and animal fat, while *Prevotella* was linked to carbs. The authors of that study also looked into controlled feeding in 10 subjects and discovered that, after starting a high-fat or low-fiber diet, the microbiome composition changed within 24 hours and stayed stable for the duration of the 10-day study. The findings imply that enterotype partitioning and diet are especially closely related. The question then becomes whether a Western-style diet's effects on the composition of the gut microbiota are linked to a higher or lower incidence of disease, and whether stable changes in the bacterial enterotype can be achieved through long-term dietary interventions.

Another diet and enterotype study included six male and four female volunteers with ages ranging from 21 to 33 years and BMIs (in kg/m^2) between 19 and 32. The study subjects were fed either an animal-based diet consisting of meat, eggs, and cheese, or a plant-based diet rich in grains, legumes, fruits, and vegetables. For five days in a row, each diet was consumed at will, and the subjects' feces were either cultured or subjected to direct analysis using 16S rRNA gene sequencing. The authors demonstrated how changes in the microbiota in the high-fat animal-based diet may be related to changes in fecal bile acid profiles and the development of microorganisms that can cause IBD. The findings demonstrated how a diet high in fat can change gut flora, resulting in dysbiosis and eventually illness.

7.4.7. Yogurt and the Gut Microbiome

A growing amount of research indicates that obesity and chronic inflammatory diseases are significantly influenced by the gut microbiota. It has been suggested that T2D and/or obesity are chronic inflammatory disorders. As was previously mentioned, obesity and type 2 diabetes are significantly influenced by the Western diet. Yogurt consumption is thought to be healthful, in contrast to a high-fat diet. This benefit may be mediated by changes to the host's gut microbiota in addition to other nutritional aspects. Over a century ago, Elie Metchnikoff postulated that using bacteria found in yogurt, one could manipulate one's intestinal microbiome to improve one's health. For decades, there had been a paucity of research on the relationship between the composition of the gut microbiota and yogurt consumption.

However, in the last eight years or so, there has been a sharp increase in the number of studies, including clinical trials that have looked into the effects of yogurt consumption on changes in the gut microbiota and its positive effects on health, thanks to advanced technology, particularly high-throughput sequencing technology and adequate bioinformatical computation. Yogurt, in particular, and other fermented foods can have up to 10^9 CFU live bacteria/g of yogurt. *Lactobacillus* Delbruck subspecies bulgaricus and *Streptococcus* salivarius subspecies thermophiles are combined to ferment yogurt.

Bifidobacteria, are often added to yogurt as well. In a phase-I clinical trial, Merenstein et al. showed that the consumption of yogurt containing *Bifidobacterium lactis* activated an array of immune genes in the immune cells from the peripheral blood of study subjects. It is interesting that both anti-inflammatory and inflammatory genes were activated, including genes that regulate interferons and T cell differentiation. It is also interesting that there were no major changes in the composition of the gut microbiota in the subjects in the 10-d trial period. However, the consumption of yogurt that contained *Lactobacilli* ($6 \times 10^7 - 2.4 \times 10^8$/g yogurt) in a Japanese study showed a significant effect on alteration in the gut microbiota.

Yogurt and other fermented milk products carry a lot of LAB that enter the digestive system. By strengthening gut epithelium tight junctions and reducing potentially hazardous enzymes generated by the residing bacteria, LAB can alter the intestinal environment. Olivares et al. found that eliminating fermented foods from the diet lowered the level of the innate immune response, which included a decline in leukocyte phagocytosis and a reduction in the immune response to infections. However, the impaired immune response could be restored by LAB. Throughout at least the last 20 years, childhood obesity has been a serious public health concern. Based on the findings of several clinical trials, Marette and Picard-Deland recently examined yogurt consumption and its effects on childhood obesity and the risk of cardiometabolic diseases. The authors came to the conclusion that yogurt helped the children and adolescents in their study maintain healthy cardiovascular systems and regulate their body weight in addition to providing essential nutrients.

The alteration in the gut microbiota played a role in the positive outcome. According to studies, people who are obese and/or have type 2 diabetes (T2D) have higher levels of endotoxins, like LPS, in their blood, which stimulates immune cells and causes inflammation. The altered gut microbiota associated with obesity is probably the cause of the rise in LPS. Odamaki et al., yogurt containing *Bifidobacterium longum* BB536 dramatically reduces enterotoxigenic *Bacteroides fragilis* in the gut microbiota, according to research done on 420 healthy participants. The results support the idea that eating some types of yogurt suppresses gut bacteria that produce endotoxins, which may lessen inflammation and endotoxemia even though this study was not intended to look into obesity and inflammation. Consuming yogurt containing *Bifidobacterium animalis subspecies lactis* BB-12 also decreased the expression of Toll-like receptor 2 (TLR2) in immune cells and decreased the levels of inflammatory cytokines. However, not all yogurt consumption has beneficial effects. Some yogurts have no effect on the gut microbiota and/or weight. Moreover, long-term yogurt consumption in immune-deficient patients can cause fatal *Lactobacillus rhamnosus* septicemia. Although yogurt consumption, in most cases, has beneficial effects on the health of hosts, we need to be careful in applying this popular biotherapy in different clinical settings.

7.5. DYSBIOSIS OF THE GUT MICROBIOTA

Trillions of non-pathogenic bacteria and viruses make up the majority of the microorganisms that comprise the human intestinal microbiota. To prevent pathogen colonization and invasion, the microbiota works in concert with the immune system and the host's defenses. Additionally, it carries out a crucial metabolic role by serving as a source of vitamins and nutrients and helping to extract energy and nutrients from food, including amino acids and SCFA. In the end, the host's intestinal microbiota is necessary for many essential processes, and as such, it may improve health. However, it is challenging to pinpoint the exact role that the intestinal microbiota plays in human disease and how it affects human health.

Exposure to a variety of environmental factors, such as food, toxins, drugs, and pathogens, can change the microbiota. As evidenced by experimental animal models, foodborne viral pathogens can cause both local and systemic inflammation, changing the composition of the microbiota and barrier function. This can lead to the development of autoimmunity, as demonstrated by type

1 diabetes and T-cell-mediated destruction of insulin-producing pancreatic β-cells. Of these, enteric pathogens have the greatest potential to cause microbial dysbiosis. The ability to culture pure isolates for identification and classification—a capability that is inevitably restricted to "culturable" microorganisms—and classical microbiological techniques have historically been the foundation for documenting dysbiosis.

A quick and thorough way to profile complex communities of microorganisms is now possible thanks to the development of high-throughput DNA based pyrosequencing technology, which allows bacteria and archaea to be classified based on individual 16S rRNA sequences directly from human samples (mostly faucal in origin) without the need for culturing. It has been demonstrated since this technology was first used that there are significant individual differences in the intestinal microbiota's composition. This can be partially explained by genetic differences between hosts, with positive correlations found between the genetic relatedness of the host under observation and similarity in the dominant faucal microbial communities. *Proteobacteria*, *Actinobacteria*, *Fusobacteria*, *Spirochaetes*, *Verrucomicrobia*, and *Leptosphaeria* are also present, but *Bacteroidetes* and *Firmicutes* are the dominant phylum. Using metagenomic analysis to investigate the functional capability of the intestinal microbiota genome (microbiome), it has been shown that almost 40% of the microbial genes present in each individual are shared with at least half the general population providing evidence for the existence of a functional core, or core microbiome. The main approach to studying changes in composition of the intestinal microbiota in relation to disease has relied primarily on the phylogenetic characterization of the microbiota of diseased individuals in comparison with apparently healthy individuals. However, since there are substantial inter-individual and intra-individual variations in addition to age-related changes in the composition of the intestinal microbiota, it is difficult to establish precise relations between human health and the presence and relative abundance of specific microbial communities. It may be possible in the future to use specific changes in compositional diversity, or even functional diversity, as biomarkers for health or specific diseases. It is important to note, however, that it is questionable whether changes in phylogenetic composition are a cause or consequence of a given disease.

Studies using germ-free mouse models of human autoimmune disease provide perhaps the strongest evidence of the direct involvement in or requirement for the intestinal microbiota in disease pathogenesis. These models allow researchers to determine the impact of exposure to and colonization by environmental microorganisms on the onset and progression of the disease (Table 7.3). The severity and/or incidence of disease are lower in germ-free environments in the majority of disease models, though not all of them, which is consistent with the microbiota acting as a "trigger" for the progression of the disease. Nevertheless, up until now, efforts to pinpoint the "pathogenic" microbiota members (pathobionts) capable of replicating the microbiota's overall impact have been unsuccessful.

Table 7.3. Summary of Gut Microbiota–Immune Axis in Various Diseases

Diseases	Findings	Reference
Gastrointestinal Infections	Provide resistance against colonization and invasion by pathobiont.	Singer et al. (2019) [222]
	Human gut microbiome metabolites induce expression of flagellin (a bacterial protein) increases EHEC motility and epithelial injury.	Tovaglieri et al. (2019)
IBD	Gut microbiota dysbiosis of IBD patients is consistently marked by an overgrowth in Proteobacteria.	Lee and Chang (2021)
	SCFA confers protection against IBD by maintaining gut barrier integrity, promoting Treg cell differentiation, and inhibiting histone deacetylases.	Furusawa et al. (2013)
Colorectal carcinoma	Generation of genotoxin such as *Bacteroides* fragilis toxin (Bft), cytolethal distending toxin (CDT), and colibactin.	Sepich-Poore et al. (2021)
	Bacterial-derived secondary bile acids and hydrogen sulfide promote proinflammatory milieu that increases CRC risk.	Hale et al. (2017)
	Bacteria such as *F. nucleatum* can adhere to colon tumors and aggravate tumorigenesis.	Yeoh et al. (2020)
Hepatocellular carcinoma	Systemic translocation of LPS promotes chronic liver injury and predisposes to HCC.	Lin et al. (1995)
	Excess butyrate production promotes HCC progression.	Singh et al. (2018)
	Secondary bile acids promote carcinogenesis and impede anti-tumor immunosurveillance in the liver.	Yoshimoto et al. (2013)

Cardiometabolic disease	LPS and other microbial ligands drive low-grade chronic inflammation and predispose to CVD.	Cani et al. (2007); Guasch-Ferré et al. (2017); Millard et al. (2018)
	Bacterial trimethylamine and its conversion to trimethylamine-N-oxide in the liver increases the risk of coronary artery disease, metabolic syndrome, stroke, and vascular inflammation.	
Rheumatoid Arthritis	*Prevotella* spp. Abundance is positively associated with new-onset rheumatoid arthritis.	Scher et al. (2013)
Allergic Diseases	Gut microbiota dysbiosis increases risk for allergic disease, e.g., food allergy and asthma.	Fazlollahi et al. (2018); Bunyavanich et al. (2016)

Source: Campbell C, Kandalgaonkar MR, Golonka RM, Yeoh BS, Vijay-Kumar M, Saha P. (2023). Crosstalk between gut microbiota and host immunity: Impact on inflammation and immunotherapy. Biomedicines; 11(2):294. https://doi.org/10.3390/biomedicines11020294.

The fact that intestinal dysbiosis is most frequently linked to GI-related illnesses, where changes in the host's (immune system's) interaction with antigens and stimuli derived from the lumen cause and/or prolong uncontrollably high levels of inflammation in the intestinal mucosa and, in certain situations, beyond, should come as no surprise.

7.5.1. What Is Dysbiosis?

A reduction in microbial diversity and an imbalance of microbial species within specific body microbiomes (i.e., the group of microbes found in a particular area of the human body). As a result, beneficial bacteria (e.g., Firmicutes) are typically reduced, but other potentially dangerous bacteria (e.g., Proteobacteria) proliferate more frequently. Impaired gut barrier function and immune-mediated inflammation—a consequence of an overreaction by the immune system—are frequently associated with dysbiosis. Thus, there may be a connection between intestinal dysbiosis and a number of illnesses, such as CRC, diabetes mellitus, and IBD.

Trillions of microorganisms, including bacteria, fungi, and viruses, live in the GI tract. The gut flora is made up of the genetic material that all microorganisms in the community have combined to form. One could consider the intestinal microbiota to be a real bodily organ that supports human health. Both local and systemic processes, including immunity, vitamin supply, nutrient transformation, gut-to-brain communication, and even tumor progression, may be impacted by gut dysbiosis. However, dysbiosis can also exist in other microbiomes, including those found in the oral cavity, vagina, and skin.

7.5.2. Is Dysbiosis Contagious?

Dysbiosis cannot be spread. On the other hand, under specific conditions, some infections may eventually lead to dysbiosis due to an imbalance in their native microbiome.

7.5.3. What Causes Dysbiosis?

The primary determinants of the microbiota's composition that can lead to dysbiosis are dietary modifications, medications (particularly antibiotics), and physical and mental stress. The most frequent and important factor leading to significant changes in the body's native microbiota is the use of antibiotics. The range of activity, dosage, and duration of an antimicrobial agent's administration all affect its capacity to affect the microbiome. Antimicrobial agents, such as ceftriaxone and meropenem, have the ability to eradicate nearly all beneficial bacteria in the body. This results in a reduction of microbial diversity and immune system suppression. Although they may also affect the skin and vaginal flora, they usually have a greater effect on the intestinal flora. Additionally, the impact on the microbiota increases with dosage and duration of administration.

Eating foods high in sugar, protein, and sulfur compounds (beef, spinach, broccoli, nuts, and chickpeas), for example, can upset the microbial balance in the gut microbiota and cause inflammation and disruption of mucus. Similar to this, stress can cause major changes in the microflora of the GI tract and skin, which can lead to an increase in potentially pathogenic microorganisms like *E. coli* and a significant decrease in beneficial bacteria like Bifidobacteria and *Lactobacilli*, respectively, and atopic dermatitis. The growth-promoting properties of norepinephrine on gram-negative bacteria or changes in GI tract motility and secretions brought on by stress could be the source of these alterations.

Modest changes in the makeup of microorganisms can create a window of opportunity for other exacerbating factors to intensify alterations and result in an imbalance. Exacerbating factors that are known to cause shifts in the microbiota and ultimately lead to dysbiosis include increased oxidative stress, as observed in immune-mediated inflammation, bacteriophages (viruses that kill the bacteria they infect), and bacteriocins (toxins produced by bacteria that inhibit the growth of other bacteria). Often, dysbiosis cannot be brought on by a single cause. The microbiota is naturally resilient and possesses a special ability to adjust to changes in environmental conditions and nutrient availability.

On the other hand, a number of interrelated events may eventually result in pathologically significant changes in the microbiome. The bacterial groups that are impacted determine the threshold that must be reached in order to cause dysbiosis. Wide changes within the *Firmicutes*, *Bacteroidetes*, and main phyla may not have any negative pathological effects. Conversely, however, a growing number of marginal groups i.e., bacteria like *Clostridium perfringens*, which are already present in the GI system in smaller amounts, have the potential to disrupt normal intestinal function (Falony et al., 2016; Mohajeri et al., 2018).

7.5.4. Signs and Symptoms of Dysbiosis

The majority of dysbiotic patients exhibit localized symptoms, which vary depending on the microbiome that is impacted. The symptoms of oral dysbiosis, which include tooth decay, cavities, and sore and bleeding gums, are typically mild when the GI tract's microbiome is disturbed. Gut dysbiosis can be accompanied by bloating, discomfort in the abdomen, belching, nausea, and changes in bowel movements (such as diarrhea or constipation). Consequently, painful urination, strange vaginal or penile discharge, rashes, or itching of the genitalia and rectum may result from modifications in the microbiota of the urogenital region. Atopic dermatitis (eczema), psoriasis, or acne are examples of skin rashes that can be a symptom of skin dysbiosis. In more extreme situations, when the immune system is compromised, weariness, trouble focusing, mood swings, irritability, anxiety, and depression are examples of systemic symptoms that may arise (Falony et al., 2016; Mohajeri et al., 2018).

7.5.5. Metabolomic Impact of the Interaction Between Diet and the Microbiome on Human Health

Food particles that are not fully broken down in the small intestine, along with endogenous substances like digestive enzymes and shed epithelial cells and mucus, find their way into the colon where they are fermented by the colonic microbiota. These substances are converted by bacteria into a wide range of metabolites that are in close proximity to the host's cells. These metabolites have the ability to impact the host's metabolic phenotype and thus the risk of disease.

Proteins and undigested carbohydrates make up the majority of the substrates available to the microbiota. A variety of metabolites, such as branched chain fatty acids (BCFA), ammonia, amines, phenolic compounds, and gases like hydrogen, methane, and hydrogen sulfide, are produced during the fermentation of these substrates. Furthermore, the intestinal microbiota plays a role in vitamin synthesis, the conversion of prodrugs to their bioactive forms, the alteration of bile acids and xenobiotics, and the activation or inactivation of bioactive food components like isoflavanoids and plant lignans.

7.5.5.1. Mechanistic Effect of Metabolites on Host Health

The primary anions in the colon are the SCFAs acetate, propionate, and butyrate, which are primarily created by bacteria fermenting undigested carbohydrates. Colonocytes can easily absorb up to 95% of the generated SCFA and use it as an energy substrate. About 10% of the daily caloric requirements in humans are met by SCFAs, as colonocytes can oxidize up to 60–70% of their energy needs. The portion that the colonocytes do not eat is carried to the liver by the portal blood stream, which passes through the basolateral membrane. In addition to serving as local energy substrates in the colon, SCFA are signaling molecules that influence insulin and glucose regulation as well as systemic lipid metabolism. Two distinct G-protein-coupled receptors, GPR41 and GPR43 (later renamed to FFAR3 and FFAR2, respectively), which are widely distributed throughout the human body, including the small intestine and colon, are at least partially responsible for mediating these effects. In the cell, SCFA can function as histone deacetylase (HDAC) inhibitors to cause hyperacetylation of histones, which modifies gene expression and induces growth arrest, apoptosis, and anti-inflammatory effects.

Plant polyphenols have been linked to a number of health advantages, such as neuroprotective, cardioprotective, chemoprotective, anti-inflammatory, and antiestrogenic properties. The *in vivo* mechanistic evidence is still unclear, though. To become biologically active, most plant polyphenols must undergo metabolic transformation, such as hydrolysis and deglycation. The microbiota in the colon breaks them down into a variety of small phenolic compounds, the physiological significance of which is unknown. Furthermore, consuming polyphenols appears to selectively alter the composition of the microbiota, according to recent research. For example, in healthy humans, the numbers of *Enterococcus*, *Prevotella*, *Bacteroides*, *Bifidobacterium*, *Bacteroides uniformis*, *Eggerthella lenta*, and *Blautia coccoides-Eubacterium rectale* are significantly increased when red wine polyphenols are consumed. Therefore, the modulation of the intestinal microbiota as well as the bioactive metabolites of polyphenols should be credited for the health benefits associated with them.

Other products of bacterial metabolism have been associated with diseases affecting the liver, cardiovascular system and the kidneys.

The gut-liver axis and the effect of intestinal microbiota on liver function have drawn more attention in recent years. The liver receives 70% of its blood supply from the intestine through the portal vein, which exposes it to metabolites produced during intracolonic fermentation. Early in the 1980s, it was proposed that the microbiota may play a causal role in the onset of non-alcoholic fatty liver disease (NAFLD). Hepatic steatosis coexisted with bacterial overgrowth in patients who had intestinal bypass surgery. It's interesting to note that the steatosis improved after receiving metronidazole treatment. A mechanism linking the microbiota to NAFLD is the bacterial metabolism of choline. When mice with NAFLD were given a high-fat diet, choline was gradually converted to methylamines, which led to a high excretion of dimethylamine (DMA) and trimethylamine (TMA) in the urine and a corresponding decrease in serum phosphatidylcholine levels.

The microbiota converts choline into methylamines, which lowers choline's bioavailability and prevents the synthesis of phosphatidylcholine, which leads to the buildup of triglycerides in the liver. This is similar to diets low in choline, which have been repeatedly linked to hepatic steatosis. As a result, the intestinal mucosa absorbs the bacterial metabolite TMA, which is then carried to the liver through the portal vein by the flavin mono-oxygenase (FMO) enzyme complex, where it is converted to trimethylamine N-oxide (TMAO). TMAO was found and validated as a cardiovascular disease (CVD) predictor in a metabolomics study profiling the plasma of patients undergoing elective cardiac evaluation. Further studies on mice demonstrated the pro-atherogenic nature of TMAO by increasing the formation of foam cells and cholesterol-loaded macrophages, thereby confirming the essential role of the intestinal microbiota in the formation of TMAO. Similarly, mice's atherosclerosis was accelerated and TMAO was produced when dietary L-carnitine, a TMA found in large amounts in red meat, was metabolized by the intestinal microbiota (Falony et al., 2016; Mohajeri et al., 2018).

7.5.6. Dysbiosis in Disease

7.5.6.1. Dysbiosis and GI-Tract-Related Disorders

7.5.6.1.1. Inflammatory Bowel Disease

Crohn's disease (CD) and UC, being the most common types, make up the majority of IBD. These conditions are marked by persistent, recurring inflammation that impacts the lining of the intestines. Despite the fact that the causes of both diseases remain unexplained, research is accumulating to suggest that imbalances in intestinal microbial populations, or dysbiosis, contribute to the development of IBD. Overall, patients exhibit a decrease in microbial population and functional diversity and stability of their intestinal microbiota with decreases in specific *Firmicutes* and a concomitant increase in *Bacteroidetes* and facultative anaerobes such as *Enterobacteriaceae*. Significant differences in the microbiota of CD versus UC patients have also been noted.

The five bacterial species that have been linked to the predominant dysbiosis in CD have been identified. *Faecalibacterium prausnitzii* is one of these species, and it has been shown to have therapeutic effects in experimental models of colitis. Changes in its abundance are associated with the prolongation of disease remission. In contrast, although a causal relationship between adherent-invasive E coli and Mycobacterium paratuberculosis and CD pathogenesis has not yet been established, they have been linked. Currently, it remains uncertain whether intestinal microbial dysbiosis is the primary cause of the inflammation in IBD or if it is only the outcome of an unbalanced GI-tract environment. The study by Gevers et al. aimed to ascertain the microbiota's status in CD cases with an early diagnosis. This study distinguished between the bacterial populations of patients with CD and healthy control patients by analyzing the microbiota of a sizable cohort of pediatric CD patients who had just received a diagnosis. Compared to healthy individuals, CD patients had lower abundances of *Erysipelotrichales*, *Bacteroidales*, and *Clostridiales* and higher abundances of *Enterobacteriaceae*, *Pasteurellaceae*, *Veillonellaceae*, and *Fusobacteriaceae*.

7.5.6.1.2. Dysbiosis and Other GI-Tract Disorders

The intestinal microbiota has also been linked to a number of other (chronic) GI-related illnesses and disorders, including IBS, celiac disease, and CRC, in addition to IBD, metabolic disorders, obesity, and type 2 diabetes (T2D). There have been documented variations in the microbiota composition between healthy individuals and various IBS subtypes, albeit not all subtypes exhibit the same changes. Changes in the microbiota composition have also been linked to celiac disease and CRC, with greater diversity and richness seen in comparison to control subjects. However, no consistent pattern of changes in the microbiota has been found in any of these diseases thus far. However, a recent study on celiac disease has illuminated the relationship between host genetics and microbiota composition and the development of the disease. The development of celiac disease is strongly associated with the expression of the leukocyte antigen DQ2. Before a disease manifests itself clinically, children with this haplotype have a different microbiota composition than non-HLA DQ2 individuals. Dietar-induced CD4 T-cell reactivity causes celiac disease.

7.5.7. Dysbiosis in Systemic Disease

7.5.7.1. Metabolic Disorders

Obese mice and humans have been shown to have higher relative abundances of Firmicutes and lower levels of *Bacteroidetes*, though not all studies have confirmed these observations. Remarkably, intestinal dysbiosis is not presently regarded as a risk factor for the diagnosis of metabolic diseases like obesity or type 2 diabetes. When compared to lean individuals, obese people have been shown to have more subtle changes in the composition of their intestinal microbiota, including a lower diversity of microbes in their composition. The intestinal microbiota's role in obesity is further supported by data from obese (ob/ob) mice devoid of leptin expression, which increases feelings of fullness. Antibiotic treatment caused alterations in the gut microbiota, which in turn decreased the incidence of metabolic endotoxemia, inflammation, and a number of obesity-related parameters, providing evidence that the microbiota played a role in the development of obesity in these mice. It is clear from studying human populations that the higher incidence of obesity and type 2 diabetes in the West can be attributed to high-fat diets and overindulgence in food, which together modify host metabolism.

The ability of the microbiota to obtain energy from food and its role in metabolism do, in fact, point to a major environmental factor influencing the risk of metabolic disease. Studies on germ-free mice have demonstrated a direct correlation between the composition of the intestinal microbiota and body weight, indicating that the microbiota plays a role in regulating fat storage and that the absence of intestinal microbes protects against diet-induced obesity. It has been suggested by these and related studies that obese people are more adept than lean people at utilizing food as fuel and storing it as fat. This may be related to or a result of the intestinal microbiota's functionality. Therefore, studying functional microbial diversity in addition to phylogenetic diversity should yield important insights into differences between various physiological states of the host, such as in obese versus lean individuals. It has been discovered that obesity is associated with a changed representation of bacterial genes and metabolic pathways, including those involved in nutrient harvesting. Also, the amount of SCFA produced by the intestinal microbiota, rather than the changes in the composition of the microbiota, is important in the development of obesity. Perhaps unsurprisingly, shifts in microbiota phyla have also been described in T2D, with metagenomics-based studies identifying discriminant metagenomic markers that may differ between different ethnicities of patients. The question remains whether dysbiosis of the intestinal microbiota is a direct cause for any metabolism-related disorder, or whether changes in the intestinal microbial communities in affected and obese individuals are an adaptation to a change in the host's diet. Two observations relevant to answering this question are one, that the transfer of microbiota from lean donors into individuals with metabolic syndrome can increase insulin sensitivity and overall amelioration of symptoms of metabolic disease and two, dietary changes in humans leads to rapid and reversible changes in the relative abundance of dominant members of the intestinal microbiota.

Studies on both humans and animals that show quick changes in the microbiota composition following Roux-en-Y gastric bypass surgery (RYGB) suggest that host physiology, behavior, the microbiome, and diet may interact, though the effect on metabolite levels has not been as thoroughly investigated. Nevertheless, RYGB surgery caused significant metabolic disruptions in a model of non-obese rats. Fecal samples taken after surgery had higher levels of colonic protein fermentation metabolites in addition to lower concentrations of oligosaccharides and higher concentrations of SCFA. These findings may indicate that there was insufficient protein digestion in the small intestine due to the bypass, which resulted in an increased protein supply to the colon and increased protein fermentation. Fecal water samples taken two and eight weeks following the procedure, interestingly, showed noticeably higher levels of cytotoxicity than samples taken from animals that were not actually operated on. It is necessary to look into whether there is a causal relationship involved in the correlation between elevated levels of fermentation metabolites of amino acids and elevated cytotoxicity in a subject with a normal weight and health.

Additionally, there is a bidirectional functional relationship between the kidney and the large intestine. While metabolites related to microbes play a role in the development of kidney disease, uremia affects the metabolism of colonic microbes. The two compounds that have been studied the most, p-cresyl sulfate and indoxyl sulfate, are thought to be the ancestors of the so-called uremic toxins. They originate from the bacterial fermentation of tryptophan and tyrosine, respectively, which is followed by sulphation in the liver or colonic mucosa. They are strongly protein-bound within the plasma and build up when renal function is compromised. Due to competition for binding sites on plasma proteins, the free, unbound levels of these solutes rise more than their total plasma levels. Both p-cresyl sulphate and indoxyl sulfate levels have been associated with overall mortality, CVD, and kidney disease progression in patients with chronic kidney disease (Falony et al., 2016; Mohajeri et al., 2018).

7.5.8. Dysbiosis and CNS-Related Disorders

Additionally, intestinal microbial dysbiosis has been linked to extraintestinal diseases, particularly those that may have an effect on the "gut–brain–axis," which influences behavior, cognition, and the CNS.

Numerous investigations have concentrated on the idea that the gut microbiota could directly rewire the hypothalamus-pituitary-adrenal (HPA) axis, a common pathway that is triggered by infections and disrupted by psychological stressors, to affect behavior and cognitive function. Enteric infections have been linked to anxiety, depression, and cognitive impairment; in contrast to conventionalized mice, germ-free mice, lacking an intestinal microbiota, exhibit changes in stress-responsivity, central neurochemistry, and behavior that suggest a decrease in anxiety. For instance, alterations in the synthesis of neurotrophic factors and hormones as well as the expression of their receptors have been linked to elevated anxiety-like behavior in germ-free mice. *Campylobacter jejuni*, a common cause of gastroenteritis, can cause brainstem activation (the nucleus tractus solitarius and lateral parabrachial nucleus) and anxiety-like behavior in pathogen-infected mice. Commensal bacteria may alter the structure of the brain by influencing GABA, which directly affects immune and neural receptors in the central nervous system (ENS and CNS). Main inhibitory neurotransmitter in the CNS is GABA. It is well recognized that the postnatal development of the ENS depends on the early microbial colonization of the intestinal tract. Thus, intestinal microbiota may affect how the CNS develops and functions.

In a maternal immune activation (MIA) mouse model, where pregnant animals are given the viral mimetic ploy (I:C), they exhibit increased intestinal permeability and develop stereotypical abnormalities in behavior, social ability, and communication that resemble ASD. This model provides evidence of a potential causal role of the intestinal microbiota in the development of autism spectrum disorder (ASD). Compared to control offspring, MIA offspring exhibit intestinal dysbiosis and a changed serum metabolomic profile marked by elevated levels of 4-ethylphenylsulphate (4EPS) derived from the microbiota. Probiotic bacteria administration restored intestinal barrier function and reduced symptoms resembling ASD.

Perhaps it is no coincidence that GI tract disorders, such as IBD and IBS, are common co-morbidities in crippling stress-related disorders, such as depression and anxiety, given the emerging preclinical data and indications in developmental disorders. According to recent research, intestinal permeability and bacterial translocation may be involved in the pathophysiology of depression by driving the immuno-inflammatory and oxidative and nitrosative stress (IOandNS) pathways. Increased immune responses (serum IgM and IgA responses) against LPS products of gram-negative gut enterobacteria—*Hafnia alvei*, *Pseudomonas aeruginosa*, *Morganella morganii*, *Pseudomonas putida*, *Citrobacter koseri*, and *Klebsiella pneumonia*—have been linked to chronic depression in humans. In both pathogen-infected and germ-free rodents, attempts have been made to investigate the possible effects of bacteriotherapy on the CNS and behavior.

Compared to mice free of specific pathogens, germ-free mice show hyperresponsive HPA axis activity in response to stress, and *Bifidobacterium infantis* reversed this hyperreaction. *B. infantis* lowered dopamine metabolite concentrations in the amygdaloid cortex and raised plasma tryptophan levels. Depression is linked to both of these metabolite concentrations. Probiotics have been shown to be effective in regulating mood in humans. A study using *Lactobacillus casei* revealed that individuals who had the lowest baseline scores in the depressed/elated dimension improved significantly in their mood scores when taking the probiotic in comparison to the placebo group. In healthy human volunteers, the combination of *L. helveticus* and *B. longum* lowered anxiety and had positive psychological effects along with a drop in serum cortisol.

A probiotic formulation decreased brain intrinsic connectivity, response to emotive stimuli, and changes in midbrain connectivity, according to functional magnetic resonance imaging (fMRI) measurements of functional brain activity (Falony et al., 2016; Mohajeri et al., 2018).

REFERENCES

1. Arboleya, S., Binetti, A., Salazar, N., Fernández, N., Solís, G., Hernández-Barranco, A., Margolles, A., de Los Reyes-Gavilán, C. G., & Gueimonde, M. (2012). Establishment and development of intestinal microbiota in preterm neonates. *FEMS Microbiology Ecology, 79*, 763–772.
2. Arumugam, M., Raes, J., Pelletier, E., Le Paslier, D., Yamada, T., Mende, D. R., Fernandes, G. R., Tap, J., Bruls, T., Batto, J. M., et al. (2011). Enterotypes of the human gut microbiome. *Nature, 473*, 174–180.
3. Azad, M. B., Konya, T., Maughan, H., Guttman, D. S., Field, C. J., Chari, R. S., Sears, M. R., Becker, A. B., Scott, J. A., & Kozyrskyj, A. L. (2013). Gut microbiota of healthy Canadian infants: Profiles by mode of delivery and infant diet at 4 months. *CMAJ, 185*, 385–394.
4. Bering, S. B. (2018). Human milk oligosaccharides to prevent gut dysfunction and necrotizing enterocolitis in preterm neonates. *Nutrients, 10*, 1461.
5. Burcelin, R. (2016). Gut microbiota and immune crosstalk in metabolic disease. *Molecular Metabolism, 5*, 771–781.
6. Compare, D., Rocco, A., Sanduzzi Zamparelli, M., & Nardone, G. (2016). The gut bacteria-driven obesity development. *Digestive Diseases, 34*, 221–229.
7. Falony, G., et al. (2016). Population-level analysis of gut microbiome variation. *Science*, https://doi.org/10.1126/science.aad3503.
8. Goodrich, J. K., Waters, J. L., Poole, A. C., Sutter, J. L., Koren, O., Blekhman, R., Beaumont, M., Van Treuren, W., Knight, R., Bell, J. T., et al. (2014). Human genetics shape the gut microbiome. *Cell, 159*, 789–799.
9. Hughes, R. L. (2020). A review of the role of the gut microbiome in personalized sports nutrition. *Frontiers in Nutrition, 6*, 191. https://doi.org/10.3389/fnut.2019.00191.
10. Human Microbiome Jumpstart Reference Strains Consortium, Nelson, K. E., Weinstock, G. M., Highlander, S. K., Worley, K. C., Creasy, H. H., Wortman, J. R., Rusch, D. B., Mitreva, M., Sodergren, E., et al. (2010). A catalog of reference genomes from the human microbiome. *Science, 328*, 994–999.
11. Kamada, N., Seo, S. U., Chen, G. Y., & Nunez, G. (2013). Role of the gut microbiota in immunity and inflammatory disease. *Nature Reviews Immunology, 13*, 321–335.
12. Khosravi, A., & Mazmanian, S. K. (2013). Disruption of the gut microbiome as a risk factor for microbial infections. *Current Opinion in Microbiology, 16*, 221–227.

13. Laterza, L., Rizzatti, G., Gaetani, E., Chiusolo, P., & Gasbarrini, A. (2016). The gut microbiota and immune system relationship in human graft-versus-host disease. *Mediterranean Journal of Hematology and Infectious Diseases, 8*, e2016025.
14. Ley, R. E., Turnbaugh, P. J., Klein, S., & Gordon, J. I. (2006). Microbial ecology: Human gut microbes associated with obesity. *Nature, 444*, 1022–1023.
15. Lloyd-Price, J., et al. (2016). The healthy human microbiome. *Nature Genetics*, https://doi.org/10.1186/s13073-016-0307-y.
16. Luckey, T. D. (1972). Introduction to intestinal microecology. *American Journal of Clinical Nutrition, 25*, 1292–1294.
17. Mohajeri, M. H., et al. (2018). The role of the microbiome for human health: From basic science to clinical applications. *European Journal of Nutrition, 57*, 1–10. https://doi.org/1007/s00394-018-1703-4.
18. Moore, W. E., & Holdeman, L. V. (1974). Human fecal flora: The normal flora of 20 Japanese-Hawaiians. *Applied Microbiology, 27*, 961–979.
19. Pantoja-Feliciano, I. G., Clemente, J. C., Costello, E. K., Perez, M. E., Blaser, M. J., Knight, R., & Dominguez-Bello, M. G. (2013). Biphasic assembly of the murine intestinal microbiota during early development. *ISME Journal, 7*, 1112–1115.
20. Qin, J., Li, Y., Cai, Z., Li, S., Zhu, J., Zhang, F., Liang, S., Zhang, W., Guan, Y., Shen, D., et al. (2012). A metagenome-wide association study of gut microbiota in type 2 diabetes. *Nature, 490*, 55–60.
21. Rajilic-Stojanovic, M., Smidt, H., & de Vos, W. M. (2007). Diversity of the human gastrointestinal tract microbiota revisited. *Environmental Microbiology, 9*, 2125–2136.
22. Rinninella, E., Raoul, P., Cintoni, M., Franceschi, F., Miggiano, G. A. D., Gasbarrini, A., & Mele, M. C. (2019). What is the healthy gut microbiota composition? A changing ecosystem across age, environment, diet, and diseases. *Microorganisms, 7*(1), 14. https://doi.org/10.3390/microorganisms7010014.
23. Roger, L. C., Costabile, A., Holland, D. T., Hoyles, L., & McCartney, A. L. (2010). Examination of faecal Bifidobacterium populations in breast- and formula-fed infants during the first 18 months of life. *Microbiology, 156*, 3329–3341.
24. Rothe, M., & Blaut, M. (2013). Evolution of the gut microbiota and the influence of diet. *Beneficial Microbes, 4*, 31–37.
25. Salminen, S., Gibson, G. R., McCartney, A. L., & Isolauri, E. (2004). Influence of mode of delivery on gut microbiota composition in seven-year-old children. *Gut, 53*, 1388–1389.

26. Tanoue, T., Umesaki, Y., & Honda, K. (2010). Immune responses to gut microbiota-commensals and pathogens. *Gut Microbes, 1*, 224–233.
27. Thursby, E., & Juge, N. (2017). Introduction to the human gut microbiota. *Biochemical Journal, 474*, 1823–1836.
28. Tidjani Alou, M., Lagier, J. C., & Raoult, D. (2016). Diet influence on the gut microbiota and dysbiosis related to nutritional disorders. *Human Microbiome Journal, 1*, 3–11.
29. Valdes, A. M., Walter, J., Segal, E., & Spector, T. D. (2018). Role of the gut microbiota in nutrition and health. *BMJ, 361*, 36–44, k2179. https://doi.org/10.1136/bmj.k2179.

CHAPTER 8

PROBIOTICS AND THEIR IMPACT ON NUTRIENT ABSORPTION

Contents

INTRODUCTION

When taken in sufficient quantities, probiotics—live microorganisms—offer the host a number of health benefits. Probiotics are frequently taken as dietary supplements or included in foods like yogurt. Although probiotics are primarily linked to bettering gut health, they also affect a person's ability to absorb nutrients. The kind and strain of probiotics that a person takes will determine how well they absorb nutrients (Markowiak et al., 2017). Probiotics can increase the synthesis of certain digestive enzymes, strengthen the lining of the gut, reduce inflammation, and boost the production of SCFAs, all of which can improve the bio accessibility of specific nutrients and phytonutrients.

Probiotics can also aid in preserving the intestinal barrier's integrity, which is necessary for adequate nutrient absorption. The proteolytic potential of specific *Lactobacillus* species. helps to produce bioactive soluble peptides that are smaller and more easily absorbed than the original protein. This can enhance the nutritional value of some proteins (Manus et al., 2021; Wang et al., 2019). It has been demonstrated that the fermentation of plant proteins with specific *Lactobacillus* species improves protein absorption by reducing antinutritional substances such as tannins, phytate, and trypsin inhibitor. According to Jager et al. (2020), fermentation can assist in lowering the pH, trypsin inhibitor activity, and phytate and tannin content of pulses. Apart from their absorption, specific probiotics are crucial for the gut's synthesis of B-complex vitamins like thiamine, B12, and folate (Hill, 1997). Probiotics can be very helpful in increasing the bio accessibility of nutrients and phytonutrients in humans, but it is important to recognize that their effectiveness varies.

8.1. PROBIOTICS AND NUTRIENT ABSORPTION

Live bacteria and yeasts are known as probiotics, and they are good for your health in general and your digestive system in particular. They are frequently taken as dietary supplements or included in some fermented foods such as kimchi, kefir, yogurt, and sauerkraut. Probiotics may affect how well nutrients are absorbed, even though their benefits for the immune system and gut health are their main areas of recognition. Maintaining a healthy balance of gut bacteria is essential for optimal digestion and absorption, as the digestive system plays a critical role in food breakdown and nutrient absorption.

Probiotics can help in several ways:

- **Enhancing Gut Health:** A healthy gut microbiota, or the community of microorganisms residing in your digestive tract, can be maintained with the aid of probiotics. Proper digestion and nutrient absorption depend on a balanced gut microbiota. An unbalanced gut microbiota can cause digestive problems and impair the absorption of nutrients. Probiotics aid in the development of good bacteria, which can enhance nutrient absorption and gut health in general (Roberfroid et al., 2010; Lynch & Pedersen, 2016; Barone et al., 2022).

- **Producing Enzymes:** Enzymes that facilitate the breakdown and digestion of specific nutrients can be produced by certain probiotic strains. Lactase, an enzyme required for the digestion of lactose (the sugar present in milk and dairy products), is produced by specific strains of *Lactobacillus* and *Bifidobacterium*. Because it helps them better digest lactose and absorb the nutrients from dairy products, this may be advantageous for people who are lactose intolerant.
- **Improving Gut Barrier Function:** The lining of your stomach serves as a barrier to regulate the flow of chemicals from the digestive system into the blood. Leaky gut, also referred to as increased intestinal permeability, can result from a compromised gut barrier. This may prevent the body from absorbing nutrients and permit toxic substances to enter the bloodstream. Probiotics can lessen intestinal permeability and increase nutrient absorption by strengthening the gut barrier.
- **Enhancing Nutrient Bioavailability:** Probiotics can also increase some nutrients' bioavailability. The term "bioavailability" describes how much a nutrient can be absorbed and utilized by your body. Probiotics can help your body digest substances that would otherwise be difficult, such as fiber, complex carbohydrates, and other substances. They do this in order to release more nutrients that your body can absorb and use.

It's important to remember that different probiotic strains and formulations have different effects on how well nutrients are absorbed, and that individual factors like diet and gut health can also affect these effects. It is advisable to speak with a healthcare provider if you're thinking about taking probiotic supplements to enhance nutrient absorption. They can provide you with guidance specific to your needs and health conditions (Barone et al., 2022; ; Lynch & Pedersen, 2016; Roberfroid et al., 2010).

8.1.1. Probiotics Improve Nutrient Absorption

While some probiotic bacterial species break down antinutrients like oxalates, others directly produce nutrients like folate.

8.1.1.1. Probiotics and Nutrient Status

Even though some foods are incredibly nutrient-dense, our bodies won't be able to use them if we can't absorb them from the stomach. In this post, we'll examine the impact of probiotics on nutrient absorption and speculate as to what role beneficial gut bacteria might play in this process as well as which essential nutrients are directly produced by said gut bacteria. Keep in mind that not everyone is a good fit for probiotic supplements. Consult your physician prior to beginning any new probiotic regimen.

8.1.1.2. Insufficient Evidence for

The following claimed benefits are only corroborated by a small number of subpar clinical studies. The use of probiotics for any of the purposes listed below is not supported by enough evidence. Never take probiotic supplements in place of anything your doctor recommends or prescribes, and always consult a physician before beginning any new regimen.

- 1. **Vitamin Levels:** Children's levels of folate and vitamin B12 were considerably raised by *L. acidophilus* consumption daily. It has been demonstrated that *L. reuteri* isolated from sourdough can produce cobalamin (vitamin B12) and that *L. plantarum* isolated from raw cow milk can produce folate (B9) and riboflavin (B2). In human studies, the vitamin status (B12, folate, and D3) was improved by certain *Lactobacillus* probiotics. Certain species also generate nutrients such as folate and B vitamins.
- 2. **Iron Levels:** In young women in south India, low levels of *Lactobacilli* were linked to iron deficiency. *L. plantarum* was also linked to a 50% increase in women's absorption of iron from fruit drinks and a 100% increase from oat bases.
- When given *L. acidophilus* supplements, preschoolers showed improved red blood cell counts and a marked decrease in the incidence of anemia. Using *B. animalis* ssp. in milk. In children aged 1–4, lactis and prebiotic oligosaccharides increased weight gain by 0.13 kg/year and decreased the risk of anemia and iron deficiency by 45%.
- 3. **Other Minerals:** *L. helveticus* raised the levels of calcium in the blood in postmenopausal women and older volunteers. Fermented goat milk containing *L. plantarum* demonstrated higher calcium retention, whereas fermented milk containing *L. fermentum* may increase the bioavailability of calcium, phosphorus, and zinc. *S. cerevisiae* may enhance the absorption of iron, zinc, magnesium, and phosphorus by breaking down phytate.

8.1.2. Probiotics Help in Improving Nutrient Absorption for Malnourished People

We can't help but be concerned about the health and nutritional conditions in India, a nation that is home to one-fifth of the world's population. More than 50% of children, young girls, and women are anemic (the body lacks iron and other vital minerals and vitamins), and more than 30% of children under five are stunted. As a risk factor for non-communicable diseases like cancer, heart disease, and type 2 diabetes, changes in diet and lifestyle have also contributed to obesity in addition to undernutrition. It is estimated that 73 million adults suffer from diabetes, and 19% of men and 21% of women are obese.

This is a serious issue that cannot be solved quickly. To guarantee better food quality and quantity as well as a healthy lifestyle, it would simply entail taking a more comprehensive approach to health and nutrition. But if our food is not being absorbed by our bodies, nutrition becomes meaningless. Thus, optimal nutrient absorption from our diets is contingent upon intestinal health. It is not surprising that over half of Indians suffer from poor intestinal health because this vital organ, which is the site of food digestion and nutrient absorption, is frequently the most neglected. This raises the question of whether diet and intestinal health are related.

As the largest immune organ in the body, the intestine depends on these bacteria to remain healthy, which is essential to the intestine's function, the significance of intestinal health has also entered the scientific discourse. Because of their extraordinary ability to gauge intestinal health, having a higher concentration of beneficial bacteria is essential for a healthy gut.

Foods containing probiotics are specially made to boost and maintain the good bacteria in the intestines. Iron and calcium absorption from ingested foods has been linked to the consumption of certain probiotic foods. It's interesting to note that research on young Indian women and children revealed that consuming probiotic- and prebiotic-enriched milk considerably improved the subjects' iron status. Certain probiotic bacteria have the ability to produce B group vitamins, which are easily destroyed by heat and for which a significant portion of the population is vitamin B deficient. Vitamin B producers that are probiotics may aid in the fight against the deficiency. About 13 million children under the age of five suffer from severe acute malnutrition, and 1–2 million of them pass away every year. This is an unfortunate fact. These kids have weak immune systems, unhealthy intestines, and long-term health consequences. Certain probiotics can lower the incidence of diarrhea and thereby prevent vitamin and mineral loss, according to studies conducted on Indian children. Despite being one of the most ignored groups, pregnant women have the highest nutrient needs. They frequently have digestive issues that prevent them from getting the most out of food, leaving them iron, zinc, folic acid, and iodine deficient. These women frequently give birth to premature and underweight children. It is commonly known that certain probiotics that have undergone scientific testing can enhance general health during pregnancy and may be a helpful supplement to a diet plan. It is predicted that the population of many countries will be higher in adults over 65 than in children under five. Immunity, digestive health, and gastrointestinal problems are all brought on by aging. According to studies, taking certain probiotics can help the elderly population's health by lowering inflammation, boosting good bacteria, and enhancing immunity and digestion (Roberfroid et al., 2010; Lynch & Pedersen, 2016; Barone et al., 2022).

8.2. INFLUENCE OF PROBIOTICS ON NUTRIENT DIGESTION AND ABSORPTION

Probiotics are live microorganisms that give the host health benefits when taken in sufficient amounts. Probiotics can affect the digestion and absorption of nutrients, even though they are primarily linked to gut health and the regulation of the gut microbiota. Here's how probiotics can influence nutrient digestion and absorption (Figure 8.1):

- **Enhanced Breakdown of Complex Carbohydrates:** Certain strains of *Lactobacillus* and *Bifidobacterium*, which are probiotics, generate amylases, which are enzymes that aid in the breakdown of complex carbohydrates into simpler sugars. This may make it easier for carbohydrates to be absorbed and digested.
- **Improved Protein Digestion:** Probiotics have the ability to produce proteases and other proteolytic enzymes that help break down proteins into smaller peptides and amino acids. The small intestine's ability to absorb amino acids may be improved by this improved protein digestion.
- **Synthesis of Vitamins:** Probiotic strains that possess this ability can synthesize B vitamins (including folate, biotin, and vitamin B12), vitamin K, and precursors of vitamin A. The total amount of these vitamins that are available for the host to absorb can be influenced by this microbial synthesis.
- **Enhanced Mineral Absorption:** Probiotics have the ability to affect how well minerals like calcium, iron, and magnesium are absorbed. They can generate organic acids, like lactic acid, which can improve the absorption of minerals in the digestive system and make them more soluble.

- **Gut Barrier Function:** Probiotics can contribute to the integrity of the gut barrier by increasing the production of mucins and tight junction proteins. The optimal absorption of nutrients is ensured while the absorption of harmful substances is minimized by a healthy gut barrier, which stops the leakage of undigested nutrients and toxins.
- **Interaction with the Gut-Microbiota:** The composition and activity of the gut microbiota can be altered by probiotics, which may have an indirect impact on the digestion and absorption of nutrients. To process and absorb nutrients efficiently, one needs a balanced gut microbiota.

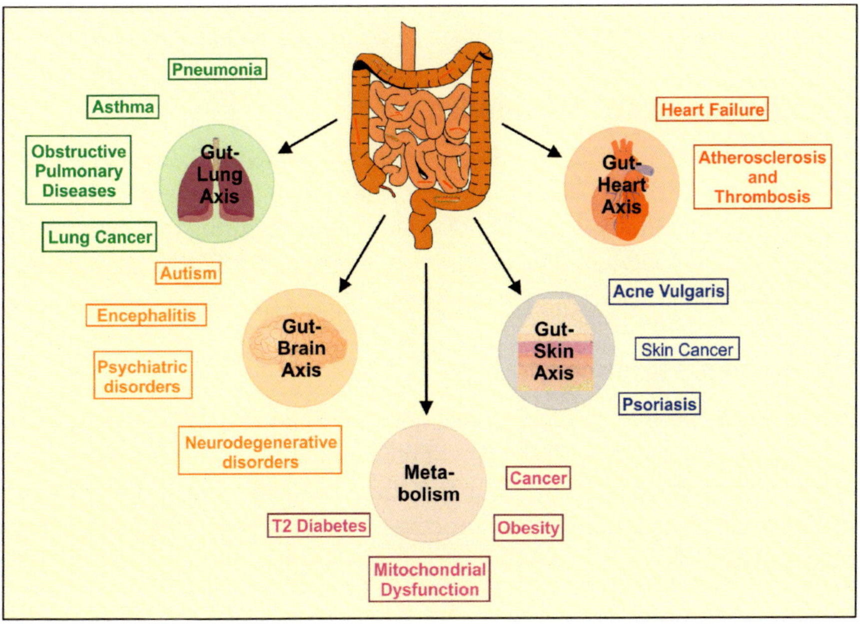

Figure 8.1. *Probiotics interaction with the gut microbiota.*

Source: Chandrasekaran P, Weiskirchen S, Weiskirchen R. (2024). Effects of probiotics on gut microbiota: An overview. International journal of molecular sciences; 25(11):6022. https://doi.org/10.3390/ijms25116022.

It's crucial to remember that probiotic effects on nutritional digestion and absorption can differ based on the precise strains of probiotics used, the dosage, and the makeup of the person's gut microbiota at the time of use. Furthermore, different people may react differently to probiotics, and more study is required to completely comprehend the mechanisms and particular effects on various nutrients.

8.3. PROBIOTICS AND THE METABOLISM OF DIETARY COMPOUNDS

Probiotics are live microorganisms that give the host health benefits when taken in sufficient amounts. You can find them in fermented foods like kefir, sauerkraut, and yogurt or you can take them as dietary supplements. Although probiotics are best known for helping digestion and promoting gut health, newer studies indicate they may also have an impact on how food ingredients are metabolized. The processes by which dietary compounds are metabolized, absorbed, and used by the body are referred to as metabolism. This covers the different micronutrients and bioactive substances found in food, in addition to macronutrients like proteins, fats, and carbs.

Probiotics can affect the metabolism of dietary compounds in several ways:

8.3.1. Carbohydrate Metabolism

Enzymes produced by certain probiotic strains aid in the breakdown of complex carbohydrates, like dietary fiber, which the body would not otherwise be able to process. Short-chain fatty acids (SCFAs) are produced during this fermentation process. These include butyrate, propionate, and acetate. SCFAs have multiple metabolic advantages and provide energy to colon cells.

8.3.2. Protein Metabolism

Enzymes produced by some probiotics help break down proteins into smaller peptides and amino acids. This can improve how well dietary proteins are absorbed and digested, guaranteeing that there is a sufficient supply of amino acids for a range of physiological processes, such as tissue regeneration and muscle growth.

8.3.3. Fat Metabolism

Certain probiotic strains may have an impact on how dietary fats are metabolized, according to some research. They might facilitate fat metabolism, enhance lipid absorption, and control fat storage. This may have consequences for cardiovascular health and weight control.

8.3.4. Bile Acid Metabolism

Probiotics can interact with bile acids, which are produced by the liver and play a crucial role in fat digestion and absorption. Some probiotic strains have been shown to modify bile acid metabolism, leading to changes in fat absorption and cholesterol metabolism.

8.3.5. Modulation of Nutrient Uptake

Probiotics have the ability to affect the expression and function of transporters, which are involved in the gut's ability to absorb different nutrients. Probiotics may alter how well vitamins, minerals, and other vital nutrients are absorbed from the diet by modifying these transporters. It's crucial to remember that probiotics' effects on metabolism can differ based on the strain or strains combined with the individual's gut microbiota composition and general health. Furthermore, investigations into this area are still being conducted, and more research is required to completely comprehend the mechanisms and therapeutic ramifications of probiotic-mediated effects on the metabolism of dietary compounds. It's always best to speak with a healthcare provider if you're thinking about taking probiotics or if you have specific questions about how they might affect your metabolism. They can offer tailored advice based on your unique needs and health status (Barone et al., 2022; Lynch & Pedersen, 2016; Roberfroid et al., 2010).

8.4. PROBIOTICS AS FUNCTIONAL FOODS: POTENTIAL EFFECTS ON HUMAN HEALTH AND ITS IMPACT ON NEUROLOGICAL DISEASES

Probiotics are currently the main focus due to their extensive health potential and represent the new buzzword in human dietary selection globally. Classically speaking, a probiotic is a viable dietary microbial supplement that acts on the intestinal tract to benefit the host. Supplements or other substances that are taken to achieve a particular goal are known as functional foods. Biotherapeutics and nutriceuticals are other terms for functional foods. Functional foods are best exemplified by probiotics.

At the moment, the most popular way to consume probiotic cells is through food products. The majority of foods high in probiotics fall into the category of functional foods. From $33 billion in 2000 to $176.77 billion in 2013, the global market for functional foods has grown, making up 5% of the overall food market. Probiotic foods are expected to account for between 60 and 70% of the whole functional food market. Dairy products containing probiotic bacteria, including ice cream, flavored liquid milk, fermented milk, baby food, milk powder, frozen dairy desserts, cheese, buttermilk, whey-based beverages, and normal and sour cream, have been produced with notable success in the last few decades. In response to the growing incidence of lactose intolerance, a number of nondiary probiotic products have emerged in recent years. These include fruit juices, soy-based products, oat-based desserts, baby foods, and breakfast cereal.

The majority of research topics remain unexplored, and it is unclear how probiotics achieve their beneficial effects on health. Probiotics, on the other hand, work to prevent the growth of pathogens through the synthesis of antimicrobial compounds, the production of lactase, the alteration of gut pH, the disruption of pathogen-receptor site binding, the disruption of existing nutrients and growth factors, and the stimulation of immunomodulatory cells.

The fact that probiotics are safe, inexpensive, and effective at preventing microbial infections is one of their most important features. Probiotics were regarded by the World Health Organization (WHO) in 1994 as the next most important immune defense system in the event that antibiotic resistance renders regularly prescribed antibiotics ineffective. We refer to the application of probiotics as microbial interference therapy when treating antibiotic resistance.

A compelling possibility is presented by the possible use of probiotics in the management and prevention of neurological disorders. To assign them the credit of enhancing neurological manifestations or reducing the incidence of neurodevelopmental disorders, however, requires a deeper comprehension of their mode of action. Up until recently, the majority of this topic's research focused on the microbiota—gut—brain axis mechanisms, with encouraging findings in children with autism spectrum disorder (ASD). Extensive research on how well probiotics modify these associations will help identify new targets that require intervention and clarify the etiopathogenesis of ASD and a few other neurological disorders.

8.4.1. The Role of Probiotics as a Functional Food Ingredient

Over the last few years, over 500 probiotic food products have been introduced to the global market, and the number of probiotic foods available is still rising. Both consumers and scientists are becoming more aware of the probiotic foods that are created when grains, fruits, and vegetables, as well as meat products, are fermented. Commercial probiotics can now be found in ice cream, milk, sour milk, fruit juices, and oat-based products. Recently developed foods containing probiotics include cheese, meat-based products, spreads made with mayonnaise, and dips made with cheese. Probiotics are becoming increasingly popular in modern society due to their unique qualities, actions, and significance, but they have a long history. Pliney the Elder, a Roman naturalist, recommended drinking fermented milk to cure digestive issues. According to research by Henry Tissler of the Pasteur Institute in Paris, infants who had bifidobacteria in their digestive systems experienced fewer gastrointestinal issues, as shown by a decrease in diarrheal problems.

According to Eli Metchnikoff's theory, lactic acid bacteria (LAB) provide a certain level of defense against intestinal autointoxication. Metchnikoff was awarded the Nobel Prize in medicine in 1908 for his work demonstrating that intestinal diseases could be treated by replacing harmful bacteria with beneficial ones. Metchnikoff had noticed that despite extreme poverty and unfavorable weather, Bulgarian rural dwellers lived to very old ages. They reportedly drank fermented milk products and lived longer on average than the wealthier Europeans. Metchnikoff came to the conclusion that the fermented milk products' LAB offered health advantages against aging.

The probiotic microbes are added to the food artificially during the probiotic food preparation process. The majority of culture preparations are synthesized for various applications and are commercially available in highly concentrated form as freeze-dried powders or highly concentrated freeze-cultures. Because *Bifidobacterium* spp. produce several metabolic components, including acetic acid synthesis, the addition of probiotic microbes may alter the food products' aroma and taste during the fermentation process. Probiotic microbes should therefore be added to food products with caution so as not to negatively impact the sensory qualities or quality of the final product.

The active cells or total viable cells per milliliter or gram of the product at the time of individual consumption determines the medicinal efficacy of probiotic food products. We must ensure that a high probiotic survival rate is maintained both during production and throughout the product's shelf life in order to maintain consumer confidence in probiotic food products.

The viability of probiotic microbes in food products during production, processing, and storage is influenced by a number of factors. Chemicals like hydrogen peroxide, coloring agents, artificial flavors, and bacteriocins are among the factors that have been identified. Food parameters include titratable acidity, water activity, molecular oxygen, and the presence of sugar and salt. Processing parameters include incubation temperature, food product cooling rate, heat treatment, storage techniques, and packing materials and methods. Microbiological parameters include the percentage of inoculums and the probiotic strains. The fermentation temperature is one of the key variables influencing qualitative parameters and the viability of probiotic microbes. The majority of probiotic bacteria can grow best in temperatures between 37 and 43°C.

A probiotic bacterial strain of *Lactobacillus casei* supplemented to infants has been shown to increase circulating immunoglobulin A levels, which is correlated with a shorter duration of diarrhea caused by rotavirus. Furthermore, *Bifidobacterium bifidum* and *Lactobacillus acidophilus* supplements to probiotic foods significantly increase the nonspecific immune phagocytic activity of circulating blood granulocytes. Consuming yogurt containing probiotic bacteria has been shown to increase cytokine production from blood mononuclear cells.

Probiotics have been shown to have health benefits, but only when they contain the minimum viable microorganism count required at the time of consumption. The minimum level of 106 CFU ml^{-1} at the time of consumption has been widely adopted by the food industry.

The United States Food and Drug Administration (US FDA) recommends that probiotic food products have a minimum probiotic count of 106 CFU ml^{-1}. A daily dosage of 108 to 109 probiotic microorganisms is very important to achieve probiotic action in humans, depending on the quantity ingested and taking into consideration the effect of storage on the viability of probiotics. It has also been suggested that probiotic supplements be taken on a daily basis. Approximately 100 g of probiotics should be taken per day to supply the intestine with 109 live cells. Probiotic foods are available in the US as tablets, powders, liquids, and capsules. Typically, they contain one or more probiotic bacterial strains. Not all probiotics have the same qualities and functions, so choosing a particular strain or species of probiotic bacteria is important for maintaining good health. While a minimum consistent dosage of a probiotic is required to achieve a positive health effect, it's important to remember that there is a safe upper limit based on animal studies.

8.4.2. Probiotics for Human Health Improvement

Probiotics are best taken by healthy individuals first because they improve overall health and protect against a variety of diseases. Improving our health is the most shrewd and preventive move we can take to shield ourselves against various diseases. However, how exactly do probiotics enhance health? The idea that probiotics either directly or indirectly improve our health will be emphasized in the paragraphs that follow. Numerous microorganism groups exist in human bodies, and these groups cooperate to carry out a variety of tasks. The most important ones are those that our digestive system performs. These microbes improve digestion and food intake. Our digestive systems are not burdened as much by these microbes.

These microbes reduce the steps required by the human body to modify complex food structures to simpler ones. The relation of probiotics to human health improvement can be summarized by the following facts and points:

- Probiotic microbes are functional and user-friendly.
- Probiotic microbes help to ferment our food from a complex form to simpler byproducts and can promote our health by several diverse methods.
- They can compete with the pathogenic microbes and take possession of our digestive system.
- The viable count of probiotic microbes could be decreased because

of many factors such as inappropriate diet, age, alcohol, and several other factors. That is why probiotics should be supplemented through our daily diet.

- In specific cases, such as after consumption of antibiotics, they are supposed to be affected severely; hence, they should be consumed orally in significant amounts along with food.
- They provide the human body with necessary useful byproducts.
- They eliminate the side effects of the pathogenic microbes.
- They trim down the quantity of food required by the human body by enabling correct digestion and metabolism of any amount of food.
- They decrease the additional burden on the human digestive system.
- As they form biofilms, they eliminate the effect of the first attack of harmful compounds on our cells.
- In many cases, probiotics could balance the deficiency in our genetic materials by serving to make use of the products of their genes such as in the case of lactose insufficiency.

8.4.3. Probiotics in Infection Control

According to reports, probiotics are essential for both preventing and controlling diarrhea following antibiotic treatment. Bifidobacteria species, *Saccharomyces boulardii*. For the treatment of diarrhea, *Lactobacillus* GG and *Lactobacillus reuteri* are typically used. Probiotics are highly successful in preventing traveler's diarrhea and other rotavirus-caused diarrheal illnesses in young children. By either generating bacteriocins or by outcompeting harmful bacteria or viruses and preventing them from adhering to the epithelial cells, probiotics suppress the microorganisms that cause diarrhea. Few studies have demonstrated the ability of probiotic bacteria to inhibit *Helicobacter pylori* colonization and functions in the stomach, which causes gastritis, peptic ulcers, and stomach cancers. Furthermore, *Lactobacillus salivarius* was demonstrated to be capable of producing large volumes of lactic acid and inhibiting the growth of *H. pylori* in mice and *in vitro*. It has been suggested that using probiotics to combat *H. pylori* will improve the antibiotic therapies used to treat the infection in terms of tolerability, eradication rate, and compliance. There has been a lot of interest in the potential role of probiotics in the treatment of influenza (Barone et al., 2022; Lynch & Pedersen, 2016; Roberfroid et al., 2010).

8.4.4. Probiotics in Maternal Health

Every developmental stage, from conception to the early postnatal period, is greatly influenced by the interaction of hereditary and environmental factors, which has long-term implications for the health of both the mother and the child. As an environmental factor, the gut microbiota is crucial because it regulates the development of immune and metabolic systems before and after neonatal life.

A few studies added a fresh perspective to the use of probiotic bacteria by shifting the emphasis from gut health to vaginal health. Another area of novel research that interests me is the use of probiotics primarily for the management of vaginal infections in general. One of the primary causes of bacterial vaginosis may be the decrease or disappearance of *Lactobacilli* in the vagina. Affected individuals are at a markedly higher risk of contracting gonorrhea, herpes simplex, chlamydia, and viral infections caused by the human immunodeficiency virus. One of the most successful and reliable methods of preventing vaginal infections is the use of probiotic *Lactobacilli*; despite the fact that only a small number of strains have been clinically proven to be effective, they have an excellent safety record. Numerous researchers stressed how crucial it is to characterize and evaluate probiotic strains clinically in relation to vaginal infections using various delivery methods, including oral suspension or tablets, dried powder, or vaginal pessaries.

One effective method of enhancing the flora and preventing certain pathogens or lessening their ability to rule is the introduction of *Lactobacilli* into the vagina through capsule or pessary. This approach appears to be a workable treatment for bacterial vaginosis and urinary tract infections. It has been demonstrated by science that a combination of *Lactobacilli* strains reduces vaginal yeast and bacterial pathogens even when taken orally. When metronidazole is used in place of an antibiotic alone, it also offers a higher cure rate. A consistent oral dosage of 108 viable probiotic *Lactobacilli* can help women maintain and restore the health of their urogenital tracts (Barone et al., 2022; Lynch & Pedersen, 2016; Roberfroid et al., 2010).

8.4.5. Probiotics in Child Health

Probiotic bacteria, mainly *S. boulardii*, *L. reuteri*, and *Lactobacillus rhamnosus* GG, have been shown to significantly reduce the duration of acute diarrhea, especially acute rotaviral diarrhea, by one day and the number of stools with a potential effect starting after three days, according to statistics obtained by the meta-analyses of a few clinical investigations related to many diarrheal diseases. Probiotics have also been shown to reduce the incidence of antibiotic-associated diarrhea (AAD) by an average of 60% when taken as a preventive measure.

Despite significant advancements in neonatal care, the mortality rate from necrotizing enterocolitis (NEC), a serious gastrointestinal disorder that affects newborns, has not decreased over the past three decades, as evidenced by persistent rates between 10 and 30%. Only a few cutting-edge therapies show promise in the fight against this fatal illness. In this case, it has been shown that breast milk is the best option for preventing NEC. Nonetheless, there is growing evidence that probiotic supplements to preterm infants in the Neonatal Intensive Care Unit may provide sufficient relief for the afflicted infants. Probiotic therapy is a novel and exciting approach at the moment. The evidence showing differences in the intestinal microbiota's organization in preterm infants is what drives the decision to supplement preterm infants with probiotics. Giving probiotics to vulnerable infants may be a useful way to alter the beneficial and healthy bacteria colonizing their stomachs, and it may also encourage research into using probiotics to treat preterm illnesses. One of the most recent meta-analysis investigations conducted in India provided adequate scientific proof to establish the effectiveness of probiotics in decreasing death and disease in preterm neonates.

Probiotic supplementation has shown to be a promising strategy for preventing very low birth weight babies from developing NEC; however, more research is needed to address a few issues. To ensure the safety and quality of probiotic preparations sold commercially, there are currently no clear regulatory guidelines in place. It is also necessary to clearly explain the best probiotic combination and dosage method (Barone et al., 2022; Lynch & Pedersen, 2016; Roberfroid et al., 2010).

8.4.6. Probiotics as Future Biotherapeutics to Control Diabetes

In order to control type 2 diabetes, some researchers are attempting to determine the potential of novel dietary-based strategies, such as probiotic interventions, such as by modifying gut hormones. More has been learned about the gut-brain connection to type 2 diabetes than just peripheral insulin resistance and beta cell failure, thanks to the identification of gut hormones like glucose-dependent insulinotropic polypeptide (GIP) and glucagon-like peptide-1 (GLP-1) incretins and their implications in glucose homeostasis. Certain reports indicate that there are differences in the microbial populations found in the gut between individuals who are lean and obese, as well as between those who have diabetes and those who do not, implying that microbial composition may differ in obesity and diabetes.

Researchers state that probiotic interventions may function as potential modulators of gut flora that alter the gut composition in a favorable way and offer a variety of health benefits to the host, given the link between changes in metabolic diseases and intestinal microbiota. A few strains of *Lactobacillus* species may have a positive impact by decreasing body fat percentage in people with diet-induced obesity, especially by shrinking the size of white adipose tissue cells. Researchers are working to safely and effectively alter the human gut microbiota. Given the great potential and possibility of using these new dietary-based approaches to effectively control type 2 diabetes, future research efforts should be concentrated on highlighting the immunomodulatory, intricate hormonal, and metabolic mechanisms underlying the microbiota–host interaction in various tissues. Furthermore, it is necessary for the clinical treatments to be assessed in carefully planned trials with patient-focused endpoints.

8.4.7. The Possibility to Prevent Colon Cancer with Probiotics

Across the world, colon cancer is one of the most prevalent cancers. The present pathway of carcinogenesis includes initial gene mutations, including tumor suppressor and oncogenes, followed by the development of dysplastic aberrant crypt foci (ACF), carcinoma or adenoma formation, and metastases.

Different possible mechanisms for the beneficial effects of probiotics on colon cancer include the following:

- Attaching and removing potential carcinogens by the probiotics;
- Synthesizing antimutagenic or antitumorigenic compounds in the colon;
- Supporting the host's immune response;
- Modifying the intestinal microflora qualitatively and quantitatively;
- Altering the metabolic functions of the intestinal microflora;
- Beneficial effects on the physiology of the host; and
- Changing the physicochemical environment in the colon.

It has been demonstrated that the probiotics *Lactobacilli*, bifidobacteria, and *Streptococcus thermophilus* are highly effective at avoiding gene mutation in the Ames assay and at halting the mutation process in carcinogenesis. The mutagenicity of *Salmonella typhimurium* strains His is tested using the Ames assay. It is capable of identifying about 90% of all known carcinogens. According to some reports, probiotic-fed rodents may be able to lower the incidence of colon ACF and mammary lesions caused by chemical carcinogens. Few scientific investigations proved that the supplementation of *L. acidophilus* could reduce the mutagenicity of urine and the feces of human individuals who intake a fried meat diet.

It has previously been shown that 2-amino-3-methyl-imidazo [4,5-f] quinoline, a mutagenic, heterocyclic aromatic amine compound that can cause colon lesions in rodents, is present in broiled meat or fish. The colon adenoma patients' upper colonic crypt mucosal cell proliferation was inhibited by supplementing them with *L acidophilus* and *B. bifidum*. Although the scientific evidence mentioned above generally points to the significant impact of probiotics on colon cancer, specific findings from human research are still lacking.

8.4.8. The Task of Probiotics in Disease Treatment

Probiotics have the potential to treat and manage actual diseases in addition to improving our health and reducing harmful infections. It is debatable, though, how probiotics could aid in the actual management and treatment of disease. Understanding the behavior of the illness and the factors that drive it are the most important aspects.

For example, illnesses associated with genetic disorders can cause specific types of deficiencies, like lactose intolerance. In these kinds of situations, probiotics will help to eliminate these deficiencies through a variety of strategies, such as giving our bodies the products of the missed gene or appropriate substitutes. When we get older, probiotics will be a great support system for us. It will lessen the strain on our biological system and allow us to engage in other activities, primarily those related to improving our capacity to utilize the food we eat. Probiotics may help a pathway that is weak or defective because of a single allele error as opposed to an error in both alleles. Just as accurately as in the case of retinoblastoma patients. In such a case, the significant basis for the Knudson hypnosis's will be totally interfered with while a single gene will not be a theme to too much stress that could show the way to a mutation.

Some of the crucial roles of probiotics in maintaining human health, disease treatment, and management are the following:

- Maintaining good condition of the genitourinary tract.
- Checking the serum cholesterol levels.
- Control of hypertension.
- In curing diarrheal diseases. *Saccharomyces cerevisiae* var *boulardii* is very effective in treating different diarrheal disorders.
- Probiotics are supplemented to trim down the antibiotic negative effect and to restore any type of loss in the beneficial microflora.
- Inhibition of the putrefactive-type fermentation, which is one of the Metchnikoff postulations regarding the benefits of probiotics.
- Probiotics support the defensive functions of the intestinal mucosa including the production and secretion of antibacterial peptides.
- Enhanced digestion of lactose against food possessing lactose.
- Decreasing the occurrence of allergy in susceptible individuals.
- Supporting bioavailability and synthesis of nutrients and enhancing the immune system.

- Probiotics are very effective in treating diarrheal disorders.
- Improving intestinal tract health.

8.5. PROBIOTICS AND THEIR EFFECTS ON METABOLIC DISEASES

LAB, or probiotics, are frequently added to foods and used in medicinal preparations. Numerous studies have shown how probiotics can help with metabolic illnesses. A study has demonstrated the impact of fermented dairy products, particularly when combined with specific strains of LAB, on serum cholesterol levels. For example, it has been discovered that a very small amount of *Lactobacillus fermentum* KC4b dry culture can extract 14.8 mg of cholesterol from the culture medium. Additionally, *lactobacilli* are crucial for the breakdown of bile salts into bile acids in the intestine, which prevents the formation of micelles. *In vivo* lipid metabolism is enhanced and lipid peroxidation is decreased by probiotics. Probiotic supplementation to the diet for several weeks enhanced immunological response without causing the release of inflammatory cytokines, delaying the onset of systemic inflammatory diabetes. There is evidence to suggest that variations in the gut microbiota's composition may exist before childhood obesity develops.

The concept of beneficial bacteria was first proposed by Russian scientist and Nobel laureate Metchnikoff at the beginning of the 1900s. He wrote in his book "The Prolongation of Life" that a person should consume a lot of beneficial bacteria because they offer the host numerous health advantages. He proposed that the gut microflora may be altered and that probiotics, or beneficial bacteria, may take the place of harmful microbes. Probiotics are typically taken as dietary supplements or as part of fermented foods that have had specially added active live cultures, like yogurt and soy yogurt. According to a working group established by the WHO and the Joint Food and Culture Organization (FAO), probiotics need to contain live microorganisms and provide a quantifiable physiological benefit. Probiotic organisms used in food must be able to survive in gastric juices and be resistant to exposure to bile in the human gut, as per the guidelines set forth by the WHO and the FAO. They also need to be able to proliferate and populate the human digestive system. In addition to these requirements, they also need to be safe, efficient, and maintain their effectiveness for the duration of the product's shelf life. With the growing understanding of the practical impact of microbes on human physiology and illness, new perspectives on therapy have surfaced, offering innovative methods for treating acute and chronic conditions. Probiotics currently fall between food and medicine, but this gap will close as research into genetically modifying beneficial bacterial species for pharmaceuticals and the production of nutrients in the human gut continues. Probiotics are essentially LAB (LAB), which are added to foods and widely used in medicinal preparations.

Currently under investigation and documentation are particular health outcomes, such as the reduction of chronic intestinal inflammatory diseases as reported by Mach and the avoidance and management of diarrhea caused by pathogens as reported by Yan et al. as demonstrated by Vanderhoff, urogenital infections as investigated by Reid, and atopic illnesses. Furthermore, Gilliland discovered that regular probiotic consumption reduced cholesterol, the risk of colon cancer, the frequency of diarrhea, and the immune system's ability to be stimulated.

Testing a supplement does not suggest that any other strain of the same species is beneficial; rather, all observed effects are exclusive to the strain or strains that are tested. The probiotic strains belonging to the genera *Lactobacillus*, *Bifidobacterium*, and *Enterococcus* show the greatest promise. According to Fuller and Gordin, the representative species are *Bifidobacterium longum*, *Bifidobacterium breve*, *Bifidobacterium bifidum*, *Bifidobacterium infantis*, *Lactobacillus acidophilus*, *L johnsonii*, *L. casei*, *L. gasseri*, *L. plantarum*, and *L. rhamnosus*. Scientists, physicians, and consumers all agree that probiotics have health benefits, but there is still debate over the underlying molecular mechanism (Roberfroid et al., 2010; Lynch & Pedersen, 2016; Barone et al., 2022).

8.5.1. Hyperlipidemia

Murray found that the two main global causes of cardiac diseases and death are hyperlipidemia and atherosclerosis. McGill established a correlation between blood cholesterol levels and the chance of developing coronary heart disease (HD). Around the world, hyperlipidemia is responsible for roughly twice as many deaths as cancer and 10 times as many as accidents. Many studies on humans and experimental animals were conducted after Mann discovered that the fermented milk of the wild type starters was the cause of the hypocholesterolemic effect in the Maasai tribesmen. St. Onge demonstrated the effect of fermented dairy products on the serum cholesterol, especially with selected strains of LAB, by increasing the utilization of cholesterol for the de novo synthesis of the bile acids.

Micelles are formed during lipid metabolism, which facilitates the intestinal absorption of cholesterol. An intermediary stage in the absorption of fat is the emulsification of the dietary fat. *Lactobacilli* are involved in the inhibition of micelle formation by deconjugating bile salts in the intestine to form bile acids. As a result, there is less absorption of cholesterol. Similar measures are taken for cholesterol that enters the colon by the enterohepatic circulation. The bile salts and the enzyme Hydroxyl Steroid Dehydrogenase (HSDH) are hydrolyzed by the *lactobacilli*-produced conjugate bile acid hydrolase (CBH), which degrades the bile salts and ultimately stops the enterohepatic circulation of bile acids. It is also believed that *lactobacilli* are responsible for the inhibition of hydroxymethyl glutarate coA (HMG CoA). There is a strong correlation between coronary heart disease and elevated serum cholesterol levels. Rao and companions showed that feeding laboratory rats fermented milk combined with animal feed reduced their serum cholesterol levels compared to feeding them skim milk in preclinical studies.

This variation cannot just be a result of cholesterol being redistributed among the liver, plasma, and other bodily pools. Hypocholesterolemia-causing factors include orotic acid, lactose, casein, and hydroxymethyl glutaric acid. 20 mice were used in a placebo-controlled study to assess how probiotic-fermented food affected their serum cholesterol levels. Over the course of 42 days, the control group was fed only cholesterol, while the experimental group was fed a food mixture that included probiotics and 1% cholesterol. *Saccharomyces boulardii* (10 9 CFU) and *L. casei* NCDC-19 (109 CFU*) were found to generate a 19% reduction in total serum cholesterol and a 37% decrease in LDL-cholesterol levels. In another study, probiotic-fed pigs (Yorkshire barrows) with hypercholesterolemia-induced blood cholesterol had an 11.8% reduction in total blood cholesterol. This study was also placebo-controlled.

Another study carried out by Yong Zhang and colleagues to study the antioxidative effects of *Lactobacillus casei* on hyperlipidemic rats found that supplementing with *L. casei* significantly decreased the levels of malondialdehyde (MDA), while the serum and liver of these rats showed increased levels of glutathione peroxidase and superoxide dismutase (SOD). Therefore, it can be inferred that probiotics enhance lipid metabolism *in vivo* and decrease lipid peroxidation. In a study on the effects of probiotics on *in vitro* cholesterol reduction, Pereira et al. found that the *L. fermentum* KC5b strain could remove up to 14.8 mg of cholesterol per gram (dry weight) of cells from the culture medium while also maintaining viability for two hours at pH 2 in bile acids. It was therefore considered a noble probiotic. Fermented milk containing *Lactobacillus acidophilus* L1 was found to lower serum cholesterol levels by 2.4% ($p<0.05$) during the first treatment period in a controlled study. Human trials on dairy products fermented with specific LAB have produced a variety of results, but they all highlighted the bacteria's ability to reduce total and LDL cholesterol levels to a modest extent. Probiotics, however, may be a good way to start the therapy, according to the research. As demonstrated in 112 male Golden Syrian Hamsters, genetically modified soy milk, with or without a probiotic fermentation, can improve hypercholesterolemia and lower the risk of atherosclerosis.

8.5.2. Diabetes

According to Arauz, the incidence of hypertension in patients with diabetes was 1.5 to 3 times higher than in those without the disease. According to a study, people with hypertension had an almost 2.5 times higher risk of developing type II diabetes than people with normotension.

Irving affirmed the link between high blood pressure, insulin resistance, and hyperglycemia during fasting. Additionally, it was discovered that the combination of high blood pressure and fasting hyperglycemia changes the structure of the dermal microvascular system. In a different investigation, it was found that giving fructose-induced diabetic rats Dahi, an Indian fermented product containing *Lactobacillus acidophilus*, *L. casei*, and *L. lactis*, reduced the accumulation of glycogen in rats' livers. Additionally, it was discovered that untreated, essential hypertensive patients had higher insulin levels during fasting and after meals than normotensive people. It is urgent to evaluate novel approaches in order to identify effective, low-risk strategies for preventing or lowering the incidence of diabetes and hypertension. Nearly 75% of diabetic patients do not achieve adequate blood pressure control, despite the paucity of anti-hypertensive medications on the market and the overwhelming evidence supporting the benefits of blood pressure control. As a result, those with essential hypertension have an increased chance of developing diabetes.

Probiotic consumption may represent a novel approach to the prevention or management of diabetes and the consequent decrease in the prevalence of hypertension. Numerous studies indicate that long-term high-fat diet consumption is associated with poor inflammatory status, which may precede the onset of diabetes in these individuals. A negative correlation was discovered between the *Bifidobacterium* spp population and the lipopolysaccharide (LPS) content of plasma. According to reports, when diabetes is brought on by a high-fat diet, the amount of *Bifidobacterium* spp. was positively connected with reduced diabetes endotoxemia, plasma and adipose tissue inflammatory cytokines, as well as enhanced glucose tolerance and glucose-induced insulin secretion. Additionally, a study has demonstrated that Bifidobacteria can lower intestinal endotoxin levels, enhance the mucosal barrier, lessen systemic inflammation, and ultimately lower the incidence of diabetes.

Al-Salami showed that probiotics (75 mg/kg body weight) pretreated diabetic and healthy Wistar rats (n = 10) before gliclazide suspension was administered produced optimal control over hyperglycemia and improved signs and symptoms in those diabetic animals. Probiotics induced a significant increase in gliclazide uptake, as reported by the authors. This has been attributed to two factors: first, probiotic therapy may help restore drug transporter activity; second, it may help restore the disrupted gut motility linked to diabetes. These results suggest that probiotics can help treat diabetes and work in concert with other diabetes medications to lower the risk of diabetes-related hypertension.

In one study, participants in the 60–83 age range took 1.5 × 1011 CFU of *Lactobacillus lactis* twice a day for six weeks. This improved immune response without causing the release of inflammatory cytokines, delaying the onset of systemic inflammatory-induced diabetes. Models of insulin resistance were generated by introducing the Ay gene into KK mice. Boden observed that a decrease in insulin sensitivity results in the regulation of blood glucose and fat metabolism due to dysfunction of the pancreatic β-cells. According to Matsuzaki, the delivery of a diet supplemented with lyophilized *Lactobacillus casei* decreased the hyperglycemia-related insulin insufficiency. In a related study, it was noted that the inhibition of the autoimmune destruction of the pancreatic β-cells was reduced in 4-week-old female, non-obese, insulin-dependent, diabetic (nod) rats fed a diet containing 0.05% lyophilized *L. casei.* The parent protein's sequence contains inactive ACE inhibitory peptides, which can be liberated through fermentation by microbial activity to form bioactive.

Of the 70 identified enzymes, about 20 enzymes that are present at the highest levels in bovine milk, have been isolated from milk and they have been characterized. The presence of a proteolytic system aids the probiotics to grow in milk products, as it degrades casein along with the lactose hydrolyzing enzymes, as was observed by Saito.

In a different investigation, it was found that feeding sour milk to spontaneously hypertensive rats (SHR) containing 2.5–3.5 mg/kg/day of VAL-PRO-PRO and ILEPRO-PRO, which was fermented by *L. helveticus* LBK-16H (10%), resulted in a 17 mm Hg decrease in SBP. In an investigation involving human subjects, the consumption of sour milk containing *Lactobacillus helveticus* and *Saccharomyces cerevisiae* resulted in a noteworthy reduction in diastolic blood pressure (DBP) ($p< 0.01$) after eight weeks. Conversely, the placebo group did not exhibit any noteworthy effects. In a study which was done on hypertensive subjects, it was observed that the consumption of 150 ml milk which was fermented by *L. helveticus* b.i.d., decreased the SBP and DBP by $p = 0.001$ and $p = 0.048$ respectively. We observed that probiotics are the potential modulators of the gut flora that change the gut flora composition in a beneficial manner and exert various health effects i.e., anti-oxidant, anti-inflammatory and anti-hyperlipidemic in type 2 diabetes mellitus patients. Thus, the consumption of probiotics shields the impairment of the pancreatic β-cells, which plays a major role in the production of insulin molecules and thus, prevents the onset of insulin dependent diabetes as well as diabetes-related hypertension (Barone et al., 2022; Lynch & Pedersen, 2016; Roberfroid et al., 2010).

8.5.3. Obesity

According to the US Department of Health and Human Services, obesity is one of the major factors contributing to preventable, unfavorable health conditions, morbidity, and mortality, as well as the utilization of healthcare resources and associated expenses. Ley studied the gut microbiota in the feces of 12 obese volunteers who had followed a weight-loss regimen for a year. They were randomized at random to follow a low-calorie, fat-restricted (FAT-R) or carbohydrate-restricted (CARB-R) diet. The guts of each subject were compared at baseline, 12, 26, and 52 weeks. The results revealed that, in comparison to the lean controls, the obese subjects had a higher proportion of firmicutes and a lower proportion of *Bacteroidetes*. Secondly, there was a decrease in the firmicute representation and an increase in the *Bacteroidetes* in the weight-loss subjects. Thirdly, these changes occurred regardless of the diet and were proportionate to the weight loss. Moreover, it was noted that a broad range of microbes were changing up or down, rather than just a single firmicute or *Bacteroidetes* member flourishing or declining. The *Bacteroidetes* and firmicutes divisions dominated the microbiota, as demonstrated by the preclinical investigations, and the bacterial flora exhibited a remarkable intraindividual stability over time.

A study conducted on animals has indicated that consuming a high-fat diet raises the serum levels of LPS. Additionally, LPS exacerbates the symptoms of fatty liver when combined with a high-fat diet. Gram negative bacteria have LPS in their cell walls, which can lead to endotoxemia and an increase in the concentration of proinflammatory cytokines in different tissues. However, an oligofructose supplement reduces the absorption of LPS from the gut lumen and promotes the growth of bifidobacteria. An enhanced insulin sensitivity and glucose tolerance are also linked to this effect. Backhed proposed in a different study that the gut microbiota affects genes that control energy storage and expenditure as well as energy harvest, i.e., both sides of the equation for energy balance. Obesity was linked to changes in the relative abundance of the two dominant bacterial divisions, the *bacteroidetes* and the firmicutes, according to the distal gut microbiota of genetically obese mice and their lean littermates, as well as those of obese and human lean volunteers. These findings suggest that the microbiome of obese individuals is better able to absorb energy from food. Additionally, they discovered that the obese microbiota produced noticeably more total body fat than the lean microbiota after colonizing a small number of germ-free mice with the obese microbiota and another group with the lean microbiota. This demonstrated that the pathophysiology of obesity included the gut microbiota as an additional contributing factor. In a different study, Kalliomaki examined stool samples taken from infants as young as 6 and 12 months old and followed them until they were 7 years old in order to examine the composition of the gut microbiota. Researchers discovered that while leaner children had more *Staphylococcus aureus* than obese children at 6 and 12 months of age, the obese children at 7 years old had fewer *Bacteroidetes*. This provided additional proof that variations in the gut microbiota's makeup might exist before obesity develops. Lee et al. investigated the ability of *Lactobacillus rhamnosus* PL60 to prevent obesity in a study using diet-induced obese mice. Conjugated linoleic acid, produced by the human-derived bacterial species *L. hamnosus*, has been demonstrated in animal experiments to decrease body fat. Even though there were no changes in the energy intake, it was discovered that the obese mice that received *L. rhamnosus* had reductions in body weight and white adipose tissue (perirenal and epididymal). The probiotic *Lactobacillus gasseri* (LG2055) was the subject of a recent, intriguing human study that showed significant reductions in body weight, BMI, waist and hip circumferences, body fat mass, and visceral and subcutaneous fat areas in the abdomen.

8.6. STRATEGIES FOR IMPROVING GUT HEALTH

Developing a healthy, balanced gastrointestinal tract (GIT) is the main goal of a multifaceted approach to improving gut health. Above all, it is critical to include a wide range of dietary fiber. A variety of fruits, vegetables, whole grains, legumes, and nuts are included in this. While insoluble fiber, which is common in whole grains and vegetables, promotes regular bowel movements and prevents constipation, soluble fiber, which is found in foods like oats and apples, gives nourishment for good gut bacteria. Including fermented foods and probiotics in one's diet is another important way to support gut health. Probiotics are live beneficial bacteria that increase the diversity of microbes in the gut. They are frequently found in yogurt, kefir, kimchi, and sauerkraut. These microbes are essential for strengthening the immune system and aiding in digestion. To support this, prebiotics ought to be added as well. These can be obtained from foods like bananas, onions, garlic, and leeks. Prebiotics are indigestible fibers that support the growth and general health of gut flora by providing sustenance for these good bacteria.

Reducing the amount of processed and sugary foods consumed is essential to further support gut health. These products may upset the delicate balance of gut flora, which could result in inflammation and discomfort in the digestive system. Alongside this, it's critical to prioritize your intake of omega-3 fatty acids. These are present in walnuts, flaxseeds, chia seeds, and fatty fish like mackerel and salmon. Inflammation of the gut lining is significantly decreased by omega-3 fatty acids, which promotes a healthier digestive system.

Retaining adequate hydration is essential for gut health. Drinking enough water helps food pass through the digestive system more easily and makes it easier for vital nutrients to be absorbed. Additionally, a major factor in achieving optimal digestion is the adoption of mindful eating practices. In addition to promoting better nutrient breakdown and absorption, chewing food thoroughly and enjoying every bite helps reduce overeating, which can cause digestive discomfort.

Drinking alcohol and caffeine in moderation is also recommended because overindulging in these substances may upset the delicate balance of gut flora and cause irritation to the digestive system. Another essential component of improving gut health is regular physical activity. It has been demonstrated that regular exercise enhances overall digestive function and supports a healthy gut microbiome. Additionally, gut health is greatly impacted by stress management. Chronic stress can have a negative impact on the digestive system; it can be lessened by practicing yoga, deep breathing techniques, meditation, or spending time in nature. Furthermore, since antibiotics can upset the delicate balance of good bacteria in the gut, their use should be done with caution. Supplementing with probiotics can help restore microbial equilibrium in cases where antibiotics are required (Barone et al., 2022; Lynch & Pedersen, 2016; Roberfroid et al., 2010).

8.6.1. The Importance of Dietary Fiber in Gut Health

The consumption of dietary fiber is essential for achieving optimal gut health because it keeps the gastrointestinal system healthy and in balance. It includes the indigestible parts of plant-based diets that make it through the digestive system mostly undigested. Because of this special characteristic, fiber has a variety of advantageous effects on the intestines. Its main purpose is to encourage regular bowel movements. Rich in whole grains, fruits, and vegetables, insoluble fiber gives stool more volume and helps it pass through the digestive system more easily. This helps avoid constipation, which is a common issue that can cause discomfort and other digestive issues.

Furthermore, the varied community of microorganisms that live in the digestive tract, known as the gut's resident microbiota, depends critically on dietary fiber for their sustenance. Foods high in soluble fiber, such as legumes, some fruits, and oats, function as prebiotics by feeding good bacteria. These microbes then convert soluble fiber into SCFAs, including propionate, butyrate, and acetate. SCFAs are essential for preserving the integrity of the gut barrier and provide energy to the cells lining the colon. This effectively acts as a barrier to prevent toxic substances and dangerous infections from entering the bloodstream.

Additionally, dietary fiber helps control appetite and weight by promoting feelings of fullness and satiety. Foods high in fiber have fewer calories per gram due to their lower energy density. This lowers the risk of overindulging and ensuing weight gain by promoting a more controlled and balanced eating style. Furthermore, foods high in fiber have slower digestion and absorption, which results in more stable blood sugar levels. This is especially important for diabetics because it lessens the risk of severe blood glucose spikes and crashes.

Dietary fiber has been associated in the long run with a lower risk of a number of chronic diseases. Consuming it has been linked to decreased incidence of heart disease, hypertension, and a few gastrointestinal disorders, such as colorectal cancer (CRC) and diverticular disease. It is thought that a variety of mechanisms, including better blood pressure regulation, elevated bowel regularity, and elevated cholesterol levels, are responsible for this protective effect.

8.6.2. Fermented Foods and Their Role in Gut Health

Foods that have undergone fermentation are essential for fostering and preserving gut health. These foods naturally undergo a microbial transformation process in which sugars and carbohydrates are transformed into organic acids or alcohol by helpful bacteria, yeasts, or fungi. This procedure offers several health advantages in addition to improving the food's flavor and preservation. The high probiotic content of fermented foods is one of their main advantages. Probiotics are live microorganisms that give the host health benefits when taken in sufficient amounts. Fermented foods are full of these good bacteria, such as yogurt, kefir, kimchi, sauerkraut, and miso. Probiotics support the diversity and replenishment of the gut microbiota through ingestion, which is essential for a healthy digestive system.

Additionally, one of the most well-known genera of probiotic bacteria, *Lactobacillus* and *Bifidobacterium*, are abundant in different strains in fermented foods. Numerous benefits to gut health have been linked to these strains. They facilitate the digestion of complex carbohydrates, improving absorption of nutrients and lowering the possibility of gastrointestinal distress. Additionally, eating fermented foods has been connected to stronger immune system performance. Since the gut contains a sizable portion of the immune system, a balanced and varied microbiota composition is necessary for the immune system to function properly. Fermented foods contain beneficial bacteria that aid in immune response regulation, strengthening and fortifying the body's defense against pathogens. Fermented foods also help to maintain a healthy, well-functioning digestive system. They encourage the synthesis of other chemicals and enzymes that aid in nutrient absorption and digestion. Furthermore, fermented foods support the gut's acidic environment, which is essential for preventing the growth of pathogens and dangerous bacteria.

Fermented foods contain a high concentration of bioactive compounds in addition to probiotics. These consist of a variety of phytochemicals, vitamins, and minerals that have been linked to a number of health advantages. For instance, fermented soybeans in miso offer a concentrated source of protein and isoflavones, while fermented cabbage in sauerkraut is a good source of vitamin C. It's crucial to remember that not all fermented foods are made equal and that the conditions and method of fermentation can have an impact on the quantity and variety of healthy microorganisms. To optimize the potential benefits for gut health, choosing naturally fermented, unpasteurized varieties free of excessive additives or preservatives is advised (Barone et al., 2022; Lynch & Pedersen, 2016; Roberfroid et al., 2010).

8.6.3. Omega-3 Fatty Acids: A Key Player in Gut Health

Essential polyunsaturated fats, omega-3 fatty acids are important for gut health and play a vital role in overall health maintenance. These fats are well known for their anti-inflammatory qualities and are mainly present in fatty fish such as salmon, mackerel, and sardines as well as flaxseeds, chia seeds, and walnuts. Regarding gut health, omega-3 fatty acids function as strong GIT anti-inflammatory agents. Numerous digestive disorders, including inflammatory bowel diseases (IBD) like Crohn's disease and ulcerative colitis (UC), can be brought on by persistent inflammation in the gut. It has been shown that omega-3 fatty acids, in particular eicosapentaenoic acid (EPA) and docosahexaenoic acid (DHA), reduce inflammation by inhibiting the synthesis of pro-inflammatory chemicals, which helps people with gastrointestinal disorders.

Furthermore, omega-3 fatty acids support the integrity of the gut barrier. Tight junctions hold the specialized cells that make up the gut lining together to form a dynamic barrier. Increased intestinal permeability, also known as leaky gut, can result from a breached barrier and permit undesirable substances to enter the bloodstream. It has been demonstrated that omega-3 fatty acids, in particular DHA, improve the function of these tight junctions, supporting the integrity of the gut barrier. Furthermore, the composition of the gut microbiota is positively impacted by omega-3 fatty acids. They promote the growth of beneficial bacteria and inhibit the growth of pathogenic microorganisms. This promotes a diverse and well-balanced microbial community in the gut, which is essential for good digestion and general health.

Additionally, the synthesis of specific lipid mediators like protectins and resolvins is aided by omega-3 fatty acids. These substances are essential for reducing inflammation and encouraging tissue healing. Omega-3 fatty acids aid in the healing and restoration of the gut lining by promoting the resolution of inflammation, which is especially advantageous for people with chronic inflammatory conditions. Apart from their ability to reduce inflammation, omega-3 fatty acids have several other health advantages. They have been linked to decreased risk of developing some chronic diseases, enhanced cardiovascular health, and enhanced cognitive function. These wider health advantages are entwined with their impact on gut health since overall well-being is enhanced by a balanced gut microbiome and decreased intestinal inflammation.

8.6.4. Balancing Gut Bacteria

A healthy and functional gastrointestinal system depends on the balance of gut bacteria. Numerous positive health effects are linked to a diverse microbiome, which is defined by a large range of helpful microorganisms. Here are some tips to promote microbial diversity and achieve a balanced gut microbiome:

- **Dietary Diversity**: Increase your intake of a variety of complete, plant-based foods. Add a rainbow of vibrant fruits and vegetables along with whole grains, legumes, nuts, and seeds various fibers and nutrients that support various strains of beneficial bacteria are found in various types of food.
- **Fiber-Rich Foods**: Give foods high in dietary fiber priority. Good gut bacteria are fed by soluble fiber, which is present in fruits, vegetables, and legumes. It functions as a prebiotic. The insoluble fiber found in vegetables and whole grains promotes regular bowel movements and gut health in general.
- **Fermented Foods**: Add foods that are naturally fermented to your diet, such as miso, kimchi, kefir, yogurt, and sauerkraut. These foods are high in probiotics, which increase microbial diversity in the gut by introducing live beneficial bacteria.
- **Probiotic Supplements**: If you have specific concerns about your gut health or are taking antibiotics, you should think about taking probiotic supplements. Select a premium supplement that contains a range of strains for optimal effects.
- **Prebiotic-Rich Foods**: Eat foods high in prebiotics, like asparagus, bananas, leeks, garlic, and onions. Prebiotics are indigestible fibers that support overall digestive health by feeding good gut bacteria and fostering their growth.
- **Limit Sugar and Processed Foods**: Reduce the amount of processed foods and sugary snacks you eat because they can upset your gut's delicate balance and cause inflammation.
- **Include Omega-3 Fatty Acids**: Include foods high in omega-3 fatty acids, such as walnuts, flaxseeds, chia seeds, and fatty fish (salmon,

mackerel). By lowering intestinal lining inflammation, omega-3 fatty acids support a more wholesome digestive system.

- **Limit Antibiotic Use**: Antibiotics can upset the delicate balance of good bacteria in the gut, so use them sparingly and only when absolutely necessary. To aid in the restoration of microbial balance, if probiotics are prescribed, think about taking supplements.
- **Manage Stress**: Engage in stress-reducing activities like meditation, deep breathing exercises, yoga, or spending time in nature. Chronic stress can negatively impact gut health, so finding healthy ways to manage it is crucial.
- **Stay Hydrated**: Drink plenty of water throughout the day to support the smooth movement of food through the digestive tract and aid in the absorption of nutrients.
- **Limit Alcohol and Caffeine**: Caffeine and alcohol abuse can irritate the digestive system and disturb the gut microbiome. With these drinks, moderation is key.
- **Regular Exercise**: Engage in regular physical activity, which has been shown to promote a healthy gut microbiome and improve overall digestive function.

A balanced and diverse gut microbiome is essential for optimum digestive health and general well-being, and you can actively work toward achieving it by adopting these lifestyle practices. It's vital to pay attention to your body and modify your strategy as necessary, keeping in mind that each person may react differently.

8.6.5. The Gut-Brain Connection

The relationship between the central nervous system (CNS) and the GIT is intricate and complex, and the gut-brain connection illustrates how mental health can have a significant impact on digestive health. This network of bidirectional communication influences physiological as well as psychological processes through neural, hormonal, and immune pathways. The vagus nerve, a lengthy cranial nerve that connects the brainstem to several body organs, including the gut, is a crucial component of the gut-brain connection. This nerve plays a critical role in controlling processes like peristalsis, the secretion of digestive enzymes, and gut motility by facilitating the transmission of signals between the brain and the digestive tract.

Furthermore, the gut contains a vast network of enteric neurons, which are sometimes referred to as the second brain. This complex system, which functions autonomously of the CNS, is in charge of regulating blood flow, food movement, and nutrient absorption in the local gut. The gut microbiota is a diverse community of microorganisms that live in the digestive tract and collectively facilitate communication between the gut and the brain. Trillions of bacteria, viruses, fungi, and other microorganisms make up this microbiota, which actively interacts with the gut lining to affect a number of physiological processes. They aid in the synthesis of neurotransmitters that are important in controlling mood and emotion, such as gamma-aminobutyric acid (GABA) and serotonin. This sensitive communication system may be disrupted by dysbiosis, an imbalanced state of the gut microbiota. Research has indicated that changes in the composition of the gut microbiota are linked to mood disorders such as anxiety and depression. On the other hand, the gut can be affected by stress and mental health issues, which can result in symptoms like bloating, altered bowel habits, and abdominal pain. Particularly long-term stress has been connected to negative consequences on digestive health. Increased intestinal permeability may result from it, making it possible for dangerous substances to enter the bloodstream and possibly set off immune reactions. This phenomenon is linked to a number of gastrointestinal disorders and is commonly referred to as leaky gut syndrome. Knowing that there is a connection between the gut and the brain emphasizes the significance of a holistic approach to health that takes mental and digestive health into account. It has been demonstrated that regular exercise, mindfulness meditation, and relaxation techniques improve gut health as well as mental health. Furthermore, a healer can benefit from a balanced diet that includes foods high in fiber, probiotics, and omega-3 fatty acids.

8.6.6. Hydration and Digestion: How Water Impacts Gut Health

Water is essential for maintaining digestive health because it affects several processes in the digestive system depending on your level of hydration. Keep an eye on the color of your urine. Light yellow or straw-colored urine indicates adequate hydration, while darker colors may indicate a need for additional fluids.

Here, we are going to discuss the relationship between hydration and digestion and explore the importance of water drinking for a healthy gut.

8.6.6.1. Lubricating the Digestive Tract

Water serves as a lubricant throughout the digestive tract, which is one of its primary functions in digestion. Saliva, which is mostly made of water, starts to break down food particles as soon as it enters the mouth, making swallowing easier. Sufficient hydration is necessary for the food to pass easily from the esophagus to the stomach and intestines, facilitating effective digestion and nutrient absorption.

8.6.6.2. Facilitating Nutrient Absorption

Sufficient hydration is necessary for the body to absorb nutrients from food. Water facilitates the breakdown of macronutrients and helps them pass through intestinal walls so that the bloodstream can absorb them. Dehydration can make it difficult for the body to absorb nutrients from food, which can result in deficiencies and poor digestive performance.

8.6.6.3. Supporting the Gut Microbiome

Trillions of microorganisms make up the gut microbiome, which is essential for immune system performance, mental health, and digestion as a whole. By facilitating functions like intestine secretions, gastrointestinal motility, and waste elimination, hydration affects the balance of the gut microbiome. According to research, including enough water in a balanced diet may help shape a gut microbiome that is beneficial to health.

8.6.6.4. Promoting Digestive Efficiency

Maintaining adequate hydration can improve the breakdown of food in the digestive system. Water consumption before meals can speed up the breakdown of food particles and the absorption of nutrients by aiding in digestion. Better overall digestive comfort and function may arise from this.

8.6.6.5. Maintaining Gastrointestinal Health

Constipation and indigestion are two digestive disorders that can be avoided with proper hydration. In addition to promoting regular bowel movements, adequate hydration also lessens the risk of digestive discomfort.

8.6.6.6. How Much Hydration Do You Need?

There are numerous variables that can affect how much water is needed to maintain gut health. How much water a person needs can vary depending on their physiology and the surroundings. For example, you require more water when engaging in vigorous physical activity and during the warmer months when perspiration is higher. That being said, a good rule of thumb for staying properly hydrated is to aim for eight to 10 cups of water a day, or 64 to 80 ounces. In addition to water, foods high in moisture are also part of this intake.

8.6.6.7. How do you Increase Hydration Levels?

To increase hydration levels, there are several strategies that go beyond simply drinking water. Here are some tips that can help:

- **Eat Water-Rich Foods:** Drink plenty of water and include foods like watermelon and cucumbers in your diet. Apart from being hydrating, these foods also have important nutrients in them.
- **Consume Fluids Before Thirst Sets In:** It's critical to hydrate yourself before you experience thirst. Waiting until you feel thirsty may be a sign of mild dehydration, so make it a goal to stay hydrated throughout the day.
- **Flavor Your Water:** Drinking more water can be encouraged by adding fruit slices, like lemon or berries, to enhance its flavor and appeal.
- **Start Your Day with Hydration:** To ensure that you stay hydrated throughout the day, start your morning routine with a glass of water.
- **Utilize Water-Based Beverages:** Include water-based beverages like herbal teas as part of your fluid intake.
- **Include Moisture-Rich Meals:** Incorporate meals with high water content, such as soups and smoothies, into your diet to boost hydration.
- **Develop Hydration-Encouraging Habits:** Whether it's setting reminders to drink water or using a fun water bottle, find ways to encourage consistent fluid intake throughout the day.
- **Choose Hydrating Snacks:** Snack on fruits and vegetables that have high water content, such as celery and oranges.
- **Consider Rehydration Solutions:** In situations where rapid rehydration is necessary, oral rehydration solutions or sports drinks can be beneficial.
- **Hydrate Before Meals:** Drinking water before meals can not only aid digestion but also contribute to your overall fluid intake.

8.6.7. Mindful Eating Practices for Optimal Digestion

Eating mindfully entails being totally present and involved in the process, which has a big positive impact on digestion and general well-being. Here are some key mindful eating practices for optimal digestion:

- **Savor Each Bite**: Give your food enough time to fully develop its flavors, textures, and aromas. Chew slowly and mindfully, savoring every taste of your food.
- **Eliminate Distractions**: When eating, reduce the amount of time you spend staring at TV, phones, or other screens. To pay attention to your body's signals of hunger and fullness, concentrate only on the act of eating.

- **Listen to Your Body:** Pay attention to your body's hunger and fullness cues. Eat when you're hungry and stop when you're comfortably satisfied, rather than relying on external cues like portion sizes.
- **Portion Control:** Use smaller plates and bowls to help regulate portion sizes. This can prevent overeating and promote a more balanced and comfortable digestion.
- **Engage Your Senses:** Observe the colors, smells, and textures of your food before taking a bite. Engaging multiple senses enhances the eating experience and fosters a deeper connection to your food.
- **Practice Gratitude:** Take a moment to express gratitude for the food you are about to eat. Recognize the effort and resources that went into producing and preparing your meal.
- **Avoid Emotional Eating:** Be mindful of emotional triggers that may lead to eating when you're not truly hungry. Find alternative ways to cope with emotions, such as engaging in physical activity or practicing relaxation techniques.
- **Chew Thoroughly**: Eating with your teeth cleanly helps break down food particles early in the digestive process, which facilitates easier digestion later on. You can also taste the flavors more fully as a result.
- **Pause between Bites**: In between bites, set down your utensils. This small action promotes eating more slowly and lets your body tell you when it's full.
- **Be Mindful of Stress:** Avoid eating when stressed or rushed. Stress can lead to poor digestion and discomfort. If you find yourself in a stressful situation, take a few deep breaths before you start eating.
- **Stay Hydrated**: To aid in digestion, sip water both before and after eating. However, since it can dilute digestive enzymes, avoid consuming too much fluid during meals.
- **Reflect on Food Choices**: Think about the foods you're eating and their nutritional content. Select complete, nutrient-dense foods that will fuel your body.

You can improve nutrient absorption, improve your digestive experience, and develop a better relationship with food by implementing these mindful eating techniques. This method enhances overall wellbeing and meal satisfaction in addition to promoting optimal digestion (Barone et al., 2022; Lynch & Pedersen, 2016; Roberfroid et al., 2010).

8.6.8. Inflammatory Foods to Avoid for Better Gut Health

It is essential to stay away from inflammatory foods in order to preserve gut health and avoid digestive discomfort. These are compounds that have the potential to elicit an immunological response and aggravate GIT inflammation. Here are some common inflammatory foods to be cautious of:

- **Processed Sugar and Artificial Sweeteners**: Overindulgence in artificial sweeteners and refined sugars can upset the delicate balance of gut flora, encouraging the growth of pathogenic microorganisms. This disproportion may be a factor in gastrointestinal inflammation.
- **Processed and Fried Foods**: Foods high in trans fats, like fried foods, processed snacks, and fast food, can aggravate inflammation. Trans fats have the ability to change the makeup of gut flora, which can result in dysbiosis.
- **Refined Grains**: The fiber and nutrients present in whole grains are absent from white bread, white rice, and other refined grains. They may cause inflammation in the gut and cause abrupt spikes in blood sugar levels.
- **Dairy Products (for Lactose Intolerant Individuals)**: Dairy products have the potential to cause gastrointestinal distress, such as bloating, gas, and diarrhea, in people who are lactose intolerant. Making non-dairy or lactose-free food choices can help avoid these problems.
- **Gluten-Containing Grains (for Gluten-Sensitive Individuals)**: Grain consumption such as wheat, barley, and rye containing gluten can cause inflammation and damage to the intestinal lining in people with celiac disease or gluten sensitivity.
- **Saturated and Trans Fats**: Excessive intake of trans fats, which are present in processed and fried foods, and saturated fats, which are present in red meat, full-fat dairy, and some oils, can cause inflammation and harm gut health.
- **Artificial Additives and Preservatives:** Some artificial additives and preservatives in processed foods can disrupt the gut microbiota and contribute to inflammation. It's best to choose whole, minimally processed foods.
- **Excessive Alcohol and Caffeine:** Overconsumption of alcohol and caffeine can irritate the digestive tract and disrupt the gut microbiota. Moderation is key to preventing inflammation.
- **Nightshade Vegetables (for Some Individuals):** While generally nutritious, nightshade vegetables like tomatoes, peppers, and

eggplants can cause digestive issues in some people due to their alkaloid content. Those sensitive to nightshades may experience inflammation in the gut.

- **Highly Spiced Foods (for Some Individuals)**: Hot sauces and chili peppers are examples of spices that can irritate the GIT, especially in people who have IBS.

It's crucial to remember that every person reacts differently to these meals, and some may be more allergic to particular ingredients than others. Furthermore, although avoiding these inflammatory foods can help promote gut health, you should also concentrate on including nutrient-dense, anti-inflammatory foods like fruits, vegetables, whole grains, and lean proteins in your diet. An intestinal system that is healthier and more resilient may benefit from this well-rounded approach.

8.6.9. Stress Management Techniques for a Healthy Gut

Stress management strategies are essential for supporting gut health because the mind-body link has a significant impact on digestive health. Prolonged stress can negatively impact the GIT, resulting in symptoms like bloating, constipation, and pain in the abdomen. Putting into practice sensible stress-reduction techniques can lessen these impacts and promote gut health at its best. Mindfulness meditation is one of the best methods for managing stress. Developing present-moment awareness and accepting oneself, thoughts, feelings, and physical sensations without passing judgment are the goals of this practice. Regular mindfulness meditation has been linked to improved gut health, lowered stress levels, and lower levels of inflammation in the body, according to research. Mindfulness has the potential to improve overall digestive function and reduce symptoms associated with conditions like IBS by calming the nervous system.

Exercises involving deep breathing are another effective method for managing stress. The body's relaxation response is triggered by methods such as diaphragmatic breathing, which involves concentrating on taking slow, deep breaths that engage the diaphragm. This promotes calmness and eases tension in the digestive tract by inhibiting the fight-or-flight reaction brought on by stress. Deep breathing is a practical and accessible method of stress relief because it can be done anytime, anywhere. Yoga is a holistic discipline that incorporates breathing exercises, physical postures, and awareness. It has been demonstrated to successfully lower stress levels and enhance intestinal health. Yoga poses that promote blood flow to the digestive organs, such as twists and mild inversions, can facilitate better digestion. Furthermore, practicing yoga can help people develop a more mindful awareness of their bodies' cues regarding hunger, fullness, and discomfort.

Frequent exercise promotes gut health and is an effective method of stress management. Endorphins are feel-good and stress-relieving chemicals that are released into the brain when you exercise. Additionally, it promotes the digestive tract's regular contractions, which facilitate the passage of food through the system. Exercises like yoga, jogging, and walking can be especially helpful for lowering stress and maintaining digestive health. In addition to helping to manage stress and maintain gut health, joyful and relaxing activities like hobbies, time spent in nature, or relaxing music listening can also help. By providing constructive outlets for stress release, these activities help mitigate the detrimental effects of long-term stress on the digestive system. Finally, getting help from mental health specialists like therapists or counselors can be very helpful for understanding and putting into practice stress-reduction strategies. Evidence-based techniques such as cognitive-behavioral therapy (CBT) can give people useful tools to better manage stress and enhance their general well-being, which includes gut health (Barone et al., 2022; Lynch & Pedersen, 2016; Roberfroid et al., 2010).

8.7. DIETARY APPROACHES TO MANAGE GUT-RELATED DISORDERS

Dietary strategies play a critical role in the management of gut-related illnesses, offering people realistic and efficient ways to reduce symptoms and enhance their general health. Dietary modifications made to address particular GIT conditions can improve digestive health and quality of life considerably. A low-FODMAPS diet is a well-known treatment for people with IBS, a disorder marked by bloating, pain in the abdomen, and changes in bowel habits. Fermentable carbohydrates, or FODMAPs, are present in a wide range of foods. By lowering the gut's fermentation of high-FODMAP foods like some fruits, vegetables, grains, and dairy products, restricting these foods can help ease symptoms. Foods that do not worsen symptoms can be identified and tolerated by individuals through the gradual reintroduction and careful observation of particular FODMAP groups.

Dietary techniques for IBD, such as UC and Crohn's disease, are intended to control inflammation and encourage intestinal healing. Limiting high-fiber foods in a low-residue diet can help minimize irritation and reduce bowel movements during flare-ups. Conversely, a well-rounded diet high in nutrients, such as fruits, vegetables, whole grains, and lean proteins, supports general health and preserves nutritional status during remission. Tight adherence to a gluten-free diet is required for people with celiac disease, an autoimmune condition brought on by gluten consumption. This means staying away from wheat, barley, and rye, as well as any other source of gluten. Celiac disease sufferers need to carefully read labels and look for gluten-free substitutes for common foods like bread, pasta, and baked goods. To manage this condition, it is imperative to maintain a well-balanced diet rich in nutrient-dense fruits, vegetables, and proteins, as well as gluten-free grains like rice, quinoa, and corn.

Bloating, diarrhea, and discomfort in the abdomen can be reduced in people with lactose intolerance by avoiding dairy products or using lactase supplements. Choosing lactose-free or dairy-free substitutes, like plant-based milk or lactose-free milk, enables people to get enough calcium without aggravating gastrointestinal distress.

Dietary changes are a common part of managing gastroesophageal reflux disease (GERD) in order to reduce acid reflux and relieve heartburn symptoms. Avoiding trigger foods such as chocolate, tomatoes, citrus fruits, caffeine, and spicy or fatty foods is part of this. Reflux episodes can also be avoided by eating smaller, more frequent meals and by not sleeping right after. Moreover, a high-fiber diet is essential for people with diverticular disease, which causes tiny pouches to form in the colon wall. By doing so, you can encourage regular bowel movements and avoid problems like diverticulitis, which is an inflammation of these pouches. Dietary fiber is abundant in foods like fruits, vegetables, whole grains, and legumes.

8.7.1. Gluten-Free Living: Navigating a Celiac Disease Diagnosis

Living a gluten-free lifestyle is crucial for those with celiac disease, an autoimmune condition brought on by gluten consumption. For people who have celiac disease, the protein gluten, which is present in wheat, barley, and rye, can cause serious harm to the lining of the small intestine. A strict commitment to avoiding all gluten sources is necessary to navigate this diagnosis. This includes hidden sources like sauces, dressings, and processed foods, as well as obvious ones like bread and pasta. Making this dietary change is essential for reducing celiac disease symptoms and avoiding long-term consequences.

The choice of naturally gluten-free foods is the cornerstone of a gluten-free lifestyle. These consist of whole grains like rice, quinoa, and corn; lean meats; dairy products; and unprocessed fruits and vegetables. It is crucial to carefully read food labels to look for any indication of gluten-containing ingredients. Furthermore, certified gluten-free goods go through a rigorous testing process to guarantee that they adhere to gluten-free standards, giving people with celiac disease a trustworthy and safe option for packaged goods. Following a gluten-free diet goes beyond cooking at home to include eating out and social events. When eating out, people with celiac disease need to let the kitchen staff know exactly what they need to eat. To avoid cross-contamination, it's imperative to ask about gluten-free options, request menu item modifications, and verify kitchen procedures. Having open lines of communication with restaurant employees can help guarantee a secure dining environment.

Furthermore, eating a diet rich in nutrients and well-balanced is essential for good health and wellbeing in general. Essential vitamins, minerals, and fiber can be found in whole, unprocessed foods like fruits, vegetables, lean proteins, and whole grains that are free of gluten. Because celiac disease damages the small intestine, it can make it difficult for affected individuals to absorb nutrients. For this reason, maintaining a balanced diet is especially important to prevent nutritional deficiencies. For those who have celiac disease, a gluten-free lifestyle is necessary, but it's crucial to approach dietary decisions carefully. Certain gluten-free products may be deficient in fiber and nutrients, especially if they are processed substitutes like baked goods and snack items. An effective and well-balanced gluten-free diet depends on including a range of whole, nutrient-dense foods.

People adjusting to a celiac disease diagnosis benefit from looking for resources and support in addition to dietary changes. Participating in in-person or online celiac support groups can foster a feeling of community and offer a forum for exchanging perspectives and experiences. A registered dietitian with expertise in celiac disease can provide individualized advice on meal planning, reading nutrition labels, and making sure you're getting the right amount of nutrients.

8.7.2. Nutrition Strategies for IBD Management

Nutritional approaches are essential to the treatment of IBD, which includes diseases such as UC and Crohn's disease. Tailored dietary approaches have the potential to alleviate symptoms, promote healing, and enhance overall well-being in disorders involving chronic inflammation of the digestive tract.

A nutrient-dense, well-balanced diet is a fundamental tenet of nutrition-based management of IBD. This comprises a range of whole grains, fruits, vegetables, lean meats, and healthy fats. These foods are rich in fiber, vitamins, minerals, and antioxidants, all of which support tissue repair, lower inflammation, and strengthen the immune system. It's critical to concentrate on nutrient-rich foods because people with IBD may experience malabsorption and nutrient deficiencies. Essential vitamins and minerals like calcium, magnesium, and zinc can be found in foods like leafy greens, fish, lean meats, nuts, and seeds. Furthermore, the anti-inflammatory qualities of omega-3 fatty acids, which are present in fatty fish like salmon and flaxseeds, can be especially helpful for people with IBD.

When considering fiber consumption in the context of IBD, care should be taken. While soluble fiber, which can be found in foods like fruits, vegetables, and oats, can help with regularity and gut health in general, some IBD patients may find it more difficult to tolerate insoluble fiber, particularly during flare-ups. Reducing the amount of fiber consumed and selecting forms that are easily digested can help stop symptoms from getting worse. People with IBD may benefit in some circumstances from particular dietary adjustments. For example, during times of active inflammation, some people may find relief by adhering to a low-residue diet. This entails cutting back on high-fiber foods, which can be more difficult to digest, such as raw fruits and vegetables, seeds, and nuts. On the other hand, progressively reintroducing foods high in fiber during remissions can support long-term digestive health and help sustain a healthy gut microbiome.

For those with IBD, especially those whose small intestine is affected by Crohn's disease, lactose intolerance is a common concern. Supplements containing lactase or dairy products without lactose can supply calcium without aggravating gastrointestinal distress. Furthermore, foods high in probiotics, such as yogurt with live cultures, can be tolerated well and may aid in maintaining a healthy gut flora. Since each person has different triggers and tolerances, individualized dietary approaches are crucial to the management of IBD. Maintaining a food journal to monitor how particular foods affect symptoms can be a useful method for figuring out individual dietary patterns. A registered dietitian with expertise in gastrointestinal health can offer customized advice on meal planning, recognizing trigger foods, and making sure you're getting the right amount of nutrients in your diet.

8.7.3. Lactose Intolerance

A common digestive ailment called lactose intolerance is caused by the body's incapacity to completely digest lactose, a naturally occurring sugar present in milk and dairy products. The inability of the enzyme lactase, which breaks down lactose in the small intestine, to function properly causes this condition. Bloating, diarrhea, and abdominal pain are just a few of the digestive discomforts that can result from lactose that has not been fully broken down. The goal of managing lactose intolerance is to reduce symptoms while maintaining a sufficient calcium intake through dietary strategies. Avoiding or consuming foods and beverages high in lactose is a crucial part of managing lactose intolerance. This includes some processed foods that have lactose added to them, as well as milk, cheese, yogurt, ice cream, and so on. Thankfully, lactose-free substitutes for numerous dairy products are easily accessible, such as lactose-free milk, cheese, and yogurt. For those who are lactose intolerant, these products have undergone special processing to eliminate or break down the lactose. Furthermore, plant-based milk substitutes like rice, almond, soy, or coconut milk can be used in place of regular milk.

Lactase supplements can be a useful help for people who still want to enjoy dairy products. The lactase enzyme found in these over-the-counter supplements helps the body break down lactose. When taken prior to dairy consumption, these supplements can help reduce gastrointestinal distress and enable people who are lactose intolerant to occasionally enjoy dairy products. Some people with lactose intolerance may tolerate fermented dairy products like yogurt better. Yogurt's live bacterial cultures aid in the digestion of lactose by breaking it down. Choosing plain, unsweetened yogurt with live cultures will minimize lactose content and offer the probiotic benefits.

Some people can also benefit from gradually introducing small amounts of lactose-containing foods over time to help them develop tolerance. To help the body adjust to lactose, a process called lactose desensitization entails introducing tiny amounts of dairy at first and progressively increasing it. It's critical to carefully read food labels because lactose can be found in a variety of unexpected foods and products, including baked goods, salad dressings, and processed foods. On ingredient lists, look for words like "milk solids," "milk by-products," or "whey," as these denote the presence of lactose. Preventing nutrient deficiencies requires incorporating foods high in calcium, low in lactose, or free of lactose. Almonds, canned fish with bones, dark leafy greens, and fortified plant-based milk are all excellent sources of calcium. To guarantee appropriate calcium intake, supplements may also be taken, under the advice of a medical professional.

8.7.4. Diverticular Disease: Fiber-Rich Diets for Colon Health

Diverticula, or tiny pouches that form in the colon wall, are a defining feature of diverticular disease. Diverticulitis is a condition caused by inflammation or infection of these pouches. Maintaining colon health and controlling diverticular disease both depend heavily on a diet high in fiber. Plant-based foods contain dietary fiber, which is crucial for both managing and preventing diverticular disease. Constipation is avoided, regular bowel movements are maintained, and the general health of the digestive system is enhanced. Fruits, oats, and legumes are rich sources of soluble fiber, which helps to facilitate softer and more effortless bowel movements by absorbing water and forming a gel-like substance in the intestines. Whole grains, vegetables, and seeds are rich sources of insoluble fiber, which helps maintain regular bowel movements by giving stool more volume.

It's imperative that people with diverticular disease include a range of high-fiber foods in their diet. This includes veggies like broccoli, carrots, spinach, and sweet potatoes as well as fruits like apples, pears, berries, and oranges. Additionally great sources of fiber are whole grains such as oats, brown rice, quinoa, and whole wheat bread. Legumes, including black beans, chickpeas, and lentils, are also a good source of soluble and insoluble fiber. It's crucial to gradually increase fiber consumption to give the digestive system time to adjust. Abrupt and substantial shifts in the amount of fiber consumed may cause discomfort, bloating, and gas. When increasing fiber intake, it's also critical to drink lots of water throughout the day because it facilitates the passage of stool through the intestines and helps it become softer.

Despite their nutritional value, nuts and seeds may need to be treated with extra caution if a patient has diverticular disease. Certain medical professionals advise against consuming hard, tiny seeds and nuts as this may lower the chance of causing irritation or obstruction in the diverticula. Nut and seed powders or finely chopped versions can also be a good substitute. Healthcare professionals might occasionally advise taking extra fiber, especially if dietary sources are inadequate. For those who find it difficult to get enough fiber from their diet alone, supplements such as methylcellulose, wheat dextrin, or psyllium husk may be helpful. Nonetheless, it's imperative to speak with a doctor before beginning any new supplement regimen. A low-fiber or clear liquid diet may be suggested during acute episodes of diverticulitis, even though a high-fiber diet is typically advised for the condition to give the colon time to recover. Reintroducing foods high in fiber gradually after symptoms have subsided can help avoid flare-ups in the future.

8.7.5. The Role of Probiotics and Prebiotics in Gut Disorder Management

Prebiotics and probiotics are important in the treatment of gut disorders and provide a safe, all-natural means of promoting digestive health. Probiotics are live, beneficial bacteria that give the host health benefits when taken in sufficient quantities. Prebiotics, on the other hand, are indigestible fibers that provide good gut bacteria with nourishment and support as they grow. Collectively, they support a resilient and well-balanced gut microbiome, which is essential for overall digestive health. Different strains of probiotics are available, and each strain offers special advantages. They function by bringing good bacteria into the gut, aiding in the restoration of microbial balance, and stopping the spread of dangerous bacteria. Some probiotic strains have demonstrated promise in symptom relief and inflammation reduction for people suffering from IBD and IBS.

Probiotic bacteria belong to two of the most prevalent genera: *Lactobacillus* and *Bifidobacterium*. They are present in supplements as well as fermented foods like kimchi, kefir, yogurt, and sauerkraut. Furthermore, some strains of the beneficial yeast *Saccharomyces boulardii* have shown promise in treating illnesses like *Clostridium difficile* infection and diarrhea brought on by antibiotics. In contrast, probiotic bacteria in the stomach receive nutrition from prebiotics. They are mostly indigestible fibers that can be found in some foods, such as bananas, chicory root, garlic, onions, leeks, and asparagus. Prebiotics nourish the cells lining the colon and support a healthy gut environment by providing a substrate for beneficial bacteria to ferment when they reach the colon undigested. This produces SCFAs like butyrate.

Synbiotics, or the combination of probiotics and prebiotics, have been shown to support gut health in a synergistic way. Synbiotics function by supplying advantageous probiotic strains with the prebiotic fibers that they prefer, which fosters the growth and activity of the probiotics. The probiotic bacteria's ability to survive and colonize the gut can be improved by this combination. Although prebiotics and probiotics seem to have promising effects on gut health, it's crucial to remember that everyone reacts differently to these supplements. Certain strains or combinations may be more or less effective depending on the type of gut disorder, the makeup of the individual's gut microbiome, and other specific health considerations. Before adding probiotics or prebiotics to a treatment plan, it is best to speak with a healthcare professional or registered dietitian, particularly for people who are taking medication or have pre-existing medical conditions. To address specific management goals for gut disorders, a healthcare professional can offer personalized recommendations and assist in choosing the most appropriate strains and dosages.

8.7.6. Plant-Based Diets for Gut Health

Diets based mostly on plants have drawn a lot of attention because they may be good for gut health. These dietary patterns reduce or eliminate animal products in favor of whole, plant-derived foods such as fruits, vegetables, legumes, nuts, seeds, and whole grains. This diet plan offers an abundance of fiber, phytonutrients, and antioxidants that can improve the health of the digestive system overall and the gut microbiome in particular. The high fiber content of a plant-based diet is one of its main benefits for gut health. In order to maintain optimal gut motility and regular bowel movements, fiber is necessary. It softens and gives stool more volume, preventing constipation and facilitating a smoother passage through the digestive system. Additionally, some forms of fiber—referred to as prebiotics—act as a food source for good gut bacteria, encouraging the development and activity of these microorganisms. Vegetables and fruits offer a variety of vital vitamins, minerals, and antioxidants in addition to being high dietary fiber sources. These nutrients are essential for immune system function in general and for maintaining the integrity and health of the gut lining. Furthermore, the wide variety of phytonutrients.

Plant-based diets often consist of legumes, such as beans, lentils, and chickpeas, which are rich in protein and fiber. Because they provide a mix of soluble and insoluble fiber, they support gut health when part of a well-balanced diet. Additionally, legumes contain resistant starch, a prebiotic that promotes the growth of good bacteria in the colon. Additional sources of fiber and vital nutrients include nuts, seeds, and whole grains like brown rice, quinoa, and oats. By fostering regularity and supporting a diverse gut microbiome, they offer prolonged energy and improve gut health overall. Diets based on plants are naturally high in heart-healthy unsaturated fats and low in saturated fats. Given that diets high in saturated fats have been linked to an increased risk of certain gut-related conditions, this may have a positive effect on gut health. Furthermore, the digestive tract's inflammation can be lessened by unsaturated fats' anti-inflammatory qualities. While plant-based diets have many advantages for gut health, it's crucial to remember that careful planning is necessary to ensure adequate nutrient intake, especially for vitamins and minerals.

8.7.7. The Gut Microbiome and Digestive Disorders

An essential part of preserving digestive health is the gut microbiome, a complex community of trillions of microorganisms living in the digestive tract. This complex ecosystem of bacteria, fungi, viruses, and other microbes affects the immune system, gut barrier integrity, and nutrient absorption, among other aspects of digestion. A variety of digestive disorders have been associated with disruptions in the composition of the gut microbiome, and diet is a key factor in regulating this complex community of microorganisms. There is mounting evidence that the gut microbiome and the severity of symptoms in conditions such as IBD and IBS interact significantly. Studies suggest that changes in the diversity and abundance of specific bacterial species could be responsible for intestinal permeability, inflammation, and increased susceptibility to gastrointestinal symptoms. Dietary therapies that focus on the gut microbiota have therefore become more well-known as a possible treatment strategy for these illnesses.

Diets high in fiber are essential for modifying the gut microbiota. Fruits, vegetables, and oats contain soluble fiber, which acts as a fermentable substrate for good bacteria, encouraging their development and activity. Whole grains, nuts, and seeds are rich sources of insoluble fiber, which helps maintain regular bowel movements by giving stool more volume. A high-fiber diet helps to create a more balanced and diverse gut microbiome, which is linked to better digestive health by supporting helpful microbes. Live beneficial bacteria, or probiotics, are present in fermented foods like yogurt, kimchi, and kefir. Certain strains of probiotics have been shown to improve gut health. These probiotics can improve the function of the gut barrier, help restore the balance of microorganisms, and reduce symptoms associated with disorders like IBD and IBS. Furthermore, probiotic supplements containing *Bifidobacterium* and *Lactobacillus* strains have demonstrated promise in treating digestive disorder symptoms.

Beneficial gut bacteria eat prebiotics, which are non-digestible fibers present in foods like onions, garlic, and bananas. Prebiotics support the development and activity of these microorganisms by feeding them, which strengthens and balances the gut microbiome. Synbiotics, or the combination of prebiotics and probiotics, have the potential to produce a synergistic effect that improves the colonization and survival of beneficial bacteria. On the other hand, diets heavy in processed foods, added sugars, and saturated fats have been linked to a gut microbiome that is less resilient and less diverse. These eating habits may cause microbial population imbalances, which may exacerbate inflammation and cause discomfort in the digestive tract. Additionally, it has been demonstrated that some dietary elements, such as polyphenols, which are present in foods like berries, green tea, and dark chocolate, are beneficial to gut health. Because of their anti-inflammatory and antioxidant qualities, polyphenols can promote gut health and have a beneficial impact on the makeup of the gut microbiome.

8.7.8. Dietary Interventions for Managing Crohn's Disease

Nutritional therapies play a vital role in the management of Crohn's disease, a long-term inflammatory gastrointestinal disorder. Although there is not a single treatment plan that works for everyone, certain dietary approaches can help people with Crohn's disease feel better overall by lowering inflammation and relieving symptoms. Concentrating on easily digestible foods is one of the main factors to take into account when using diet to manage Crohn's disease. This entails choosing lean proteins, properly cooked fruits and vegetables, and easily digested carbs like potatoes and rice. Vegetables can be easier on the digestive system by steaming or cooking. Furthermore, limiting raw and high-fiber food consumption during flare-ups can help reduce gastrointestinal distress. It's important for people with Crohn's disease to be aware of foods that trigger symptoms. These are particular foods that have the potential to worsen symptoms or cause flare-ups. Foods high in insoluble fiber, dairy products (especially if lactose intolerant), spicy foods, and high fat foods are common culprits, though trigger foods can vary from person to person. Maintaining a food diary can assist with pinpointing specific trigger foods and provide dietary guidance.

During flare-ups, a low-residue diet is frequently advised in order to minimize the quantity of undigested food that travels through the digestive tract.

High-fiber foods such as whole grains, nuts, seeds, and raw fruits and vegetables are restricted in this diet. Rather, it places more emphasis on lean proteins, dairy substitutes, well-cooked fruits and vegetables, and refined grains. This may help alleviate symptoms such as diarrhea and stomach pain. It is imperative for people with Crohn's disease to include foods high in nutrients and anti-inflammatory compounds. It has been demonstrated that foods high in omega-3 fatty acids, such as walnuts, flaxseeds, and fatty fish like salmon and mackerel, have anti-inflammatory qualities. Consuming foods high in antioxidants, such as kale, spinach, and berries, can also aid in the reduction of inflammation and promote overall digestive health.

Diets such as the Low FODMAP Diet or the Specific Carbohydrate Diet (SCD) may also be beneficial for people with Crohn's disease. The Low FODMAP Diet focuses on particular types of carbohydrates that can be poorly absorbed in the small intestine, while the SCD concentrates on removing complex carbohydrates that are more difficult to digest. These diets ought to be followed under the supervision of a medical professional or licensed dietitian.

Enteral nutrition or nutritional supplements may be advised in specific circumstances, particularly for people who are at risk of malnutrition or have severe inflammation. This entails taking liquid formulas or specific diets that offer all the nutrients required in an easily digested format. It's critical that people with Crohn's disease collaborate closely with a medical professional or registered dietitian to create a customized eating plan that takes into account their unique requirements and preferences. For people with Crohn's disease, this expert advice can help guarantee that dietary interventions support symptom management, lower inflammation, and improve general well-being (Barone et al., 2022; Lynch & Pedersen, 2016; Roberfroid et al., 2010).

REFERENCES

1. Alagawany, M., Abd El-Hack, M. E., Farag, M. R., Sachan, S., Karthik, K., & Dhama, K. (2018). The use of probiotics as eco-friendly alternatives for antibiotics in poultry nutrition. *Environmental Science and Pollution Research International, 25*(11), 10611–10618.
2. Arauz-Pacheco, C., Parrott, M. A., & Raskin, P. (2004). The treatment of hypertension in adult patients with diabetes. *Diabetes Care, 25*(1), 134–147.
3. Barone, M., D'Amico, F., Brigidi, P., & Turroni, S. (2022). Gut microbiome-micronutrient interaction: The key to controlling the bioavailability of minerals and vitamins? *Biofactors, 48*(2), 307–314.
4. Cani, P., Delzenne, N. M., Amar, J., & Burcelin, R. (2008). Role of gut microflora in the development of obesity and insulin resistance following high-fat diet feeding. *Pathologie Biologie, 56*, 305–309.
5. David, L. A., Maurice, C. F., Carmody, R. N., Gootenberg, D. B., Button, J. E., Wolfe, B. E., & Turnbaugh, P. J. (2014). Diet rapidly and reproducibly alters the human gut microbiome. *Nature, 505*(7484), 559–563.
6. Gebrayel, P., Nicco, C., Al Khodor, S., Bilinski, J., Caselli, E., Comelli, E. M., Egert, M., Giaroni, C., Karpinski, T. M., Loniewski, I., Mulak, A., Reygner, J., Samczuk, P., Serino, M., Sikora, M., Terranegra, A., Ufnal, M., Villeger, R., Pichon, C., Konturek, P., & Edeas, M. (2022). Microbiota medicine: Towards clinical revolution. *Journal of Translational Medicine, 20*(1), 111.
7. Gibson, G., Hutkins, R., Sanders, M., et al. (2017). Expert consensus document: The International Scientific Association for Probiotics and Prebiotics (ISAPP) consensus statement on the definition and scope of prebiotics. *Nature Reviews Gastroenterology & Hepatology, 14*, 491–502.
8. Gress, T. W., Ress, T., Nieto, F. J., Shahar, E., Wofford, M. R., & Bramcati, F. L. (2000). The atherosclerosis risk in communities, hypertension, and antihypertensive therapy as risk factors for type 2 diabetes mellitus. *New England Journal of Medicine, 342*, 905–912.
9. Guarner, F., & Malagelada, J. R. (2003). Gut flora in health and disease. *The Lancet, 361*(9356), 512–519.
10. Hikey, M. (2004). Probiotic dairy products. In A. Tamime (Ed.), *Current Legislation of Probiotic Products* (pp. 73–97). Oxford: Blackwell Publishing.
11. Hill, C., Guarner, F., Reid, G., Gibson, G. R., Merenstein, D. J., Pot, B., & Sanders, M. E. (2014). The International Scientific Association for Probiotics and Prebiotics consensus statement on the scope and appropriate use of the term probiotic. *Nature Reviews Gastroenterology & Hepatology, 11*(8), 506–514.

12. Hill, M. J. (1997). Intestinal flora and endogenous vitamin synthesis. *European Journal of Cancer Prevention, 6*(Suppl. 1), S43–S45.
13. Huttenhower, C., Gevers, D., Knight, R., Abubucker, S., Badger, J. H., Chinwalla, A. T., & Structure, A. H. M. B. (2012). Structure, function and diversity of the healthy human microbiome. *Nature, 486*(7402), 207–214.
14. Ibnou-Zekri, N., Blum, S., Schiffrin, E. J., & von der Weid, W. T. (2003). Divergent patterns of colonization and immune response elicited from two intestinal *Lactobacillus* strains that display similar properties *in vitro*. *Infection and Immunity, 71*, 428–436.
15. Irving, R. J., Walker, B. R., Noon, J. P., Watt, J. C. M., Webb, D. J., & Shore, A. C. (2002). Microvascular correlates of blood pressure, plasma glucose, and insulin resistance in health. *Cardiovascular Research, 53*, 271–276.
16. Jäger, R., Zaragoza, J., Purpura, M., Iametti, S., Marengo, M., Tinsley, G. M., Anzalone, A. J., Oliver, J. M., Fiore, W., Biffi, A., Urbina, S., & Taylor, L. (2020). Probiotic administration increases amino acid absorption from plant protein: A placebo-controlled, randomized, double-blind, multicenter, crossover study. *Probiotics and Antimicrobial Proteins, 12*(4), 1330–1339.
17. Kho, Z. Y., & Lal, S. K. (2018). The human gut microbiome—A potential controller of wellness and disease. *Frontiers in Microbiology, 9*, 1835.
18. Lynch, S. V., & Pedersen, O. (2016). The human intestinal microbiome in health and disease. *New England Journal of Medicine, 375*(24), 2369–2379.
19. Mach, T. (2006). Clinical usefulness of probiotics in inflammatory bowel disease. *Journal of Physiology and Pharmacology: An Official Journal of the Polish Physiological Society, 57*(Suppl. 9), 23–33.
20. Manus, J., Millette, M., Uscanga, B. R. A., Salmieri, S., Maherani, B., & Lacroix, M. (2021). *In vitro* protein digestibility and physico-chemical properties of lactic acid bacteria fermented beverages enriched with plant proteins. *Journal of Food Science, 86*(9), 4172–4182.
21. Markowiak, P., & Śliżewska, K. (2017). Effects of probiotics, prebiotics, and synbiotics on human health. *Nutrients, 9*(9), 1021.
22. Mayer, E. A., Knight, R., Mazmanian, S. K., Cryan, J. F., & Tillisch, K. (2014). Gut microbes and the brain: Paradigm shift in neuroscience. *Journal of Neuroscience, 34*(46), 15490–15496.
23. Pereira, D. I. A., & Gibson, G. R. (2002). Cholesterol assimilation by lactic acid bacteria and bifidobacteria isolated from the human gut. *Applied and Environmental Microbiology, 68*(9), 4689–4693.

24. Reid, G. (2008). Probiotic *Lactobacilli* for urogenital health in women. *Journal of Clinical Gastroenterology, 42*(Suppl 3 Pt 2), S234–S236.
25. Roberfroid, M., Gibson, G. R., Hoyles, L., McCartney, A. L., Rastall, R., Rowland, I., & Meheust, A. (2010). Prebiotic effects: Metabolic and health benefits. *British Journal of Nutrition, 104*(S2), S1–S63.
26. Sonnenburg, J. L., & Sonnenburg, E. D. (2014). Starving our microbial self: The deleterious consequences of a diet deficient in microbiota-accessible carbohydrates. *Cell Metabolism, 20*(5), 779–786.
27. Sowers, J., Epstein, M., & Frohlich, E. D. (2001). Diabetes, hypertension, and cardiovascular disease: An update. *Hypertension, 37*, 1053–1059.
28. St-Onge, M. P., Farnworth, E. R., & Jones, P. J. H. (2000). Consumption of fermented and nonfermented dairy products: Effects on cholesterol concentrations and metabolism. *American Journal of Clinical Nutrition, 71*, 674–681.
29. Tsai, T. Y., Chen, L. Y., & Pan, T. M. (2012). Effect of probiotic fermented, genetically modified soy milk on hypercholesterolemia in hamsters. *Journal of Microbiology, Immunology and Infection*. [Epub ahead of print] PMID: 22749666.
30. Vanderhoof, J. A. (2008). Probiotics in allergy management. *Journal of Pediatric Gastroenterology and Nutrition, 47*(Suppl 2), S38–S40.
31. Walden, K. E., Hagele, A. M., Orr, L. S., Gross, K. N., Krieger, J. M., Jäger, R., & Kerksick, C. M. (2022). Probiotic BC30 improves amino acid absorption from plant protein concentrate in older women. *Probiotics and Antimicrobial Proteins*. https://doi.org/10.1007/s12602-022-10028-4.
32. Wang, J., & Ji, H. (2019). Influence of probiotics on dietary protein digestion and utilization in the gastrointestinal tract. *Current Protein & Peptid Science, 20*(2), 125–131.
33. Wu, H. J., & Wu, E. (2012). The role of gut microbiota in immune homeostasis and autoimmunity. *Gut Microbes, 3*(1), 4–14.
34. Yan, F., & Polk, D. B. (2006). Probiotics as functional food in the treatment of diarrhea. *Current Opinion in Clinical Nutrition and Metabolic Care, 9*(6), 717–721.
35. Yong, Z., Du, R., Wang, L., & Zhang, H. (2010). The antioxidative effects of probiotic *Lactobacillus casei* on the hyperlipidemic rats. *European Food Research and Technology, 231*, 151–158.
36. Zakrzewska, Z., Zawartka, A., Schab, M., Martyniak, A., Skoczeń, S., Tomasik, P. J., & Wędrychowicz, A. (2022). Prebiotics, probiotics, and postbiotics in the prevention and treatment of anemia. *Microorganisms, 10*(7), 1330.

CHAPTER 9

GUT MICROBIOME, AGING, AND LONGEVITY

Contents

INTRODUCTION

The process of growing older, which is influenced by both genetics and environment, is referred to as aging. It entails modifications to the biological, environmental, behavioral, and social processes' dynamics. The main molecular and cellular markers of aging are loss of proteostasis, telomere attrition, genomic instability, and mitochondrial dysfunction. These markers set off compensatory mechanisms like cellular senescence, deregulated nutrient sensing, and mitochondrial dysfunction, which in turn cause stem cell exhaustion and altered intercellular communication, which account for the functional decline linked to aging. The biological and genetic underpinnings of aging and age-related diseases can be better understood thanks to the quick development of next-generation sequencing technologies. There is mounting evidence that the gut microbiota is responsible for many age-related alterations and influences lifespan in all species.

Both the host and the microbiome are physiologically affected by aging, and host–microbiota interactions may have an overall effect on aging. The immune system's reaction is largely influenced by the microbiome, and dysregulation of it can lead to pro-inflammatory conditions. Aging causes the immune system to gradually deteriorate, which throws off the balance between pro- and anti-inflammatory activity. Age-related alterations in pro-inflammatory status lead to low-level systemic inflammation, or "inflammaging," which raises the risk of developing chronic illnesses and disabilities, such as mortality, cognitive decline, metabolic disease, cardiovascular disease (CVD), and frailty. Furthermore, through the "gut–brain axis," which is mediated by immunological, hormonal, and neural mediators, gut microbes can interact with the brain and influence behavior, including higher-order cognitive processes. When taken as a whole, the microbiome provides an intriguing viewpoint for comprehending the cognitive and physical aspects of aging.

It is unclear what "healthy aging" actually entails when health is generally declining. No universally accepted definition of healthy vs. unwell aging. Heterogeneity is a feature of aging. As people age, they diverge more from one another. Even within the same body, different tissues age at distinct rates. There are various angles from which to study aging: "normal" (i.e., aging that is normal or average, pathological aging (i.e., linked to particular illnesses or additional signs of accelerated aging), and effective aging. The current review excluded pathological aging brought on by disease and concentrated on successful and normal aging. Centenarians and other exceptionally long-living people are prime examples of extremely successful aging. They have escaped or survived the majority of the illnesses that cause morbidity and death in the majority of other elderly people, but they might still exhibit some telltale symptoms of aging. These very old people can provide insightful information about the best possible aging process. Long-lived individuals (centenarians and nonagenarians) still have inflammation, but it is less than in "normal" older adults, and concomitant conditions balance their pro-inflammatory status.

9.1. OVERVIEW OF GUT HEALTH AND LONGEVITY

As new studies continue to reveal the significant impact of our gut microbiome—the community of trillions of microorganisms that live in our digestive systems—on general health and the aging process, there is an increasingly clear connection between gut health and longevity. It is becoming more and more important to comprehend the link between gut health and longevity as we try to discover the keys to living longer, healthier lives.

9.1.1. Gut Health and Longevity: A Dynamic Connection

- **Inflammation and Aging**: An important factor in age-related disorders, chronic inflammation is a defining feature of aging. An equilibrium gut microbiome contributes to the maintenance of a state of controlled inflammation. The gut is crucial for immune regulation. On the other hand, an unbalanced microbiome can cause long-term, low-grade inflammation, which heightens the risk of age-related illnesses like diabetes, neurodegenerative diseases, and CVD and speeds up aging.
- **Nutrient Absorption**: Our digestive systems can lose some of their capacity to effectively absorb vital nutrients as we age. Maintaining overall health and longevity requires the ability to extract essential nutrients from our food, which can only be achieved through a healthy gut with a variety of beneficial bacteria.
- **Gut-Brain Axis**: Through the gut-brain axis, the gut and brain interact to affect mood, thought processes, and even behavior. Better mental health, a decreased risk of age-related cognitive decline, and a lower chance of mood disorders have all been linked to a gut microbiome that is in balance. On the other hand, disorders such as anxiety and depression may be exacerbated by a weakened immune system.
- **Immune Function**: A large fraction of the body's immune cells resides in the gut. The body's defenses against infections and illnesses are aided by a strong immune system, which is supported by an ideal balance of gut bacteria. Maintaining immune function as we age is essential for long life, and the gut microbiota is essential to this process.

9.1.2. Strategies for Promoting Gut Health and Longevity

- **Dietary Fiber**: Nutrient-dense foods such as fruits, vegetables, whole grains, legumes, and nuts provide fiber to the gut microbiota. Both regular bowel movements and the maintenance of a resilient and diverse gut microbiome are facilitated by fiber.
- **Probiotics and Prebiotics**: A balanced gut microbiome can be restored and maintained by including probiotic-rich foods like yogurt, kefir, sauerkraut, and kimchi, as well as prebiotic-rich foods like garlic, onions, asparagus, and bananas.
- **Omega-3 Fatty Acids**: Omega-3 fatty acids found in fatty fish, walnuts, and flaxseeds have anti-inflammatory qualities that can help reduce inflammation caused by aging in the gut and other parts of the body.
- **Hydration**: Maintaining adequate hydration is crucial for digestive health. Water facilitates the passage of food through the digestive system and aids in the preservation of the intestinal mucous lining.
- **Limit Sugar and Processed Foods**: Consuming too much sugar can encourage the growth of bad gut flora, which can cause inflammation and age-related illnesses. It is best to stay away from processed foods for long term because they can upset the gut microbiota and are frequently loaded with unhealthy fats and additives.
- **Bone Broth**: Bone broth is rich in collagen and amino acids, which can help support and repair the gut lining, addressing issues like leaky gut syndrome.
- **Stress Management**: Prolonged stress can hasten aging and have a deleterious impact on gut health. Deep breathing exercises, yoga, and meditation are examples of stress-reduction techniques that can support gut health and longevity.
- **Regular Exercise**: Physical activity promotes regular bowel movements, reduces inflammation, and supports overall digestive health. It is also associated with a longer, healthier life.
- **Regular Health Screenings**: The elderly population should have routine screenings for health issues and should speak with medical professionals for tailored advice on age-related health issues and gut health maintenance.
- **Intermittent Fasting**: Intermittent fasting, a practice that alternates between eating and fasting, may improve gut health and lengthen life by encouraging the development of good bacteria in the gut and lowering inflammation, according to some research.

Research and interest in the complex relationship between gut health and longevity are expanding. Through thoughtful dietary choices, lifestyle modifications, and stress management, people can cultivate a diverse and balanced gut microbiome that will promote overall health and increase the likelihood of living a longer, healthier life. In order to achieve longevity, it is becoming more and more important to prioritize gut health as we learn more about the profound effects of the gut on the aging process.

9.2. AGE-RELATED INFLUENCES ON THE MICROBIOME

Loss of homeostasis, diminished function, and increased susceptibility to death are all signs of aging. Diseases associated with aging include metabolic, neoplastic, infectious, and degenerative conditions that cause frailty and cognitive decline. The basic markers of aging in mammals have been identified, both molecular and cellular, but these are correlated with alterations in the microbiome that influence the rate of age-related decline (Figure 9.1).

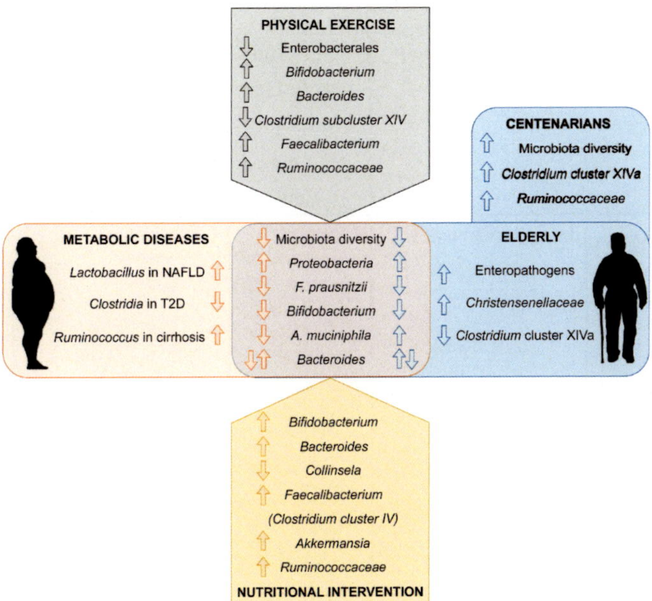

Figure 9.1. *Microorganism–host signaling as a contributor to healthy or unhealthy aging.*

Source: Juárez-Fernández M, Porras D, García-Mediavilla MV, Román-Sagüillo S, González-Gallego J, Nistal E, Sánchez-Campos S. (2021). Aging, gut microbiota and metabolic diseases: management through physical exercise and nutritional interventions. Nutrients, 13(1):16. https://doi.org/10.3390/nu13010016.

The microbiome's age-related alterations are highly variable and impacted by both internal and external environmental factors. For instance, the steady decline in the alimentary tract's physiology inevitably has an impact on the gut microbiome. Increased inflammation linked to aging, genomic instability, dysfunction of cells and mitochondria, decreased proteostasis, and epigenetic dysregulation are some of these alterations. These further contribute to the development of chronic illnesses, metabolic problems, and compromised gut-brain communication. The consequences on the host's behavior and way of life (such as increased frailty, medication use, surgery, decreased physical activity, and poor diet) may worsen the effects on the gut microbiota. The composition and function of the microbiome in older individuals are also influenced by their lifetime personal lifestyles, especially their diet, but they also present opportunities for positive behavioral changes. The impact of social interactions on the microbiota's composition and how society treats its elderly members has received less attention (Table 9.1).

Table 9.1. Gut Microbiota Composition of the Elderly (≥60 Years Old) and Centenarians (≥99 Years Old)

Subjects	**Methodological Approach**	**Main Findings in Gut Microbiota Composition**		**Reference**
84 subjects from Northern Italy (50F, 34M). Young adults (20–40 years old) **(Y)**, elderly (60–80 years old) **(E)**, centenarians group (99–104 years old) **(C)** and offspring of the centenarians (59–78 years old) **(F)**	HITChip analysis and qPCR (16S rRNA)	**Elderly**	**Centenarians**	Biagi et al. (2010)

		• ↑ *Akkermansia muciniphila*	• ↓ α-diversity index • ↑ facultative anaerobes from *Proteobacteria* phylum (*E. coli, Haemophilus, K. pneumoniae* or *Pseudomonas*) and *Bacilli* class (*Bacillus* or *Staphylococcus*) • ↓ *Clostridium* cluster XIVa • ↓ bifidobacteria • Rearrangement of *Clostridium* cluster IV (↓ *Faecalibacterium prausnitzii* and ↑ *Clostridium leptum*)	
161 **elderly** Irish subjects (82F, 79M) (>65 years old) and a **control** group (5F, 4M) (28–46 years old)	Pyrosequencing with 454 system (16S rRNA V4 region)	• ↓ *Firmicutes proportion* • ↑ *Clostridium* cluster IV (specially, *Faecalibacterium* spp.) • ↓ *Clostridium* cluster XIVa		Claesson et al. (2011)

24 volunteers from China (14F, 10M) classified in **Group RC** (100–108 years old), **Group RE** (85–99 years old) and **Group CE** (80–92 years old)	Illumina MiSeq and qPCR (16S rRNA V4 region)	• ↑ α-diversity in centenarians • ↑ *Escherichia, Roseburia* in centenarians • ↓ *Lactobacillus, Butyricimonas, Coprococcus, Parabacteroides, Akkermansia, Sutterella, Faecalibacterium* in centenarians		Wang et al. (2015)
24 semi-supercentenarians (>105 years old, **group S**) (18F, 6M) vs. 15 young adults (22–48 years old, **group Y**) (8F, 7M) from Northern Italy. Results of **C and E groups** from Biagi et al. 2010 study were incorporated.	Illumina MiSeq and qPCR (16S rRNA V3-V4 region)	**Elderly**	**Centenarians**	Biagi et al. (2016)
		• ↓ *Bifidobacterium*	• ↓ *Bifidobacterium* in C • ↑ *Bifidobacterium* in S	

		• ↓ *Bacteroidaceae, Lachnospiraceae, Ruminococcacea* with age • ↑ *Eggerthella, Bilophila, Akkermansia, Anaerotruncus, Christensenellaceae* and *Synergistaceae* with age		
168 Chinese individuals (85F, 83M) grouped into **long-living group** (≥90 years old), **elderly group** (65–83 years old) and a **younger age group** (24–64 years old)	Illumina MiSeq (16S rRNA V3-V4 region)	• ↑ *Clostridium* cluster XIVa, *Ruminococcaceae, Akkermansia* and *Christensenellaceae* in long-living group • ↑ microbial diversity in long-living group		Kong et al. (2016)

367 Japanese volunteers between 0 and 104 years old (210F, 157M)	Illumina MiSeq and qPCR (16S rRNA V3-V4 region)	**Elderly**	**Centenarians**	Odamaki et al. (2016)
		-	· ↑ microbial diversity in centenarians	
		• ↑ *Bacteroidetes* (*Bacteroides* and *Clostridiaceae*) and *Proteobacteria* (*Betaproteobacteria* and *Deltaproteobacteria*) with age • ↓ *Actinobacteria* with age • ↑ *Porphyromonas, Treponema, Fusobacterium* and *Pseudoramibacter* with age		

A total of 1095 healthy Chinese volunteers (533F, 562M) classified into eight groups according to their age (**children, adults, elderly, centenarians**)	Illumina MiSeq (16S rRNA V4 region)	**Elderly**	**Centenarians**	Bian et al. 2017)
		• ↓ *Blautia* after 60 years old • ↑ *Prevotella* and *Bacteroides* in 60–79 years old group	• ↓ *Bacteroides* and *Bifidobacterium* genera in the oldest groups vs. the youngest groups • ↓ *Prevotella* and *Bacteroides* in centenarians	
		• ↑ *Dorea, Clostridium insertae sedis, Clostridium sensu strictu 1, Marvinbryantia* and members of *Prevotella* in older subjects vs. young groups		

80 Indonesian subjects (50F, 30M): **young group** (25–45 years old) and **elderly group** (≥70 years old)	Yakult intestinal flora-scan (YIF-SCAN) (qPCR method)	• ↓ microbiota concentration • ↑ *Lactobacillus reuteri* and *Enterobacteriaceae* • ↓ *Clostridium cocoides, Bacteroides fragilis, Clostridium leptum, Bifidobacterium, Prevotella* and *Lactobacillus plantarum*		Rahayu et al. (2019)
187 elderly subjects from three groups of age (**65–70 years old**), (**90–99 years old**) and (**100+ years old**) from East China (120F, 67M)	Illumina MiSeq (16S rRNA V3, V4 and V5 regions)	**Elderly**	**Centenarians**	Wang et al. (2019)

		• ↑ *Clostridium, Parabacteroides* and *Streptococcus* in 90–99 years old group vs. 65–70 years old group • ↓ *Megamonas, Blautia* and *Coprococcus* in 90–99 years old group vs. 65–70 years old group • ↑ *Bacteroides fragilis, Parabacteroides merdae, Ruminococcus gnavus, Coprococcus* and *Clostridium perfringens* in 90–99 years old group	• ↑ community richness (Ace and Chao1 index) in centenarians (90–99 years old and 100+ years old groups) • ↑ *Ruminococcaceae, Alistipes and Barnesiella* in 100+ years old group vs. 65–70 years old group • ↓ *Lachnospira and Prevotella* in 100+ years old group vs. 65–70 years old group • ↑ *Synergistetes, Verrucomicrobia* and *Proteobacteria* in longevity group vs. younger elderly group	

56 South Korea subjects classified in **centenarians** (95–108 years old) (27F, 3M), **elderly** (67–79 years old) (7F, 10M) and **adults** (26–43 years old) (3F, 6M)	Pyrosequencing with 454 system (16S rRNA V1–V3 regions)	**Elderly**	**Centenarians**	Kim et al. (2019)
		• ↑ *Proteobacteria* in elderly vs. adults • ↓ *Bacteroidetes* in elderly vs. adults	• ↑ *Verrucomicrobia* in centenarians vs. elderly • ↑ *Proteobacteria, Actinobacteria* and *Verrucomicrobia* in centenarians vs. adults • ↑ *Akkermansia, Clostridium, Collinsella, Escherichia, Streptococcus* and *Christensenellaceae* in centenarians vs. elderly and adults • ↓ *Faecalibacterium* and *Prevotella* in centenarian's vs elderly and adults	

F: female; M: male. Changes (↑: increase; ↓: decrease) in the relative abundance of selected microbial taxa and in bacterial diversity with age. Names in bold denote each group for the corresponding study and are defined in the table.

Source: Juárez-Fernández M, Porras D, García-Mediavilla MV, Román-Sagüillo S, González-Gallego J, Nistal E, Sánchez-Campos S. (2021). Aging, gut microbiota and metabolic diseases: management through physical exercise and nutritional interventions. Nutrients; 13(1):16. https://doi.org/10.3390/nu13010016.

There is mounting evidence that the spread of microorganisms within a social group has significant positive effects on health. While strain tracking studies have focused primarily on the examination of pathogen dissemination within social groups, they have also revealed the sharing of commensals and mutualistic microorganisms within social networks. When compared to microbiomes from people in different households, those who live in the same house typically have similar compositions. Pets kept in the home can spread germs by serving as transmission vectors. Furthermore, people with more extensive social networks appear to have gut microbiomes that are more diverse. The "social microbiome," or collective microbial metacommunity of a family, is subject to change over time, with older individuals often experiencing a reduction in their ability to acquire microorganisms. Further variation is contingent upon the individuals' living arrangements—alone or in institutional care. In contrast to the past and traditional, non-industrialized societies, modern socioeconomically developed societies are characterized by a number of social changes, including a loss of group living and a tendency toward loneliness.

9.2.1. Gut Microbiome and Aging-Related Metabolic Diseases

Global aging populations have also resulted in a sharp rise in the prevalence of metabolic disorders. A number of metabolic diseases that are common in older adults, including obesity, insulin resistance, fatty liver and hepatic steatosis, atherosclerosis, CVDs, and decreased motor activity, can potentially or partially be attributed to dysbiosis in the gut microbiome and microbial metabolites, as well as abnormalities of gut barrier integrity and enhanced pro-inflammatory cytokines. Changes in the gut microbiome with age can also affect the gut-brain axis, impairing neural, endocrine, nutritional, and immunological signals via the enteric nervous system (ENS) between the gut and the brain. This may also contribute to diseases of the central nervous system (CNS), including multiple sclerosis, autism, depression, and anxiety. These aging-related changes in the gut microbiome may not always be brought on by aging; instead, they may be related to the general decline in health that is accompanied by malnutrition and an increased propensity to require medications such as antibiotics and non-steroidal anti-inflammatory drugs.

According to several studies, the decline in microbiome diversity is not so much associated with chronological age as it is with frailty related to aging. It has also been demonstrated that age-related disruptions in the gut microbiome play a significant role in the development of age-related pathological conditions like diabetes, NAFLD, chronic inflammation, and CVD. Current research is focusing on safe and efficient ways to improve/restore gut eubiosis for better health, such as dietary therapies involving probiotics, fiber ingredients, nutraceuticals, and others, while also extensively studying the mechanisms by which the gut microbiome may positively influence host aging. One of the main pathways to stabilize immunosenescence is the direct modulation of the gut microbiome, which may reduce inflammatory responses and enhance adaptive immune responses. Additional mechanisms that may be contributing to these effects include increased antioxidant activity, control over host fat deposition and metabolism, suppression of insulin resistance, enhancement of immune system and mucosal barrier integrity, increased SCFA production, alleviation of gut peptides related to lipid metabolism, and glucose homeostasis, and up-regulation of specific genes involved in the metabolism of xenobiotics.

9.2.2. Microbiome-Based Ageing Interventions

Using the complex connection between our microbiome and aging processes, microbiome-based aging interventions are becoming a ground-breaking strategy for addressing age-related health issues. Trillions of bacteria, viruses, and fungi make up the human microbiome, which is essential for both promoting and sustaining health. People's microbiomes change significantly with age, frequently leading to changes in microbial composition and a decrease in microbial diversity. These alterations may lead to altered metabolic processes, reduced immunological response, and systemic inflammation—all of which are associated with aging and age-related illnesses like diabetes, neurodegenerative diseases, and CVD. Researchers are looking into several microbiome-based interventions that aim to improve or restore microbiome health to counteract these effects. One well-known tactic is dietary interventions, which emphasize eating foods high in fiber, prebiotics, and polyphenols to promote the development of good gut bacteria and preserve a balanced microbiome. Another important area of interest is probiotics, or live beneficial bacteria. They can be used to replenish good bacteria and possibly mitigate some of the negative effects of an aging microbiome.

The potential of prebiotics, indigestible fibers that encourage the growth of beneficial bacteria, to maintain a healthy gut environment is also being studied. Fecal microbiota transplantation (FMT), another novel strategy, is the transfer of gut microbiota from a healthy donor to a recipient who has an unbalanced microbiome. While FMT is currently mainly used to treat certain types of intestinal infections, more research is being done to determine whether it can have a wider impact on aging and health. In microbiome research, personalized medicine—where treatments are customized to a person's specific microbiome profile—is gaining prominence. This strategy has the potential to maximize treatment effectiveness and enhance patient outcomes. These microbiome-based therapies have the potential to improve general well-being and quality of life while also lessening the effects of aging, provided that scientific knowledge on the subject is furthered. Further investigation into the relationship between the microbiome and aging processes may yield new and useful approaches to managing age-related illnesses and encouraging healthier aging.

9.2.3. Does the Alteration of the Gut Microbiota Affect the Health of Elderly People?

According to Metchnikoff's original theory, which was published more than a century ago, the accumulation of "putrefactive bacterial autotoxins" that leak from an elderly person's colon is what causes their disabilities. He hypothesized that this leakage causes phagocytes to become tissue-destructing agents. Centuries later, research revealed the significance of persistent inflammation on the activation and malfunction of phagocytic cells, supporting this theory. People residing in long-term care nursing facilities and community-dwelling apartments have very different fecal microbiomes, which is consistent with their roles in healthy aging. In adults who live in communities, dysbiosis, increased microbiota instability, and member loss can reduce the many advantages commensal microbes have for the host. This likely explains why there is a positive correlation between the biological, rather than chronological, age and the gut dysbiosis of community-dwelling adults, with the decrease in gut microbial diversity increasing as elderly people become more frail. The fecal microbiota of aged mice and frail elderly people is less abundant in *Lactobacillus* and *Faecalibacterium* and more abundant in the *Oscillibacter*, *Alistipes*, and *Bacteroidetes* phylum and the *Eubacteriaceae* family. In a similar vein, our team discovered that gut dysbiosis in old macaques and mice leads to increased insulin resistance due to a marked decrease in beneficial commensals and their metabolite SCFAs in the gut and circulation, as well as an increase in pro-inflammatory microbes.

The decrease in *Akkermansia* was a major factor in this process in aged mice. It led to intestinal leakage and activated CCR2+ monocytes, which transformed innate B1a B cells into pathogenic 4–1BBL+TNF+ B cells (referred to as 4BL cells). These 4BL cells then induced insulin resistance through the 4-1BBL/4-1BB axis. On the other hand, in spite of variations in lifestyle and ethnicity, the alpha diversity of the top 500 OUTs is higher in Chinese and Italian centenarians (though the Italian centenarians' Simpson reciprocal index of diversity was lower). This suggests that the preservation of the richness of commensal microbes in the gut may be beneficial for healthy aging. Overall, there may be negative effects on healthy aging and consequently lifespan due to the decreased diversity of the intestinal commensal microbes, which frequently shows up as a reduction of beneficial and enrichment of pro-inflammatory members.

9.2.4. The Importance of Microbial Metabolites Such as Short Chain Fatty Acids in Immunity

The gut microbiota's composition varies with age, but the microbiota's core remains constant. Fecal DNA sequencing and PCR analyses primarily show changes in the abundance of its constituent parts, which have an impact on the microbial community and their interactions with the host as part of a symbiotic ecosystem. As the microbial community becomes more diverse after weaning and the transition to adulthood, the abundance of commensals dominated by *Bifidobacterium*, which produces lactate and acetate, a beneficial SCFA, from milk oligosaccharides, gradually declines in the gut microbiota of breastfed infants. Other bacteria, such as their primary producer Firmicutes, that metabolize fibers to SCFAs complement their decline in young adults. However, both *Bifidobacterium* and some members of Firmicutes, such as *Clostridium* clusters IV (*Ruminococcus obeum*, *Roseburia intestinalis, E. ventriosum* et rel., *E. rectale* et rel., *E. hallii* et rel.) and of *Clostridium* cluster XIVa, (*Papillibacter cinnamovorans* et rel., and *F. prausnitzii* et rel.), are decreased in aging and in centenarians. This explains why *Bifidobacterium* abundance in centenarians and the elderly inversely correlates with inflammaging, consistent with their benefit in extending mouse longevity as inducers of mucus production and colonic tight junctions, as well as pro-inflammatory cytokines (from macrophages, for example) and colonic senescence.

Commensal microbes, immune cells, and the colonic epithelium are given energy by SCFAs, which also cause the production of mucus, for example they support the growth of different advantageous commensal microbes, safeguard the functions of the epithelial barrier, and enhance immunological tolerance and gut homeostasis. Through direct stimulation of immune cells, SCFAs also control immune responses. Increases in SCFAs like acetate, propionate, and butyrate in the sputum of patients with cystic fibrosis are positively correlated with neutrophil infiltration and negatively correlated with Pseudomonas aeruginosa expansion.

While preventing the synthesis of nitric oxide synthase, which is involved in airway inflammation, the SCFAs cause the production of IL-8/CXCL8 and the release of granulocyte-macrophage colony-stimulating factor (GM-CSF) and granulocyte colony-stimulating factor (G-CSF), which are necessary for the recruitment and persistence of neutrophils. However, increasing SCFAs—which comes from eating a high-fiber diet—inhibits neutrophil recruitment, which in turn reduces airway inflammation and improves the survival of influenza-infected mice. According to this study, SCFAs also hinder neutrophil infiltration into the lung airways by attracting bone marrow-derived Ly6c− patrolling monocytes, which when they differentiate into alternatively activated macrophages decrease the expression of the chemokine CXCL1, which recruits neutrophils, and activate CD8+ T cell effector functions.

Extracellular SCFAs are also taken up by multiple G-protein coupled receptors (GPRs) and used as substrates for β-oxidation of fatty acids in CD8+ T cell and colonocyte metabolism. Through GPR41 and GPR43, butyrate improves the memory potential and recall responses of antigen-primed CD8+ T cells by decoupling the tricarboxylic acid cycle from glycolytic input. Butyrate triggers anti-inflammatory reactions by binding to GPR109A, GPR41, and GPR43, which are expressed on the surface of neutrophils, dendritic cells, and macrophages. In a similar vein, it stimulates the differentiation of Tregs and T cells that produce IL-10 by interacting with GPR109A and promoting anti-inflammatory pathways in colonic macrophages and dendritic cells. Through GPR43, butyrate regulates colitis by upregulating the expression of the transcription factor Blimp-1 and the production of IL-10 in Th1 CD4+ T cells (without influencing the conversion of Tregs or Th17 cells). As they are necessary for the restoration of Tregs in GF mice following butyrate supplementation, other SCFA non-producer microbes also support this process. SCFAs from *Clostridium* species clusters IV and XIV induce and regulate colonic FoxP3+ Tregs.

Strong inhibitor of histone deacetylases (HDACs), butyrate induces epigenetic modifications in both hematopoietic and non-hematopoietic cells through internalization via the previously mentioned GPRs or other unidentified pathways. This promotes anti-inflammatory immune responses by inhibiting the production of several pro-inflammatory cytokines (IL-6, IL1-β, and IFN-γ) and inducing the release of anti-inflammatory IL-10 and TGF-β. Because it inhibits HDAC3, butyrate causes monocytes to differentiate into macrophages and alters the transcriptional and metabolic programs of macrophages, which improves their capacity to eradicate bacteria. Nuclear factor-κB (NF-κB) is rendered inactive by SCFAs, and TNFα production is decreased in peripheral blood mononuclear cells and neutrophils.

In order to fight infections, SCFAs boost intestinal IgA and systemic IgG responses by promoting the differentiation of B cells that produce antibodies. When SCFAs enter the bloodstream and travel to far-off organ locations like the brain and lungs, they regulate the systemic immune response. SCFAs guard against allergic airway disorders by controlling Tregs in the lungs. SCFAs have a direct impact on the maturation and function of microglia in the brain, which lowers neuroinflammation. But in some circumstances, SCFAs can also have pro-inflammatory effects. For example, SCFAs can exacerbate inflammation linked to colitis in mice by promoting T-bet and IFNγ expression in Tregs and in regular CD4+ T cells. Furthermore, in multiple sclerosis-affected mice, SCFAs can worsen inflammation in airway epithelial cells and the production of IL-17 and IFNγ-producing T cells in the CNS. SCFAs are associated with both motor deficits and an acceleration of pathogenic α-synuclein (αSyn) aggregation in the mouse model of Parkinson's disease (PD).

9.2.5. Alteration of the SCFA Levels in Elderly People

A community of commensals and several bacterial populations collaborate to produce SCFAs. For instance, *A. muciniphila* stimulates the production of butyrate from *Anaerostipes caccae*, *Eubacterium hallii*, and *Faecalibacterium prausnitzii*, as well as the growth of other commensal microbes by liberating oligosaccharides and maintaining the mucus layer. Meanwhile, *E. hallii* upregulates pseudovitamin B12 to stimulate the production of propionate from *A. muciniphila*. Through the activation of hypoxia-inducible factor 1α, butyrate enhances the barrier function of intestinal epithelial cells (IECs) and shields them from the damaging effects of *C. difficile* toxin. This likely explains why centenarians' guts continue to contain high concentrations of butyrate as well as higher concentrations of the butyrate-producing *Anaerotruncus colihominis* et rel. (from *Eubacterium limosum et rel.*) and *Clostridium* cluster IV (from *Clostridium* cluster XV), *Bifidobacterium*, and additional bacteria linked to health, like *Christensenellaceae* and *Akkermansia*.

In addition, *A. muciniphila* and its outer membrane protein Amuc_1100* stimulate IECs to produce mucus, which is required to maintain the integrity of the intestinal barrier and to support other advantageous commensals. Consequently, the reduction in this significant bacterium leads to dysbiosis in the gut and damages the intestinal epithelium, which exacerbates gut leakiness and systemic endotoxemia.

In turn, this leads to a chronic proinflammatory state that increases circulating levels of IL-6, IFNs, TNFα, and IL-1, which in turn makes older people more frail and less fit. According to a recent study by our group, the decline of *A. muciniphila* in the gut of elderly mice sets off a series of inflammatory processes that lead to an increase in insulin resistance. It begins with the loss of the colonic mucin layer, which is thought to be the result of commensal bacteria like *I. butyriciproducens*, *F. prausnitzii*, *R. faecis*, and *A. butyraticus* that produce butyrate being lost. This, in turn, causes a decrease in the amount of SCFAs, especially butyrate, in the gut lumen and circulation.

In aged mice, dysbiosis and gut leakage are further enhanced by the reduction of this vital energy source for colonocytes and suppressor of proinflammatory and pathogenic commensal bacteria, which perpetuates inflammaging. Since butyrate is also a strong TLR4 signaling inhibitor, when it is less in the bloodstream, bacterial stimuli, like endotoxin, can more easily cause inflammation in the omentum and cause BM-derived inflammatory CCR2+

monocytes to express more IFN-γ, 4-1BB, and CD40L. These monocytes use 4-1BBL, CD40, and IFNγR1 signaling to transform innate B1a B cells into 4BL cells after infiltrating the omentum. The 4BL cells then, at least partially, use the 4-1BBL/4-1BB axis to promote insulin resistance in elderly hosts. Elderly humans and macaques appear to follow a similar pathway, as their elevated insulin resistance is linked to 4BL cells produced by monocytes activated by aging.

Elderly individuals exhibit a shift from saccharolytic fermentation to unfavorable proteolytic activities, as evidenced by an increase in phenols, ammonia, and branched fatty acids—metabolites from protein fermentation—while decreasing levels of SCFAs from carbohydrate fermentation. This transition happens gradually in older adults and can be sped up by low-fiber diets or the use of antibiotics. Reduced SCFA levels increase vulnerability to pathogens and opportunistic bacterial species. For example, mice treated with the antibiotic cefoperazone have lower SCFA levels, which increases the amount of *Candida albicans* colonizing the mice. Similar to this, antibiotic-induced butyrate-producing microbial depletion lowers butyrate levels, which causes epithelial oxygenation and the growth of aerobic bacteria like *Salmonella enterica* and *Escherichia coli*. Mechanistically, reduced activity of the intracellular butyrate sensor peroxisome proliferator-activated receptor γ (PPAR-γ) is the main mechanism responsible for colonocyte oxygenation. This, in turn, causes iNOS activation and an increase in nitrate levels. The mechanisms through which age-related diseases are exacerbated by gut dysbiosis.

9.3. GUT MICROBIOME AND METABOLISM

Molecular biologist Joshua Lederberg defines the gut microbiome as the entirety of bacteria, viruses, protozoa, fungus, and other microorganisms as well as their combined genetic material that are present in the gastrointestinal tract (GIT). The bacteria living in the GIT, both commensal and pathogenic, are collectively referred to as the gut microbiota. The past 10 years have seen an investigation into the gut microbiota for possible interactions between gut microbes and hosts, including effects on immunological, metabolic, and neuroendocrine responses. Short-chain fatty acid (SCFA) synthesis, enzyme synthesis, vitamin and amino acid synthesis, and mineral absorption are all significantly influenced by the gut microbiota. The fermentation byproducts butyrate, propionate, and acetate are crucial for gut health because they protect against infections, give epithelial cells energy, and improve the integrity of the epithelial barrier. The function of resident bacterial genes and their possible relationship to human health and metabolism are the subject of ongoing research. Furthermore, research into the possibility that nonpathogenic bacterial strains can promote the immune system's ability to recover from pathogenic diseases is ongoing. The gut microbiota of humans is categorized into numerous groups known as phyla. The four primary phyla that constitute the gut microbiota are Firmicutes, Bacteroidetes, Actinobacteria, and Proteobacteria. Although bacteria are found throughout the human body—including the oral cavity, placenta, vagina, skin, and gastrointestinal tract (GIT)—the majority reside in the GIT, with the colon housing the largest population, primarily anaerobic bacteria (Figure 9.2). To understand the extent of bacterial presence and its potential effects on the host, it is worth noting that the human body expresses approximately 20,000 eukaryotic genes, whereas the gut microbiome expresses 3–3 million prokaryotic genes.

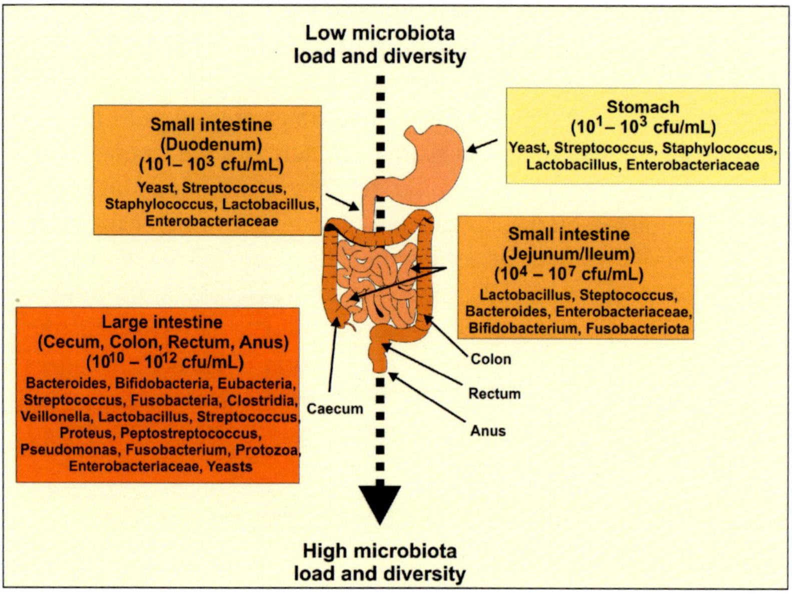

Figure 9.2. *Gut microbiota predominance.*

Source: Chandrasekaran P, Weiskirchen S, Weiskirchen R. (2024). Effects of probiotics on gut microbiota: An overview. International Journal of Molecular Sciences; 25(11):6022. https://doi.org/10.3390/ijms25116022.

Numerous factors influence the development and modification of the gut microbiome, such as the way a baby is born and fed, the environment, stress exposure, food, medications, life stage, and co-occurring diseases (Figure 9.2). The term dysbiosis refers to a change in the microbial community that lowers the diversity and quantity of commensal bacteria. Research points to connections between gut dysbiosis and long-term medical disorders like cancer, metabolic syndrome, inflammatory bowel disease (IBD), and heart disease.

9.3.1. Gut Microbiota

The term "gut microbiota" refers to the collection of bacteria, viruses, fungi, and archaea that inhabit an animal's digestive system. The total of all the genomes of the gut microbiota is known as the gastrointestinal metagenome. The majority of the human microbiome is found in the gut. The gut microbiota affects many processes, such as colonization, pathogen resistance, intestinal epithelium maintenance, metabolism of food and medication, immune function regulation, and behavior through the gut-brain axis.

The gut microbiota's microbial makeup differs depending on the digestive tract's location. With between 300 and 1000 distinct species represented, the colon has the highest microbial density of any human-associated microbial community that has been studied to date. Around 99% of gut bacteria originate from 30 to 40 different species, making bacteria the largest and most studied component to date. Bacteria can make up as much as 60% of the dry mass of excrement. Anaerobic bacteria make up more than 99% of the bacteria in the gut, but high densities of aerobic bacteria can be found in the cecum.

When compared to other parts of the body, the gut microbiota contains the highest concentration and variety of bacteria. One to two years after birth is when the gut flora in humans is established. By then, the intestinal mucosal barrier and the intestinal epithelium have co-developed in a way that is both protective of the gut flora and tolerant of it, acting as a barrier against pathogenic organisms.

Certain gut microbiota and humans have mutualistic relationships rather than just a commensal one (a non-harmful coexistence). Certain microorganisms found in the human gut assist the host by breaking down dietary fiber into SCFAs, like butyric and acetic acids, which the host can then absorb. Moreover, intestinal bacteria are involved in the synthesis of vitamins B and K as well as the metabolism of xenobiotics, cholesterols, and bile acids.

9.3.2. Functions of the Gut Microbiota

The immune system's direct defense against pathogens, the intestinal epithelium's development and maintenance, the induction of antibody production there, and the metabolism of otherwise indigestible food components are all examples of its functions. Later research also revealed its involvement in immune system development, and still more research concentrated on the gut-brain axis.

9.3.2.1. Direct Inhibition of Pathogens

The gut flora community actively fights pathogens by completely colonizing the area, utilizing all nutrients, and secreting chemicals called cytokines that either kill or inhibit unwanted organisms that would compete with it for nutrients. Distinct cytokine production is induced by distinct strains of gut bacteria. Chemical substances known as cytokines are created by our immune system to start the inflammatory response against infections. Competing organisms like *Clostridium difficile*, which are otherwise kept in check, can become established when the gut flora is disturbed.

9.3.2.2. Development of Enteric Protection and Immune System

Within one to two years of birth, humans develop a gut flora that is comparable to that of an adult. The intestinal lining, including the intestinal epithelium and the intestinal mucosal barrier it secretes, develops in tandem with the establishment of the gut flora. This results in the intestinal lining being somewhat tolerant of communalistic microorganisms and even supportive of them, while also acting as a barrier against pathogenic ones. In particular, the mucosa-producing goblet cells multiply and the mucosa layer thickens, creating an outer mucosal layer that is accessible to benevolent microorganisms for adherence and feeding, as well as an inner layer that is impenetrable to even these organisms. Furthermore, during the period that the gut flora is developing and establishing, gut-associated lymphoid tissue (GALT), a component of the intestinal epithelium that recognizes and responds to pathogens, appears and develops. The GALT that forms is resistant to other microbes but tolerant to species of the gut flora. Additionally, GALT typically develops tolerance to food that the baby is exposed to, food-derived digestive products, and metabolites—molecules formed by metabolism—produced by the gut flora.

The human immune system produces cytokines, which can suppress the immune response to preserve homeostasis and promote healing following insult or injury, as well as trigger the immune system to produce inflammation to defend itself. It has been demonstrated that certain bacterial species that are found in the gut flora can selectively stimulate the immune system to produce cytokines. For instance, certain species of *Clostridium* and *Bacteroides fragilis* seem to trigger an anti-inflammatory response, while other segmented filamentous bacteria stimulate the production of inflammatory cytokines. Additionally, the immune system's generation of antibodies can be controlled by gut flora. This regulation has the effect of making B cells change their class to IgA. In the majority of cases, IECs trigger NF-kB signaling by gut flora, which in turn causes the secretion of additional signaling molecules. Nevertheless, gut flora can also activate B cells through T helper cells. By interacting with B cells, these signaling molecules cause IgA class switching. In mucosal environments such as the gut, IgA is a crucial class of antibody. IgA has been demonstrated to contribute to the gut community's diversity. The immune system can also be altered due to the gut bacteria's ability to produce metabolites that can affect cells in the immune system.

9.3.3. Metabolism

The gut microbiota can perform a range of metabolic tasks that humans cannot or can only perform to a limited extent because it encodes a significantly higher number of genes than its human host. The bacteria in the gut can synthesize all essential and nonessential amino acids, make a wide range of vitamins, and bio transform bile. Furthermore, the microbiome supplies the essential metabolic pathways for the breakdown of nondigestible carbohydrates. These include unabsorbed sugars and alcohols from the diet, host-derived mucins, some oligosaccharides that evade digestion, and large polysaccharides like gums, cellulose, hemicellulose, and resistant starches. This function provides the host with absorbable substrates and energy recovery, while also providing nutrients and energy for the growth and multiplication of bacteria. In the colon, one of the main sources of energy is the metabolism of carbohydrates.

Because some types of gut flora contain enzymes that human cells lack for breaking down certain polysaccharides, the human body would not be able to utilize some of the undigested carbohydrates it consumes without them. It takes 30% more calories for rats raised in sterile conditions with no gut flora to maintain their body weight compared to their normal counterparts. Certain carbohydrates, fiber, oligosaccharides, and sugars that the body was unable to process and absorb—such as lactose in the case of lactose intolerance—as well as sugar alcohols, mucus secreted by the gut, and proteins are examples of carbohydrates that humans cannot digest without the assistance of bacteria.

A process known as saccharolytic fermentation is how bacteria convert the carbohydrates they ferment into SCFAs. Among the products are butyric acid, propionic acid, and acetic acid. Host cells can utilize these substances as a major source of nutrition and energy. Fermentation also yields organic acids like lactic acid and gases that are used in signaling and can induce flatulence. Butyric acid gives gut cells energy, butanoic acid helps the liver produce ATP, and acetic acid is utilized by muscle.

In addition, gut flora produces vitamins like biotin and folate and aid in the absorption of dietary minerals like calcium, iron, and magnesium. *Methanobrevibacter smithii* is distinct from other bacteria species because it belongs to the domain Archaea instead of being a species itself. It is also the most prevalent methane-producing archaeal species in the human gastrointestinal microbiota. Additionally, the gut microbiota provides the body with vitamins K and B12, which it either cannot produce or only produces in small amounts.

9.3.3.1. Pharmacomicrobiomics

The genome of humans (i.e., the genetic makeup of each person and every microbe that lives on or inside their body) differs greatly from person to person. There is a great deal of potential for interactions between drugs and an individual's microbiome because the total number of microbial cells in the human body is over 100 trillion, which is a significant number greater than the number of cells in Homo sapiens, which is tens of trillion. Examples of these interactions include drugs changing the composition of the human microbiome, drug metabolism through microbial enzymes changing the drug's pharmacokinetic profile, and microbial drug metabolism affecting a drug's clinical efficacy and toxicity profile.

In addition to carbohydrates, medications, phytochemicals, and food poisons are among the other xenobiotics that the gut microbiota can metabolize. It has been demonstrated that the gut microbiota metabolizes over 30 medications. Sometimes a drug's microbial metabolism renders it inactive.

9.3.3.2. Contribution to Drug Metabolism

The gut microbiota is a diverse community of enriched genes with enormous biochemical potential to alter drugs, particularly oral ones. Drug metabolism can be impacted by gut microbiota through both direct and indirect mechanisms. Microbial enzymes that have the ability to alter the chemical structure of drugs administered mediate the direct mechanism. On the other hand, the microbial metabolites that influence the expression of host metabolizing enzymes like cytochrome P450 mediate the indirect pathway. A few decades ago, research was done on the impact of the gut microbiota on the drug's pharmacokinetics and bioavailability. These effects can take many different forms; for example, it can cause drug toxicity in the case of irinotecan or activate inactive drugs like lovastatin or inactivate active drugs like digoxin. Numerous studies have examined the effects of gut microbiota on the pharmacokinetics of various medications.

The human gut microbiota is essential for regulating how medications are absorbed by the body. Direct drug metabolism is possible through the synthesis and release of a number of enzymes by the gut microbiota, including the microbial biotransformation of L-dopa by the enzyme's decarboxylase and dehydroxylase. Conversely, by modifying the host's drug metabolism, gut microbiota may also change how the medications are metabolized. Microbial metabolites or altered host metabolites, which alter host metabolizing enzyme expression, can mediate this mechanism.

Numerous studies have shown that the gut microbiota is involved in the metabolism of more than 50 medications. For instance, the human gut microbiota partially activates the lactone prodrug lovastatin, a cholesterol-lowering medication, by producing active acid hydroxylated metabolites. On the other hand, a component of the gut microbiota, such as *Eggerthella lanterna*, inactivates digoxin, a medication used to treat congestive heart failure. A cytochrome-encoding operon found in *Eggerthella lanta* is upregulated by digoxin and linked to digoxin-inactivation. Additionally, the toxicity and effectiveness of chemotherapeutic drugs like irinotecan can be affected by gut microbiota. The β-glucuronidase enzymes, which are encoded by the microbiome, are responsible for this effect as they enable the recovery of the irinotecan's active form that causes gastrointestinal toxicity.

9.3.3.3. Secondary Metabolites

The gut microbiota is composed of approximately 10^{13}–10^{14} bacteria, each of which has the genetic capacity to carry out far more complex chemical reactions than those that humans can. This community possesses a vast biochemical capacity to generate unique secondary metabolites, which are occasionally generated through the metabolic conversion of dietary items like fibers and endogenous biological substances like bile acids or indole. Microbial metabolites, particularly secondary bile acids (BAs) and short chain fatty acids (SCFAs), are significant for human health and disease states.

Secondary bile acids (BAs) are among the most significant bacterial metabolites generated by the gut microbiota. Lithocholic acid (LCA) and deoxycholic acid (DCA), two secondary BAs, are created when bacteria biotransform primary bile acids like cholic acid (CA) and chenodeoxycholic acid (CDCA) into these metabolites. Hepatocytes produce primary bile acids, which are hydrophobic substances kept in the gall bladder. The gut microbiota then transforms these metabolites into secondary metabolites that are more hydrophobic. The enzymes known as bile salt hydrolases (BSH), which are conserved in several phyla of gut microbiota, including Actinobacteria, Firmicutes, and *Bacteroides*, are in charge of the initial metabolic step of secondary bile acids. It has been shown that secondary BAs like DCA and LCA prevent *Clostridium difficile* from germinating and growing.

9.3.3.4. Dysbiosis

The preservation of intestinal homeostasis depends on the gut microbiota. Dysbiosis, or an imbalance in the natural microflora, is linked to the development of intestinal cancer. Increased intestinal carcinogenesis is linked to changes in the microbial community caused by the secondary bile acid deoxycholic acid. Due to dysbiosis, the colon is exposed to more secondary bile acids, which can damage DNA and result in mutations that can cause cancer in colon cells. The colon's high bacterial density (roughly 10^{12} per milliliter) that are susceptible to dysbiosis in contrast to the small intestine's comparatively low density (roughly 10^2 per ml).) could explain why the colon has a cancer incidence that is more than 10 times higher than that of the small intestine.

9.3.3.5. Host Protection and Immune-System Development

In the gut lining, numerous bacteria compete with pathogens for nutrients and attachment sites while also producing antimicrobial compounds. This keeps pathogens from colonizing the gut lining. This phenomenon is referred to as the competitive-exclusion effect or barrier. Pathogenic bacteria can attach to host cells in the gut wall and use these attachment sites to enter the epithelial cells. *In vitro* experiments reveal that nonpathogenic bacteria vie with one another for these attachment sites on the border of IECs, obstructing the attachment of pathogenic, enter invasive bacteria and their subsequent entry into the epithelial cells. Furthermore, the enteric microbiota can outcompete pathogenic bacteria for resources simply by virtue of their sheer numbers because bacteria maintain their collective habitat by managing and consuming all available resources and compete for nutrients in their immediate surroundings. Furthermore, bacteria possess the capacity to synthesize antimicrobial substances called bacteriocins, which can impede the growth of their rivals. The ability to produce these bacteriocins is commonly found in gastrointestinal bacteria.

The primary barrier separating the immune system from the outside world is the intestinal epithelium. The gut microbiota and its metabolites have ongoing, dynamic interactions that impact how a host's immune system develops. Both in terms of its structural elements and its functionality, bacteria are essential to the gut-mucosal immune system's early development. They also continue to be important to the system's function as it matures. In order to prevent infections, IECs use specific receptors to communicate with the innate immune system. These receptors identify and bind to specific molecules linked to bacteria, triggering an immune response in the host and the release of white blood cells, cytokines, and protective peptides. The outcome may be an inflammatory reaction to pathogenic organisms, a defensive reaction to commensal bacteria, or a catalyst for the cell death of the host.

Additionally linked to preventing allergies (an excessive immune response to harmless antigens) is exposure to intestinal bacteria. Research has indicated that the intestinal bacterial composition of allergic infants and young children differs from that of non-allergic individuals. The immune system is thought to be stimulated and trained to react proportionately to all antigens by the intestinal microbiota. An immune system that is not properly trained may, and frequently does, overreact to antigens due to alterations in the composition of the intestinal microbiota during early life.

9.3.4. The Gut–Brain Axis

The intestinal microbiota and its metabolites may be able to reach the brain through the gut–brain axis, a communication system that combines immunological, hormonal, and neuronal signaling between the gut and the brain. Because of the bidirectional nature of this communication system, the brain can control immune and gastrointestinal processes, including peristalsis and mucin production. Over the past 10 years, significant progress has been made in understanding the critical connections between gut microbiota and brain function demonstrating how the gut microbiota's composition is influenced by stress and how the host's stress response is influenced by the gut microbiotas and the CNS's reciprocal interactions. Stress has been demonstrated to affect the integrity of the gut epithelium as well as peristalsis, secretions, and mucin production. These modifications to the gut microbiota's habitat have the potential to affect microbial composition and/or metabolism.

The biochemical communication between the CNS and the GIT is known as the "gut-brain axis." The term "microbiome-gut-brain axis" is occasionally used to describe paradigms that explicitly include the gut flora. However, that term has been expanded to include the role of the gut flora in the interplay. The CNS, neuroendocrine and neuroimmune systems, including the hypothalamic-pituitary-adrenal (HPA) axis, sympathetic and parasympathetic arms of the autonomic nervous system (ANS), including the ENS, the vagus nerve, and the gut microbiota, are all included in the broad definition of the gut-brain axis.

A systematic review from 2016 examined the preclinical and small human trials that have been conducted with certain commercially available strains of probiotic bacteria and found that among those tested, *Bifidobacterium* and *Lactobacillus* genera (*B. longum*, *B. breve*, *B. infantis*, *L. helveticus*, *L. rhamnosus*, *L. plantarum*, and *L. casei*), had the most potential to be useful for certain CNS disorders.

9.3.4.1. Irritable Bowel Syndrome

IBS is a category of functional bowel disorders characterized by features of disordered defecation and abdominal pain or discomfort that is linked to bowel movements or changes in habits (Table 9.2).

Table 9.2. *Principle Pathogenetic Mechanisms in IBS in Which the Microbiota is Involved*

Mechanism	Description
Impaired gut barrier	Disruption of the intestinal epithelial barrier integrity allows increased gut permeability, permitting translocation of microbes and microbial components across the epithelium. This exposure of the mucosal immune system to luminal microbes and antigens provokes aberrant inflammatory responses that mediate symptoms. Structural and functional defects of tight junction proteins contributing to this barrier impairment have been described in IBS (Barbosa and Levy, 2000; Barcena et al., 2019; Berti et al., 2018).
Aberrant immune response	Intestinal mucosa in IBS patients exhibits immune activation, detectable through increased production of pro-inflammatory cytokines. This mucosal inflammation may occur due to direct immune stimulation from translocated microbes or indirect activation by microbial antigens. Immune responses appear skewed towards pro-inflammatory Th1 and Th17 pathways in IBS. Visceral hypersensitivity, i.e., heightened pain perception from the bowel, is also associated with mucosal immune activation, contributing to abdominal pain symptoms.
Molecular mimicry	Molecular mimicry between microbial antigens and host proteins can prompt cross-reactive immune responses due to sequence or structural homology between a bacterial epitope and self-antigen. This bacteria–host mimicry induces autoimmune reactions targeting host cells and tissues, thereby perpetuating inflammation and tissue damage. Antibodies formed against bacterial cytolethal distending toxin B, for example, can cross-react with the host protein vinculin and disrupt intestinal nerve function.

Brain–gut axis	The gut microbiota interacts bidirectionally with the CNS and ENS through neuronal, endocrine, and immune signaling pathways in the gut-microbiota–brain axis. Microbial metabolites such as short-chain fatty acids can modulate neurotransmission or induce epigenetic changes that alter nerve signaling. Stress can also change microbiota composition and function through effects on gastrointestinal motility, secretions, and epithelial permeability mediated by the hypothalamic–pituitary–adrenal axis.
Epigenetic changes	Microbial metabolites, especially short-chain fatty acids like butyrate, can induce epigenetic changes by inhibiting HDACs. This inhibition causes histone hyperacetylation, thereby regulating chromatin structure and gene transcription. By modifying epigenetic processes controlling host gene expression, microbial metabolites may contribute to intestinal and neural changes underlying IBS.

Source: Napolitano M, Fasulo E, Ungaro F, Massimino L, Sinagra E, Danese S, Mandarino FV. (2023). Gut dysbiosis in irritable bowel syndrome: A narrative review on correlation with disease subtypes and novel therapeutic implications. Microorganisms; 11(10):2369. https://doi.org/10.3390/microorganisms11102369.

In adults and adolescents worldwide, IBS is estimated to affect 10–20% of cases. IBS is thought to have more than one cause, but the precise cause is unknown. Its development is believed to be influenced by multiple factors, including genetics, GIT dysfunction, visceral hypersensitivity, infection, inflammation, immunity, and psychopathology. The low-grade intestinal inflammation linked to the syndrome is believed to be facilitated by variation in the gut microbiota in addition to these other factors. Within the GIT, the gut microbiota can either directly inhibit the growth of pathogenic bacteria or hinder their adherence to the intestinal wall. Moreover, enteric pathogens that may be linked to IBS symptoms may adhere more readily to the gut lining when there is dysbiosis, or microbial imbalance. A notable, two-fold increase in the ratio of Firmicutes to *Bacteroidetes* has been observed in IBS patients, suggesting that changes in the normal microbiota's composition and disrupted colonic fermentation may be major factors in the development of digestive symptoms.

9.3.5. Inflammatory Bowel Disease

Crohn's disease (CD) and ulcerative colitis (UC) are included in IBD. Anywhere along the length of the GIT, CD can cause inflammation in a pattern resembling cobblestones, with affected areas broken up by healthy tissue. Additionally, ulcerations that can extend the whole length of the intestinal wall and cause fissures that could puncture the wall and affect other organs like the kidney or uterus are characteristic of it. Contiguous inflammation that only affects the intestinal wall's outer layers is the typical presentation of UC. It usually starts at the rectum and is mainly contained in the colon.

In chronic disease, aberration of the gut microbiome due to loss or overabundance is crucial for the persistence of inflammatory responses, even though individual microbial species may have important roles in immunomodulation. Research demonstrating the ability of antibiotics to prevent or reduce inflammation in patients as well as in disease-modeling mice has provided evidence for the role of the gut microbiota in the pathophysiology of IBD. Additionally, after receiving their fecal transfer, patients with UC who had been inoculated with stool from healthy donors showed signs of remission within a week, and full recovery was observed after four months.

Large-scale changes in the diversity and abundance of the Firmicute population—whose relative abundance has been shown to be significantly lower in IBD patients—are the focus of IBD dysbiosis, much like it is with IBS. The decrease in Firmicute populations is particularly concerning as they are recognized for producing significant amounts of SCFAs, like butyrate and acetate, which are well-known for their strong anti-inflammatory characteristics. A thorough analysis of the microbial changes seen in IBD patients was conducted by Nagalingam and Lynch. They described a wide range of microbial outcomes in various subjects and research projects, but they were unable to draw any conclusions about the microbial causes or effects of IBD or about pertinent microbial biomarkers.

9.3.6. Systemic Metabolic Diseases

Type 2 diabetes and obesity are examples of systemic metabolic diseases. The discovery that the gut microbiota of metabolically obese mice carrying a mutation in the leptin gene differed markedly from that of mice lacking the mutation provided early evidence that the gut microbiota plays a role in obesity. Subsequent analysis revealed that in the gut microbiota of obese mice, the ratio of Firmicutes to *Bacteroidetes* was shifted in favor of Firmicutes, while *Bacteroidetes* predominated in lean mice. In more recent human studies, the researchers discovered that the gut microbiota's composition was different in obese people compared to those of normal weight and that these differences were caused by variations in the host's body weight. However, subsequent research has not been able to show a clear correlation between obesity and the phylum and species-level ratios of Firmicutes to *Bacteroidetes*.

Type 2 diabetes is a complicated condition that is influenced by environmental and genetic factors. It has grown to be a global public health emergency. Genome-wide association studies, which identify genetic components in a patient's genome, are primarily used to investigate the underlying genetic contributors to type 2 diabetes. Recent studies have suggested that the makeup of the intestinal microbiota may also be associated with the development of type 2 diabetes. The gut microbiota of individuals diagnosed with type 2 diabetes showed minimal differences from the control group, despite a decrease in bacteria that produce butyrate, which could have metabolic advantages. This finding implies that the pathophysiology of type 2 diabetes may be directly linked to a state of functional dysbiosis rather than any particular microbial species. Increases in the number of opportunistically pathogenic bacterial categories were also noted, though the prevalence of these pathogen categories appeared to vary widely.

The gut microbiota generates exometabolites like SCFAs, which may control a host's metabolic processes, and is involved in the uptake of nutrients and energy. Type 2 diabetes and obesity are two metabolic disorders that have been linked to SCFAs. When compared to lean individuals, obese adults and children have been found to have higher levels of fecal SCFAs, primarily butyrate and propionate. Variations in the proportion and concentration of each individual SCFA may correspond with shifts in the types of bacteria that are present.

There is a plethora of evidence that suggests the immune system and metabolism are closely related, and the gut microbiota is becoming more and more understood as a key link between the immune system, environment, and genes. The growing paradigm that an immune imbalance is caused by a metabolic imbalance, with obesity and inflammatory diseases at one end of the spectrum and starvation and immunosuppression at the other, is one of the many reasons to maintain a healthy weight. Any dysbiosis that causes a disordered change in the makeup and activity of gut microbes, as opposed to a directed one, might contribute to a person's increased vulnerability to illness.

9.3.7. Atopic Eczema and Other Allergic Disease

Atopic eczema, atopic asthma, rhinitis, and type 1 food allergies are examples of allergic diseases that have increased in incidence globally over the past 50 years. In the developed world, this incidence is currently 20% of the population, which accounts for a significant portion of the overall disease burden. In what is known as the "atopic march," allergic disease symptoms usually start with eczema (0–2 years), followed by asthma (>5 years), and rhinitis (>8 years) as a result of atopic sensitization, which mainly affects the first two years of life and can last a lifetime. The UK government lost almost £500 million annually to atopic eczema in 1996, an inflammatory skin condition that affected up to 20% of infants under the age of one year in England and Wales in 2006.

A mutation in a specific human gene involved in skin-barrier function and the mode of birth (vaginal vs. cesarean) are known to be implicated in atopic eczema, but there are many possible causes that are not fully understood. Characterization of the gut microbiota of atopic eczema patients revealed that, at one month of age, the gut microbiota of those with the condition was substantially less diverse than that of infants without the condition, especially in terms of the *Bacteroidetes* phylum. The study also found that the atopic eczema group had less *Bacteroidetes* diversity at 12 months of age, indicating that atopic eczema patients might continue to have less bacterial diversity than healthy controls. Moreover, infants with atopic eczema showed a reduced quantity of Proteobacteria, whose cell walls contain lipopolysaccharide (LPS) molecules. Due to LPS's capacity to trigger an immune response in the host, atopic eczema risk is increased in infants who receive less LPS exposure.

The notion of decreased quantitative and qualitative exposure to the microbial world during the neonatal period, known as the hygiene hypothesis, explains the notable rise in allergic disease. It is based on the finding that smaller, especially urbanized families have higher rates of atopy as a result of reduced exposure to microbial challenge. Instead of immunological tolerance, the aberrant outcome to allergen processing of immune response is caused by this underexposure to microbial antigens.

9.3.8. Alterations in Microbiota Balance

9.3.8.1. Effects of Antibiotic Use

The host's health and capacity for food digestion may be impacted by changing the population of gut bacteria, for instance by using broad-spectrum antibiotics. Antibiotic-associated diarrhea (AAD) can be brought on by an antibiotic's direct irritation of the colon, its alteration of the microbiota, or its promotion of the growth of pathogenic bacteria. The rise in antibiotic-resistant bacteria discovered after their use, which infect hosts and cause diseases that are challenging to treat with antibiotics, is another negative consequence of antibiotic use.

The body's capacity to ferment carbohydrates and metabolize bile acids can be hampered by altering the quantity and species of gut microbiota, which may result in diarrhea. Runny stools can result from poorly broken-down carbohydrates absorbing too much water, or diarrhea can be caused by a deficiency in the SCFAs that the gut microbiota produces. Reduced native bacterial species also impair harmful species' ability to grow, like *Salmonella kedougou* and *C. difficile*, which can grow out of control even though diarrhea may not be primarily caused by these species' overgrowth. Newer methods of treating *C. difficile* infections include transplanting donor feces' fecal microbiota. Around 90% of patients reported success with minimal side effects in the early reports of the treatment. Restoring the bacterial balances of the *Bacteroides* and firmicutes classes of bacteria is thought to be the cause of efficacy.

In severe illnesses, the composition of the gut microbiome is also altered, possibly as a result of immune compromise, failure to eat, or intestinal ischemia in addition to the use of antibiotics. The adverse consequences of this have sparked interest in selective decontamination of the digestive tract, a therapy that eliminates pathogenic bacteria while promoting the regrowth of healthy ones. Antibiotics modify the microbiota population in the GIT, which can impact the host's metabolic, hormonal, and immune homeostasis globally, as well as intra-community metabolic interactions and calorie intake through the use of carbohydrates. There is solid proof that taking probiotics containing *Lactobacillus* species can help avoid diarrhea brought on by antibiotics, and that taking probiotics containing *Saccharomyces* (e.g., after systemic antibiotic treatment, *Saccharomyces boulardii*) may aid in preventing *Clostridium difficile* infection.

9.3.8.1. Pregnancy

A woman's gut microbiota changes during pregnancy; these changes are comparable to those observed in metabolic syndromes like diabetes. There are no negative effects from the altered gut microbiota. The mother's first-trimester samples are similar to the newborn's gut microbiota. From the first to the third trimester, the microbiome's diversity declines as the population of some species increases.

9.3.8.2. Probiotics, Prebiotics, Synbiotics, and Pharmabiotics

Microorganisms known as probiotics are thought to have health benefits when ingested. Prebiotics, in terms of gut microbiota, are generally indigestible fiber compounds that pass through the upper portion of the digestive tract undigested. They function as a substrate for beneficial gut flora, promoting their growth or activity. Food ingredients or dietary supplements that combine probiotics and prebiotics in a way that promotes synergism are referred to as synbiotics. The term "pharmabiotics" can refer to a variety of products, including probiotics that have been genetically engineered or otherwise optimized for optimal performance (shelf life, survival in the digestive tract, etc.) or pharmaceutical formulations (standardized manufacturing that can obtain regulatory approval as a drug) of probiotics, prebiotics, or synbiotics. The end products of the metabolism of gut flora, such as vitamins, etc. Research suggests that treating certain probiotic bacterial strains may help treat chronic idiopathic constipation and irritable bowel syndrome (IBS). Those organisms most likely to result in a decrease in symptoms have included:

- Bifidobacterium breve
- Bifidobacterium infantis
- Enterococcus faecium
- Lactobacillus plantarum
- Lactobacillus reuteri
- Lactobacillus rhamnosus
- Lactobacillus salivarius
- Propionibacterium freudenreichii
- Streptococcus thermophilus

9.3.8.3. Fecal Floatation

About 10–15% of people regularly produce feces that float in toilet water (known as "floaters"); the remaining people produce feces that sink (known as "sinkers"), and the production of gas is what makes the floaters. Conventional mice typically produce "floaters," but gnotobiotic germfree mice (bred in a germ-free isolator) produce "sinkers." The colonization of the gut microbiota in germfree mice results in the transformation of food into microbial biomass and the enrichment of various gasogenic bacterial species, which transforms the "sinkers" into "floaters."

9.3.8.4. Effects of Exercise

Exercise and gut microbiota have recently been demonstrated to be related. Although endurance athletes usually incorporate both moderate and intense exercise into their training regimen, the effects on health are not the same. The relationship between gut microbiota and endurance sports is contingent on training status and exercise intensity.

9.4. BIOLOGICAL AGEING DRIVES CHANGES TO THE GUT MICROBIOME

9.4.1. The Gut Microbiome During Aging

The gut microbiome's composition is extremely dynamic after birth and progressively becomes more stable as the immune system matures, the body transitions to a solid food diet, and early adulthood brings about a standardization of lifestyle factors (Koenig et al., 2011). While the gut microbiome's composition is relatively stable if undisturbed during early to mid-adulthood, as people age, their gut microbiota experiences a period of increased volatility, marked by changes in the diversity of genera and functional capacity (Badal et al., 2020; Claesson et al., 2011). Aging-associated alterations to microbial diversity and composition of the gut microbiome have been observed in multiple species, including flies (e.g., Drosophila), worms (e.g., *Caenorhabditis elegans*), rodents, dogs, monkeys (Ambrosini et al., 2019; Dinić et al., 2021; Kubinyi et al., 2020; Lee et al., 2019; Pallikkuth et al., 2021; Scott et al., 2017; van der Lugt et al., 2018; Wei et al., 2021) and in humans (Biagi et al., 2010; Claesson et al., 2012; Jeffery et al., 2016; Leite et al., 2021). This may be due to a multitude of biological and environmental factors, including immunosenescence (Bosco and Noti, 2021), altered physiology of the GIT (Soenen et al., 2016), development of age-related diseases, and in humans, increased exposure to medication and altered diet associated with long-term care facilities (Claesson et al., 2012).

In humans, mid-to-late adulthood is when signs of gut microbiome aging first emerge (Wilmanski et al., 2021), while the precise moment of this transition may differ depending on a person's particular genetic, environmental, and lifestyle factors. Aging is generally linked to changes in the gut microbiota's functional capacity, a decrease in core microbial genera, an increase in the uniqueness of bacterial communities, and a change in microbial diversity (Claesson et al., 2011, 2012; Ghosh et al., 2022a,b; Wilmanski et al., 2021). Critically, there is discernible variation across studies looking at the gut microbiome in various older populations. This variation could be attributed to variations in microbiome analysis methodology or to dietary, cultural, environmental, or geographic differences that can affect the gut microbiome (Biagi et al., 2010, 2017; Claesson et al., 2012; Odamaki et al., 2016; Salazar et al., 2017; Wilmanski et al., 2021). For instance, the genus *Bacteroides* and other genera within the family *Bacteroidales* including *Alistipes* and *Parabacteroides* have been observed at higher levels in people aged 65þ compared to younger, healthy adults (Claesson et al., 2011). However, other studies indicate that aging leads to a decrease in *Bacteroidaceae*, the family that contains *Bacteroides*, along with a reduction in *Faecalibacterium* and *Lachnospiraceae*, and an increased abundance of *Akkermansia*, as reviewed by Badal et al. (2020).

These variations in research findings imply that it is currently hard to determine the precise moment at which a person's gut microbiome changes to that of a "elderly" state. However, less healthy people did not exhibit as much of an age-associated drift in the gut microbiome, indicating that the microbiome shift that happens with aging may be advantageous to and indicative of host health (Wilmanski et al., 2021). Moreover, in a 4-year follow-up, maintaining a higher prevalence of *Bacteroides* or a low level of microbial uniqueness was linked to a higher risk of mortality. This study shows that certain bacterial genera may have significant roles in the microbiota's function as a novel marker for lifespan in elderly humans.

9.4.2. Extreme Aging Shows Unique Microbial Signatures

It is critical to understand the distinction between healthy aging, which promotes longevity, and biologic aging, which causes physiologic decline. As for the gut microbiome signatures linked to higher death rates found by Wilmanski et al. (2021) may be harmful, but there might also be advantageous microbes that are exclusive to or more prevalent in long-living, healthy individuals and that could promote the host's health. Indeed, distinct microbial signatures emerge with age and are particularly noticeable in centenarians (people who have lived for 100 years or longer); however, the culture and location of the assessment seem to influence the variations in bacterial genera. The gut microbiome of centenarians in a well-characterized Italian cohort contained higher abundances of a number of taxa, including *Christensenellaceae* and *Akkermansia*, some of which have been linked to health in the past, indicating that they may help maintain health as people age (Biagi et al., 2016).

Separate studies revealed that the *Bacteroidaceae*, *Lachnospiraceae*, and *Ruminococcaceae* families shrank as people aged. In a different study conducted in South Korea, centenarians' fecal microbiomes contained higher concentrations of *Akkermansia*, *Collinsella*, *Clostridium*, and *Christensenellaceae*, and lower concentrations of *Prevotella* and *Faecalibacterium*. This was linked to a higher estimated capacity of the gut microbiota to support the phosphatidylinositol signaling pathway, different N-glycan biosynthesis, and glycosphingolipid biosynthesis. Synergies between clinical studies may highlight potential health-promoting or health-degrading properties of specific bacteria, which could inspire investigations into whether these bacteria may also support cognitive health during extreme ageing. However, several studies have reported differing findings regarding the composition of the gut microbiome and ageing. It is tempting to hypothesize that bacterial taxa linked to longevity may play a role in creating an environment within the aging host that promotes longevity and, in turn, supports the extreme limits of aging in human life, possibly by protecting an individual from environmental stressors. In fact, early studies on the African turquoise killifish (*Nothobranchius furzeri*) showed that the gut microbiota is causally related to longevity; introducing the microbiota from young fish into middle-aged fish enhanced the fishes' motor behavior and lifespan, and it was also associated with unique transcriptional changes in intestinal immunity (Smith et al., 2017). Moreover, progeroid mice that received the fecal microbiota from wild-type mice demonstrated a discernible increase in lifespan and health span in a mouse model of progeria, a condition of accelerated ageing wherein gut microbiome is also altered (Barcena et al., 2019). Although whole FMT has been successful in extending life in these models, there is presently insufficient data to determine whether FMT extends life in humans. Furthermore, FMT is a thorough clinical process.

Consequently, using specific probiotic bacteria to influence the ageing microbiome may be a more effective way to promote health as one ages. In fact, single bacterial strains might be the key to extending lifespan; progeroid mice's lifespans could be similarly extended by administering the bacterium *Akkermansia muciniphila* alone (Barcena et al., 2019). These studies did not examine whether these microbiota-targeted interventions could affect brain health during advanced ageing. However, other studies have shown that Akkermansia protects against neurologic diseases, such as Alzheimer's disease (AD), multiple sclerosis, amyotrophic lateral sclerosis, and epilepsy models. These findings raise the possibility that elevated Akkermansia may be beneficial as people age.

On the other hand, some bacteria might cause harm instead of benefit. Studies conducted on germ-free mice have revealed that they have an approximate 17% longer lifespan than their pathogen-free counterparts. This implies that there could be significant interactions between bacteria and their host that hasten the aging process (Tazume et al., 1991). Thus, the eradication of these potentially harmful microbes associated with aging, possibly by means of specific antibiotic use or CRISPR knockdown of specific genes or pathogens, offers an additional avenue for intervention. These tactics are currently constrained, though, by our current lack of scientific understanding of the pathogenic behavior of the bacteria.

9.5. THE GUT MICROBIOTA ACTIVELY CONTRIBUTES TO BRAIN HEALTH DURING AGEING

9.5.1. The Microbiota and the Ageing Brain: Establishing Connections

According to Peters (2006), the aging brain is characterized by alterations on several levels, including cellular, morphological, and functional changes. Systemic low-grade inflammation can result in increased blood–brain barrier permeability, which can activate microglia, induce neuroinflammation, and raise reactive oxygen species production, all of which can worsen oxidative stress and mitochondrial dysfunction (Peters, 2006). Neuronal volume loss and changes in neurotransmitter levels, including dopamine, serotonin, and gamma-aminobutyric acid (GABA), are additional effects of aging. This may result in reduced neurogenesis and synaptic plasticity, which may lead to functional deficits like cognitive impairments (Figure 9.3). After reaching adulthood, the brain's volume gradually decreases and its volume is associated with reduced grey and white matter in specific areas of the brain that are sex-dependent, like the cortical and subcortical areas involved in cognitive processes (Murphy et al., as well as Peters (2006). Stroke and neurodegenerative illnesses may be caused by additional alterations in diffusion and vasculature (Lendahl et al., 2019).

While the impact of ageing on the mammalian (Fransen et al., 2017; Thevaranjan et al., 2017; van der Lugt et al., 2018) and human (Biagi et al., 2016; Claesson et al., 2012; Jeffery et al., 2016; Leite et al., 2022; Wilmanski et al., 2021) gut microbiome is well established, the impact of the gut microbiota on brain ageing has only recently been investigated, and relies dominantly on preclinical evidence and association analysis derived from small clinical trials.

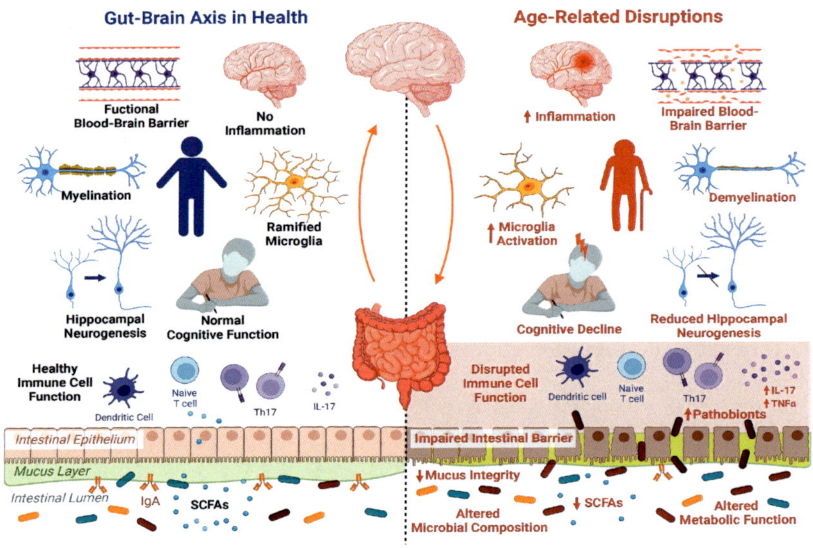

Figure 9.3. *The microbiota–gut–brain axis during aging. The process of aging initiates distinct alterations in the gut, brain and signaling pathways. These manifest in alterations in the gut microbiota composition and metabolic function, including a reduction in short-chain fatty acids (SCFAs), depleted mucus integrity and impaired intestinal barrier function, allowing for increased invasion of pathobionts from the intestinal lumen.*

Source: Boehme M, Guzzetta KE, Wasén C, Cox LM. (2023). The gut microbiota is an emerging target for improving brain health during ageing. Gut Microbiome;4:e2. doi:10.1017/gmb.2022.11

Research employing functional magnetic resonance technology (FMT) has revealed that the gut microbiota of elderly individuals can impair cognitive function and neurobiological traits when inherited by younger individuals. For example, it has been demonstrated that microbes transplanted into young mice from aged and diseased models impair learning, memory, and neuroplasticity in the recipient young mice (D'Amato et al., 2020; Kim et al., 2021; Li et al., 2020). Changes in microbial metabolites, such as short chain fatty acids, which have been demonstrated to control host microglial maturation and function, also occurred at the same time (Erny et al., 2021). Furthermore, by introducing FMT from elderly patients suffering from AD into adult rats lacking gut microbiota, humanized models were created. This resulted in a decrease in hippocampal neurogenesis and memory impairments, both of which were linked to clinical cognitive scores (Grabrucker et al., 2022). Additionally, pro-inflammatory plasma cytokine levels were elevated, motor strength was decreased, and the mortality rate from ischemic stroke was higher in young mice that received gut microbiota from elderly donors (Spychala et al., 2018). On the other hand, after being colonized with the microbiota of young donor mice, older mice recovered better from their stroke and lived longer (Spychala et al., 2018), illustrating how the microbiota from young and old people differs in function and influences how quickly the brain heals after trauma.

The recipient's age also seems to have a significant impact on the way that a person reacts to a FMT. Contrary to popular belief, young germ-free (GF) mice receiving microbiota from old mice showed increased pro-longevity markers and hippocampal neurogenesis when compared to young mice receiving FMT from young mice; however, aged GF mice did not exhibit the same effects (Kundu et al., 2019). The gut microbiota affects hippocampal neurogenesis (Guzzetta et al., 2022; Ogbonnaya et al., 2015), and is known to deteriorate as one ages (Kozareva et al., 2019), implying that age-related variables may be more important than an individual's reaction to a microbial community (Wilmanski et al., 2022). Overall, these studies show that aged people's gut microbiota differs functionally, which may be a factor in the deterioration of cognition and changes in neurobiology that come with aging.

Interestingly, when introduced into older mice, the gut microbiota from younger mice seems to be able to recover certain aspects of age-related brain degeneration. Recent research from two studies supported the same conclusions, showing that FMT from young to old mice improved memory and learning deficits associated with aging. The gut microbiota of young mice reversed age-related changes in hippocampal transcriptional profiles and metabolites and restored age-related changes in peripheral and hippocampal immune responses. These findings point to possible mechanisms by which the gut microbiota of young mice enhances cognitive function by regulating immune and metabolic pathways. Regarding this, it was demonstrated that the microbially-derived metabolite δ-Valero betaine, which is elevated in old mice and humans, directly impairs learning and memory functions. It was also found that this metabolite was decreased in old mice after feed-forward transfer from young donor mice (Mossad et al., 2021). Furthermore, certain metabolites derived from microbes have been associated with changes in the gut microbiota with age and have been found to be elevated in elderly individuals. These findings suggest that certain metabolites may have an adverse effect on cognitive function as people age. For example, isoamyl amine supplementation induced cognitive dysfunction by triggering microglia cell death in young mice (Teng et al., 2022). Meanwhile, N6-carboxymethyllysine administration lead to oxidative stress and mitochondrial damage in microglia in young and aged mice, although a starker effect was evident in young mice when N6-carboxymethyllysine was administered intraperitonially rather than by oral gavage whereas the delivery method was insignificant in aged mice, suggesting a protective role of the intestinal barrier in minimizing damage caused by harmful ageing related microbially-derived metabolites (Mossad et al., 2022). However, it is still unknown if alterations brought on by aging in the gut are beneficial or detrimental to cognitive function as we age. For instance, research by Wilmanski et al. (2021) found that a higher gut microbiome uniqueness score correlated to several health metrics in elderly individuals, whereas a lower gut microbiome uniqueness score was associated with earlier mortality. This could suggest that ageing-related changes in the gut foster increased uniqueness in the microbiome which may in turn prevent or slow the progression of ageing.

Consequently, it is possible that certain probiotic bacteria are elevated in healthy aging, and that the negative effects of aging may be caused by these bacteria's incapacity to survive and thrive in an unhealthy gut. In light of this theory, we shouldn't assume that FMT from young to old people only has positive effects because it may also have an impact on the sensitive microorganisms that currently support healthy aging. More research is required to fully investigate the potential beneficial and/or harmful effects of functional magnetic resonance therapy (FMT) on various aspects of health in the aging process.

9.5.2. The Gut Microbiome during Brain Ageing and in Age-Related Neurodegenerative Diseases

A growing body of research is showing connections between the gut microbiota and age-related neurodegenerative illnesses such as multiple sclerosis (MS), AD, PD, and mild cognitive impairment (MCI). Apart from the changes observed in the microbiota of human populations, research using preclinical disease models underscores the possible causative functions of the gut microbiota in neurodegenerative illnesses that affect hosts. There is a wealth of research on the relationship between the microbiota in AD, PD, and MS. As a result, the main points of our review will be the evidence that supports the use of gut microbiota-targeted therapies as well as the relationship between these diseases and the aging gut microbiota.

9.5.2.1. Alzheimer's Disease and Mild Cognitive Impairment

An estimated 57 million people worldwide are thought to be affected by AD, the most common age-related dementia, with estimates reaching 152 million by 2050 (Nichols et al., 2022), whereas MCI with pre-dementia states is approximately 16% common in older subjects in the USA; the most common type is amnestic MCI, which has a high risk of developing into AD (Petersen et al., 2010). Notably, it has been demonstrated that the gut microbiota of AD patients residing in the USA has higher amounts of *Bacteroidetes*, including *Bacteroides*, and lower relative abundances of Firmicutes and *Bifidobacterium* (Vogt et al., 2017).

In comparison to healthy, age-matched peers, AD patients in a Chinese cohort had higher abundances of *Prevotella* and lower levels of *Lachnospira*, *Bacteroides*, and *Ruminiclostridium*_9 (Guo, 2021). Comparing elderly patients without dementia to those with MCI, a Japanese cohort discovered that patients with cognitive decline had a higher prevalence of *Bacteroides* (Saji et al., 2019), however, some have noted that in patients with MCIs, only *Lachnospira* was significantly lower (Guo, 2021). Bacterial genera that were differentially abundant in AD were also different in MCI, indicating that changes in the gut microbiome may occur before AD onset, according to additional research on the gut microbiota of people ranging from healthy aging over MCI to AD (Li et al., 2019).

Even though there is a dearth of clinical mechanistic evidence, two case studies offer some hope regarding the gut microbiota's direct ability to reduce AD dementia symptoms. Interestingly, during his 2-month follow-up visit, the cognitive score of an 82-year-old man with AD who received FMT to treat *Clostridium difficile* infection improved from MCI levels to healthy cognition (Hazan, 2020). In a similar vein, a 90-year-old AD patient received FMT for C. difficile showed a noticeable improvement in just three months on multiple cognitive function tests (Park et al., 2021). Even with its small sample size, this clinical FMT literature validates the translation of earlier pre-clinical results showing that FMT from young male C57BL/6 wild-type mice restored some aspects of cognition when transplanted into older male C57BL/6 mice (Boehme et al., 2021).

Moreover, memory impairment, Aβ accumulation, synaptic dysfunction, and neuroinflammation were reduced by FMT from wild-type mice into male

APP/PS1 transgenic mice, a genetic mouse model of AD (Sun et al., 2019), whereas FMT into naïve C57BL/6 mice from a different genetic AD mouse model called 5xFAD mice results in decreased hippocampal neurogenesis and cognitive impairment (Kim et al., 2021). Apart from the complete microbiota transfer, efforts have also been made to modify AD pathology through the administration of particular AD-associated bacteria. A weekly administration of *Bacteroides fragilis* between the ages of 2 and 5 months increased the amyloid plaque burden in APP/PS1 mice, and increasing levels of *Bacteroides* were found in aging Tg2576 transgenic mice and correlated with the levels of amyloid plaque in their brains (Cox et al., 2019). Additionally, preclinical data involving the administration of antibiotics to rodent models of AD supports the hypothesis that the gut microbiota plays a role in the development of AD. For example, a combination of antibiotics administered from an early age to three distinct amyloid mouse models of AD (APP/PS1, 5xFAD, and App^{NL-G-F}) decreased the amyloid plaque load in male mice later in life, but no change was seen in female mice. Additionally, it was shown that early (pre-weaning) administration of the antibiotic cocktail was sufficient to observe long-lasting changes in the gut microbiome and lower amyloid pathology later in life (Minter et al., 2017).

Nevertheless, the administration of specific antibiotics had no effect on the development of amyloid plaques (Dodiya et al., 2020). Early antibiotic administration decreased gut alpha-diversity, but in the APP/PS1 model, it increased the relative abundance of *Akkermansia* (Minter et al., 2017). It's interesting to note that mice raised with antibiotics during infancy developed permanent changes to their peripheral and central immune systems, including higher numbers of regulatory T cells. Furthermore, microglia seem to be essential for triggering this response; mice treated with the antibiotic cocktail did not show decreased plaque levels in their microglia, despite the fact that microglia depletion also marginally decreased plaque burden (Dodiya et al., 2021). It was also shown in GF mice that antibiotic treatment enhanced spatial and recognition memory performance in the 5xFAD model of AD. Antibiotic therapy starting at the weaning stage also decreased anxiety in wild-type mice injected intracerebrally with amyloid peptides to mimic the onset of AD at 80 days of age (Mosaferi et al., 2021). All things considered, the increasing amount of data clearly links the gut microbiota to the pathophysiology of AD and suggests that it may be a target for new treatments.

9.5.2.2. Parkinson's Disease

The second most prevalent age-related neurodegenerative disease is PD, which may be particularly influenced by the gut microbiota, which may start the disease's pathogenesis in the gut (Yang et al., 2019). According to research on mice, misfolded alpha-synuclein propagates along the vagus nerve, which may be the fundamental mechanism behind the death of dopaminergic neurons in the striatum, a part of the brain that regulates motor behavior (Kim et al., 2019a). In fact, gastrointestinal symptoms are present in PD patients well in advance of the disease's onset (Sung et al. 2014). When vagotomy was used as a therapy to treat peptic ulcer disease in the 1970s and 1980s, a Danish study examined the risk for PD in those patients. It discovered that truncal vagotomy, which involves cutting the entire nerve, was linked to a lower risk of developing PD (Svensson et al., 2015). Two years later, a Swedish cohort confirmed this observation (Liu et al., 2017). These epidemiological data strongly suggest that the vagus nerve plays a role in the pathophysiology of PD, and they reinforce the theory that the disease first develops in the gut rather than the brain. According to a recent preprint, PD and IBD may be related, meaning that early in life, anti-inflammatory treatments that target the gut may help prevent PD (Espinosa-Oliva et al., 2022).

Alpha-synuclein overexpressing mice's motor function was deteriorated by microbiota transferred from recently diagnosed PD patients in animal models of the disease (Sampson et al., 2016), and improving motor symptoms by eradicating germs or using antibiotics to reduce the microbiota led to modifications in microglia function. This study lends credence to the hypothesis that PD may be exacerbated by the gut microbiota. Early-life *Citrobacter rodentium* infection in Pink1–/–mice increased motor dysfunction and disease pathogenesis. This increase was associated with a higher number of mitochondrial specific cytotoxic CD8 T cells in the brain. PD is thought to be an age-related condition, but this study raises the possibility that immunologic programming during infancy influences the pathophysiology of PD. Thus, interactions between the gut and microbiota over the course of a lifetime may be significant predictors of age-related neurologic disease.

9.5.2.3. Multiple Sclerosis

Ageing can have a significant impact on the course of multiple sclerosis (MS), a demyelinating neurodegenerative disease that typically manifests in people in their 20s to 40s (Dobson and Giovannoni, 2019). A smaller percentage of patients will initially present with a progressive form of the disease (primary progressive MS, PPMS), whereas the majority of MS patients develop a relapsing remitting form (RRMS) (Lassmann, 2018). Many patients with RRMS will progress to a secondary progressive disease course (SPMS) during the course of the disease, and one of the main risk factors for developing SPMS is getting older (Lassmann, 2018). It is crucial to comprehend whether host-microbiome interactions during aging impact this transition because progressive MS is more resistant to treatment and is linked to greater disability and quality of life. Patients with progressive multiple sclerosis exhibit elevated levels of

Clostridium bolteae, *Ruthenibacterium lactatiformans*, and *Akkermansia*, accompanied by decreased levels of *Blautia wexlerae*, *Dorea formicigenerans*, and *Erysipelotrichaceae* CCMM, as well as modified β-diversity in microbiota (Cox et al., 2021).

According to the Expanded Disability Status Scale, the abundance of *Clostridium* species was linked to worsening disability status, pointing to a potentially harmful role. It's interesting to note that, despite *Akkermansia* being elevated in progressive MS, higher abundance of the bacterium correlated with lower disability and was shown to improve certain aspects of disease pathology in a multiple sclerosis mouse model, possibly by lowering IL-17-producing γδ T cells and RORγtþ (Cox et al., 2021). Consequently, it's possible that some microbes prevent the development of neurodegenerative diseases, and their extinction could be harmful; however, more research is needed to confirm this theory. Antibiotic treatment improves disease and GF mice resist EAE in the experimental autoimmune encephalomyelitis (EAE) model of MS, indicating that the microbiota may play a role in the pathophysiology of the disease.

It is interesting to note that colonizing mice with the putative probiotic species *Lactobacillus reuteri* increased their susceptibility to EAE. This is probably because *Lactobacillus reuteri* can convert dietary tryptophan into active metabolites that bind to the aryl hydrocarbon receptor (AhR) on T cells and cause the production of IL17 (Montgomery et al., 2022). Additionally, EAE was made worse by introducing MS patient microbiota into GF mice (Berer et al., 2017; Cekanaviciute et al., 2017). According to recent research, mice with advanced MS exhibit a more progressive form of the disease as they age (Cahill et al., 2019), posing the query of whether aging-related alterations in the gut microbiota could be the primary cause of the advancement of MS disease.

9.5.3. Potential Mechanisms for Microbiota–Gut–Brain Signaling in Aging

The mechanism or mechanisms underlying the communication between the microbiota, gut, and brain in age-related cognitive decline are still unknown. Differences in host brain aging are likely due in part to changes in microbiota-immune signaling, given the increased immunosenescence and inflammation that accompany aging. Furthermore, research from the past few years has shown that FMT from young to old mice can reverse some age-induced changes to gut-associated immunity. Moreover, this FMT reduced the size of the microglia soma and counteracted the age-related rise in proinflammatory-associated microglial genes, indicating a reduction in immunological activation in the brain after FMT from young mice to old mice. On the other hand, when aged mice received microbiota from aged-matched mice, their plasma levels of the anti-inflammatory cytokine IL-10 were higher than when they received FMT from young mice, according to the same study. This variation in plasma IL-10 may indicate that the transfer of microbiota from young to old mice does not necessarily result in improved health, or it may instead point to the inflammatory phenotype seen in elderly recipient mice.

Previous research has shown that the gut microbiota associated with aging is crucial for host inflammation, raising the possibility of microbiota-driven alterations in gut–brain axis communication as people age. GF mice that are older than conventional aged mice exhibit reduced levels of systemic inflammatory cytokines such as TNFα and IL-6, along with an enhanced capacity of macrophages to phagocytose bacteria. These findings suggest that the microbiome is a significant factor in both immunosenescence and inflammation. Additionally, feeding young mice FMT from older mice transferred the age-related phenotypes of inflammation (higher systemic IL-6 and TNFα), immune dysfunction, and compromised gut barrier integrity. It was also associated with a decrease in certain bacterial taxa like *Akkermansia*, indicating that the gut microbiota is a direct cause of inflammation. Moreover, the ageing-induced deficit in the Peyer's patch germinal cell response, a crucial component of intestinal immunity, was restored when microbes were transferred from young to old mice via heterochronic FMT (Stebegg et al., 2019). This study provided evidence for the age-related decrease in intestinal stem cell activity.

Microbe-associated molecular patterns (MAMPs) and other small molecules secreted metabolites activate the stimulatory receptors expressed by monocytes and microglia. According to Chu and Mazmanian (2013), these compounds are continuously produced by gut bacteria and influence host immunity. According to certain research, under pathological circumstances, bacteria that originate in the gut may be present in the human brain. LPS, for instance, has been found in the human hippocampus and neocortex in a small study involving six AD patients (Zhao et al., 2017). Trimethylamine-N-oxide, a metabolite produced by bacteria, has been detected in human cerebrospinal fluid (Rio et al., 2017).

Moreover, there is growing discussion regarding the role of the oral microbiome in initiating neurodegenerative illnesses like AD (Shoemark and Allen 2015). For example, the brains of AD patients have been found to contain toxins known as gingipains, which are produced by the oral bacteria *Porphyromonas gingivalis* (Dominy et al., 2019). Orally administered bacteria-derived SCFAs could partially restore the impaired maturation of microglia in GF wild-type mice or mice treated with high-dose antibiotics to nearly eliminate the microbiota (Erny et al., 2015). The amyloid beta (Aβ) protein levels in GF APP/PS1 mice, a model of AD, were lower, indicating that amyloid load can be influenced by metabolites derived from bacteria. When the microbiome of a conventionally raised APP/PS1 mouse was colonized in GF APP/PS1 mice, the levels of Aβ were higher than when the WT microbiota was colonized (Harach et al., 2017).

Recent research has shown that the metabolite acetate, which is produced by bacteria, controls the activity of microglia in the brain. Oral acetate administration to GF mice in the 5xFAD mouse model of AD induced a pro-inflammatory phenotype in cortical microglia, with decreased amyloid plaque phagocytosis resulting in accelerated pathology (Erny et al., 2021). However, in a different mouse model of AD where Aβ was injected into the hippocampus of SPF mice, research from a Japanese group revealed that acetate partially counteracted cognitive impairment (Kobayashi et al., 2017). It has been demonstrated that supplementing aged mice with butyrate, the second major SCFA, along with acetate, reduced neuroinflammation (Matt et al., 2018). Yetyrate supplementation enhanced associative memory in APP/PS1-21 mice, even at a late stage of AD-related pathology, according to another preclinical study (Govindarajan et al., 2011). It is unclear how precisely the gut microbiota affects AD through SCFAs and changed inflammation, although it may have something to do with changed age-related pro-inflammatory monocyte trafficking to the brain (van de Wouw et al., 2019) since microglia do not appear to express SCFA receptors (Erny et al., 2015). Additionally, butyrate has been demonstrated to influence the ENS's release of gut hormones and serotonin, which can activate the vagus nerve and cause endocrine signaling that can influence brain function (Stilling et al., 2016).

Apart from SCFAs, various metabolites derived from microorganisms have also been connected to the well-being of the brain as we age. Older animals have higher levels of the microbially produced metabolite δ-valerobetaine, which has been shown to affect memory and learning (Mossad et al., 2021). In the meantime, FMT from young mice into elderly recipient mice improved learning and memory performance and decreased systemic and brain levels of δ-valerobetaine (Mossad et al., 2021). By utilizing information from the UK Biobank, the authors demonstrated a correlation between aging and elevated levels of δ-valerobetaine in the body. It's interesting to note that GF and conventional mice with δ-valerobetaine supplementation had an increase in visceral fat mass, and in humans, systemic levels of this metabolite correlate with visceral adipose tissue mass (Liu et al., 2021), a crucial component of dementia in later life (Cereda et al., 2007), pointing to δ-valerobetaine as a possible target for upcoming treatments.

In the meantime, a variety of metabolites of dietary tryptophan produced by microorganisms, known as indoles, have been linked to aging. Indoles have been demonstrated to modulate neuroinflammation through microglia control of astrocytes, and through binding to the AhR, they can act both locally on gut inflammation and neuroinflammation in the brain. Moreover, the uniqueness of the gut microbiome has been positively correlated with indoles in blood plasma. Notably, indoles are produced by *Bacteroides* species, and in 85þ individuals, a high *Bacteroides* level has been associated with increased mortality. A growing body of clinical data indicates a possible connection between dementia and bile acids. Researchers of the AD metabolomics consortium discovered that bile acid synthesis is different in AD compared to cognitively healthy individuals by combining transcriptomics analyses with a metabolic network analysis approach. It is interesting to note that certain bile acids detected in the brain cannot be accounted for by enzymes expressed locally, indicating that the gut microbiota may be the source of these bile acids.

Preclinically, metabolomic analyses suggest that secondary bile acids may play a part in restoring the balance of the upset gut microbiota, which is linked to an improvement in progeria mice's lifespan and overall health after FMT. Remarkably, a recent study discovered that centenarians have higher levels of isoallolithocholic acid, a secondary bile acid that has been demonstrated to affect host T-cell function by promoting Treg cell differentiation by raising mitochondrial activity (Hang et al., 2019), however, it is not yet known if secondary bile acids have direct effects on the brain. Alterations in the metabolism of secondary bile acid may influence cognition as people age and may represent a novel way to control brain aging.

A key component of the microbiota-gut-brain axis signaling pathway is the vagus nerve. Interestingly, a recent study discovered that FMT from elderly mouse donors had a negative impact on young mice's hippocampal function, but that this effect could be lessened by increasing vagal activity (Rei et al., 2022). According to preliminary clinical evidence, elderly subjects' quality of life may be enhanced in certain situations by stimulating the vagus nerve (Bretherton et al., 2019), as well as enhanced motor function and walking speed in PD patients, which was linked to a drop in an inflammatory biomarker and an increase in the myokine BDNF (Mondal et al., 2021).

9.6. STRATEGIES FOR TARGETING THE GUT MICROBIOME FOR BETTER BRAIN HEALTH DURING AGING

It is crucial to keep looking into whether treatments that target the gut microbiota can lessen the effects of aging on the brain as the connections between host brain aging and microbiota become more obvious. The secret to the fountain of youth may lie in many methods for changing the gut microbiota, including pharmaceutical treatments like FMT and antibiotics, as well as dietary and lifestyle choices like probiotics, prebiotics, a Mediterranean diet, intermittent fasting, and exercise (Figure 9.4). The available evidence is reviewed in the section that follows, with an emphasis on clinical evidence and preclinical data provided as supporting evidence.

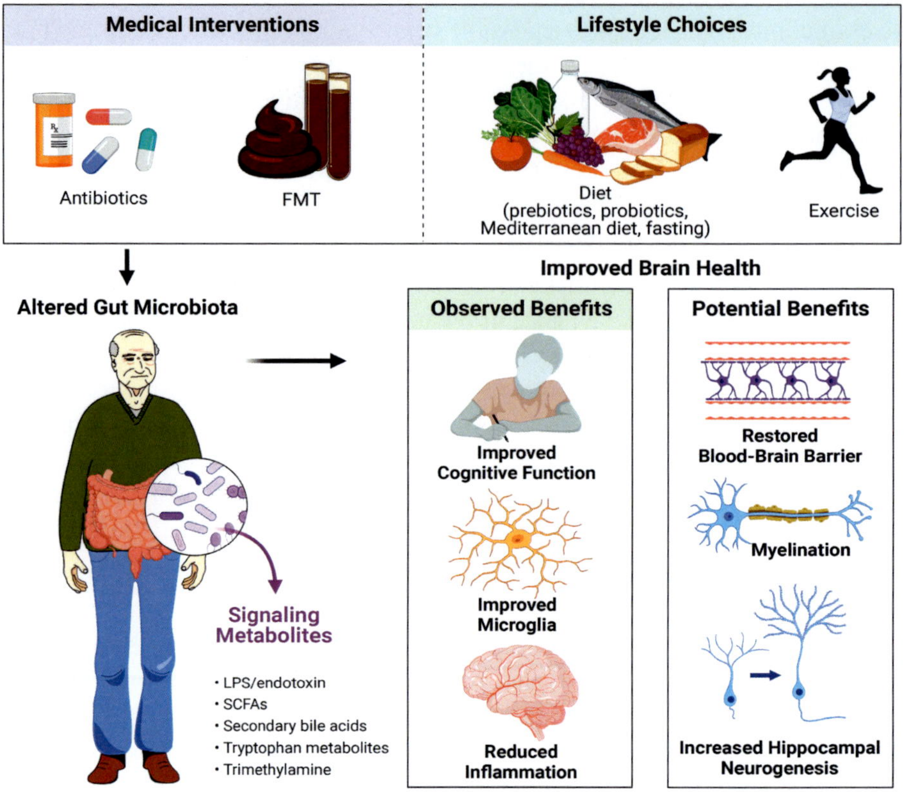

Figure 9.4. *Potential strategies for intervening on the microbiota–gut–brain axis for improved brain health in aging.*

Source: Boehme M, Guzzetta KE, Wasén C, Cox LM. (2023). The gut microbiota is an emerging target for improving brain health during ageing. Gut Microbiome;4:e2. doi:10.1017/gmb.2022.11

9.6.1. Fecal Microbiota Transplant

Nowadays, fecal microbiota transplants are frequently used in medicine to treat patients with C infections that are resistant to antibiotics. difficile, and its potential applications in various other illnesses and conditions, such as those affecting the health of the brain, are being investigated (Sorbara and Pamer, 2022). Remarkably, small-scale clinical trials have demonstrated that FMT can help PD patients with their symptoms (Kuai et al., 2021; Xue et al., 2020), notwithstanding the small sample size limitations in these research articles. Additionally, individual case studies have demonstrated that FMT was adequate for improving mobility metrics in a multiple sclerosis patients (Engen et al., 2020), and enhance cognitive functioning in two patients with AD who received FMT for their *C. difficile* infections (Hazan, 2020; Park et al., 2020). This suggests that FMT could be a viable therapeutic option for individuals with neurological diseases associated with aging.

However, there are still concerns about the safety, logistics, and viability of clinical FMT, particularly when it is used in a community where immunodeficiency and frailty are more common (Marquez et al., 2020). It would therefore be extremely helpful to develop alternative, less invasive, and risky strategies to remodel the microbiome–gut–brain axis, which would allow physicians to move beyond FMT, even though preclinical FMT studies have shown promise at improving brain health during ageing.

9.6.1.1. Antibiotics

Since antibiotics are frequently used to treat infections and their use is increasing significantly among the elderly population, it is imperative to determine whether using antibiotics as people age has unintended positive or negative effects. For example, antibiotics may have unintended consequences if they damage helpful microorganisms or cause antibiotic resistance (Barbosa and Levy, 2000). On the other hand, it is mainly unknown whether using antibiotics to get rid of a bad gut bacterium can enhance brain function. Additionally, a number of antibiotics taken orally have the ability to leak from the gut, which means that they can affect other systems directly, including the brain, independent of the microbiota-gut-brain axis. In disease states where barrier function is known to be compromised, this is more concerning (Nau et al., 2010).

Strong preclinical data suggests that, as was previously mentioned, giving antibiotics to rodent models of AD may be therapeutic for neurodegeneration. Despite a dearth of clinical evidence, a pilot open-label trial comprising 10 patients with mild to moderate probable AD dementia found that rifaximin, a minimally absorbed antibiotic, treatment for three months significantly decreased serum levels of neurofilament-light and moderately improved other serum markers for disease severity; however, no improvements to cognition were noted during this time (Suhocki et al., 2022). Remarkably, rifaximin treatment improved working memory and inhibitory control, as well as fronto-parietal and subcortical activation and connectivity, in another study that gave the medication to patients with cirrhosis and cognitive impairment (Ahluwalia et al., 2014).

However, there are many reasons why antibiotics might not be the ideal course of treatment, including the possibility that they could have detrimental neurotoxic side effects. For example, broad-spectrum antibiotics may reduce the number of helpful symbiotic gut bacteria, which may be essential for supporting the aging brain. Numerous case studies have connected the onset of delirium in the elderly with antibiotic use and elevated plasma levels of antibiotics. Additionally, following a roughly 7-year follow-up period, a recent observational study with over 14,000 women who self-reported their use of antibiotics discovered associations between chronic antibiotic use during midlife and slightly lower cognitive scores (Mehta et al., 2022), although the kind of antibiotic and the cause of long-term antibiotic use were unknown. The fact that antibiotics are used to treat particular medical conditions rather than just focusing on their effects on brain health further complicates the few clinical studies linking antibiotic use and cognitive function.

The potential use of antibiotics as a therapeutic for age-related neurodegenerative diseases may be limited because it is unclear whether using antibiotics will benefit people who are not actively fighting a bacterial infection or not, and because using antibiotics encourages the growth of bacteria that are resistant to them. Thus, conditions where specific pathogenic gut microbes have been identified may benefit from the development of novel antibiotics, such as engineered phages that could specifically eliminate individual identified pathogenic bacteria.

9.6.1.2. Probiotics

Probiotic-mediated microbiome-host modulation is becoming recognized as a potentially effective strategy to enhance cognitive function in aging humans (Eastwood et al., 2021). According to Hill et al. (2014) probiotics are live microorganisms that provide health benefits to the host when given in sufficient quantities. Probiotics have been shown to improve cognition in elderly patients with MCI, according to recent research from Japanese researchers. A 4-month intervention with a *Bifidobacterium breve* strain may be an effective strategy for improving memory functions of subjects with MCI, according to a group from Japan that demonstrated improvements in multiple cognitive domains, including immediate, delayed, and visuospatial memory, in a randomized, double-blind, placebo-controlled trial (Xiao et al., 2020). The same team previously looked into how this strain affected older adults' cognitive function with a variety of memory complaints. They discovered that, after stratifying the cohort, the probiotic only had positive effects in a subset of people who had more severe MCI (Kobayashi et al., 2019). Interestingly, recent research conducted by the group indicates that it might affect age-related brain atrophy (Asaoka et al., 2022). A 12-week supplementation with fermented soybean derived from *Lactobacillus plantarum* significantly improved a cognitive composite score in a Korean study, primarily due to an improvement in attention in MCI subjects (Hwang et al., 2019). In contrast to these, the first study in a Caucasian population demonstrated the effectiveness of *Lactobacillus rhamnosus* GG, a different strain in the *Lactobacillus* genus, in improving a cognitive composite score in older adults with MCI (Sanborn et al., 2020). Despite the intriguing nature of the results, care must be taken with the study's design. For instance, the researchers were not blind to the groups' randomization.

The beneficial effects of *L. rhamnosus* GG could be mediated through immunomodulatory pathways (Hibberd et al., 2014; Schultz et al., 2003), which have been linked to poor cognitive performance in the elderly (Baune et al., 2008; Trollor et al., 2010). In a small Japanese cohort, (Ueda et al. (2021) observed a reduced relative abundance of *Faecalibacterium prausnitzii* in elderly with MCI compared to healthy individuals. By isolating two *F. prausnitzii* strains from the healthy elderly group, the authors improved cognitive impairments in mice concomitant with changes in genes related to oxidative stress and mitochondrial function (Ueda et al., 2021). Notably, one of the *F. prausnitzii* strains was given in a pasteurized form suggesting that beneficial effects are not only restricted to live bacteria, which directly exert effects on the host but could be also mediated through effects elicited by extracellular membrane components or could act immunomodulatory. Apart from *F. prausnitzii*, another next-generation probiotic, *A. muciniphila*, demonstrated a comparable phenomenon in a small randomized controlled study (RCT) involving humans, primarily affecting certain metabolic parameters in its pasteurized form (Depommier et al., 2019) Although there is not any concrete clinical evidence to support the idea that *Akkermansia* enhances cognition in the elderly, *A. muciniphila* may affect metabolic variables that are linked to cognition, like insulin sensitivity (Willmann et al., 2020), implies that it might be able to enhance cognitive health as we age.

Probiotic effectiveness in AD was also evaluated, with varying results. An Iranian study found no beneficial effects on cognition in patients with severe AD when a multispecies mixture containing various stains and species of the genera *Lactobacillus* and *Bifidobacterium* was used for 12 weeks (Agahi et al., 2018). When brain damage is too severe, it's possible that probiotics won't improve cognition. After 12 weeks of supplementation, another Iranian study demonstrated improved insulin sensitivity, lower levels of the inflammatory marker CRP, and an improved cognitive score on the Mini-Mental State Examination (MMSE) with probiotic milk supplementation containing *Lactobacillus acidophilus*, *Lactobacillus casei*, *Bifidobacterium bifidum*, and *Lactobacillus fermentum* (Akbari et al., 2016). Additionally, a study discovered that the MMSE score of AD patients was improved by a probiotic mix that included *L. acidophilus*, and *Bifidobacterium longum* along with selenium (Tamtaji et al., 2019), whereas a different study in a Caucasian population found no advantages of a multispecies probiotic containing *L. casei* W56, *L. lactis* W19, W22, *B. acidophilus*, *B. lactis* W51, *B. bifidum* W23, *L. salivarius* W24, *L. paracasei* W20, *L. plantarum* W62, and *L. lactis* W52 on a cognitive score in AD (Leblhuber et al., 2018) perhaps as a result of the abbreviated intervention period. The recruitment criteria used in AD and MCI studies are very different from one another. The evaluation metrics for cognitive tests also vary in a similar way. To produce more trustworthy evidence, larger, better-controlled RCTs with standardized inclusion and cognitive evaluation criteria, as well as longer interval follow-ups, are required (Xiang et al., 2022).

A combination of five *Lactobacillus* strains and five *Enterococcus* strains was found to have positive effects on high-fat diet-induced disturbances of the microbiome, intestinal permeability, metabolism, and immunity, in addition to interventions with classical LAB. These studies, which used preclinical models, ultimately improved physical decline in older mice (Ahmadi et al., 2020). Remarkably, a recent preclinical study indicates that *Enterococcus* faecalis regulates social behavior and the stress response in mice that is brought on by social stress (Wu et al., 2021b), even though this study did not take age into account.

The evidence supporting the beneficial effects of probiotics on cognition in healthy older adults with memory complaints is very limited, except studies looking at older adults with dementia or pre-dementia status. Thus far, in this case, probiotic-induced improvements have only been seen in situations with additional stressors (Eastwood et al., 2021). For example, a study conducted in Korea discovered that fermented milk containing *Lactobacillus helveticus* IDCC3801 enhanced cognitive performance in cognitive fatigue assessments (Chung et al., 2014). Although more research is required to determine the factors influencing sensitivity to probiotic interventions, it appears that an individual's cognitive status and any accompanying physiological changes may be predictive of their susceptibility to cognitive decline. There are few studies on probiotic use for PD. Pilot results from an open-label, single-arm, baseline-controlled trial point to potential advantages of *L. plantarum* PS128 for improving motor abilities and quality of life (Lu et al., 2021). To corroborate these preliminary findings, more placebo-controlled research is required.

Many clinical studies are investigating the potential of microbiome-targeted strategies that make use of microbial metabolites like SCFAs to prevent neurodegenerative conditions. Duscha et al. (2020) have demonstrated that while three years of propionic acid intake positively influenced functional parameters through a reduced relapse rate associated with reduced brain atrophy, two weeks of propionic acid intake results in a sustained increase of immunoregulatory

T (Treg) cells. Furthermore, preclinical models have demonstrated that oral butyrate supplementation can be used to modulate neuroinflammation and cognitive decline (Govindarajan et al., 2011; Matt and others, 2018), although human research on these results is still pending. On the other hand, increased propionic acid in serum was linked to a higher risk of cognitive decline in a population-based cohort of over 9,000 community members in France who were 65 years of age or older (Neuffer et al., 2022). Mediation analyses revealed a strong correlation between hypercholesterolemia and glycemia and a fasting blood glucose level, indicating that metabolic disruption may be a potential mechanism in this negative association.

Intriguingly, a recent study suggests that the bacteriophages group *Caudovirales* can improve executive function and memory in both preclinical models and in humans (Mayneris-Perxachs et al., 2022). However, the impact of bacteriophages on cognitive health in the elderly remains underexplored. It is plausible that some of the apparent effects that bacteria have on the host may be in part mediated or influenced by bacteriophages (Teng et al., 2022).

9.6.1.3. Prebiotics

Prebiotics, which are substrates that host microorganisms specifically use to produce health benefits, have not been shown to enhance brain health in the aging process in clinical trials. Even so, there is some evidence that, in certain circumstances, a prebiotic mix containing fructo-oligosaccharides and inulin can improve frailty, a risk factor for cognitive decline, without having a positive effect on cognition. Preclinically, inulin supplementation reduced neuroinflammation in middle-aged mice's brains (Boehme et al. (2020) showed that inulin and *Enterococcus faecium* together demonstrated comparable outcomes in aged mice, and altered metabolites linked to brain aging, such as TMAO, in the brains of elderly mice after prolonged stress. Potential pathways could involve the regulation of brain metabolites, the stimulation of gut hormones and neurotransmitter signaling, and the modulation of peripheral and brain immune activation through microbial metabolite signaling. Additionally, in healthy elderly individuals, the combination of multiple bacterial strains (*L. acidophilus*, *L. paracasei*, *L. rhamnosus*, and *B. lactis*) and the prebiotic fructo-oligosaccharide mildly improved cognitive performance.

9.6.1.4. Diet

It has been suggested that cutting-edge dietary strategies that alter the composition of the gut microbiota can prevent age-related decline and enhance physiological well-being. Strong evidence from epidemiological studies indicates that a Mediterranean diet may help the aging human brain.

In preclinical models, elements of a Mediterranean diet, such as vitamin A and omega-3 polyunsaturated fatty acids (O3-PUFAs), have been demonstrated to have protective effects on the aging brain. Additionally, research in preclinical and clinical contexts has demonstrated the beneficial effects of polyphenols on cognition (de Vries et al., 2021). On the other hand, in preclinical models, these nutrients have been demonstrated to regulate the gut microbiota and to restore brain health after microbiome-related disruptions. Research from the NU-AGE consortium clarified the connection between altered gut microbiota brought on by a Mediterranean diet and better health outcomes in the elderly, demonstrating a lower risk of frailty and an enhanced state of health linked to an intermediate microbiome response (Ghosh et al., 2020).

Following a Mediterranean diet has been associated with higher taxonomic abundance, which has been positively correlated with better health outcomes like reduced frailty and enhanced cognition and negatively correlated with inflammatory markers like C-reactive protein and interleukin-17 (IL-17). These findings suggest that a Mediterranean diet may help encourage a healthier aging process. Growing preclinical and clinical evidence identifies IL-17 as a key cytokine implicated in neurodegenerative diseases and cognitive decline, indicating that dietary interventions that lower IL-17 may support healthy aging. Interestingly, compared to low adherence, elderly people with high adherence after a year demonstrated enhanced episodic memory and global cognition (Marseglia et al., 2018) which were linked to changes in their gut microbiome, indicating that, in comparison to FMT, broad dietary intervention may be a less intrusive method of modifying the gut microbiota in older individuals to improve cognition.

The production of short- and branch-chained fatty acids increased while secondary bile acid production decreased, according to an analysis of inferred microbial functions. Notably, metabolic network analysis in AD clinical trials suggests that secondary bile acids may be related to cognitive decline. The authors present here a correlation between altered bile acid profile, genetic variants linked to AD, and cognitive alterations derived from a large multicenter study. Further research has demonstrated that following a Mediterranean diet can slow the advancement of AD. Introducing a more comprehensive diet change has also been evaluated for its impact on dementia-type cognitive impairment, and it shows promise for a deeper modification of the gut microbiome than supplementation. The authors discovered differences in multiple taxa with distinct patterns between subjects who were cognitively impaired and those who were cognitively normal in a pilot study that used a modified Mediterranean-ketogenic diet. Conversely, diets typical of the West, which are heavy in sugar and saturated fat, may cause cognitive dysfunction in the elderly. More insight into the role of the microbiota in this process was provided by a recent study that found that the microglia of elderly mice accumulated N6-carboxymethyllysine, an advanced glycation end product, in a microbiota-dependent manner. This accumulation was linked to mitochondrial dysfunction and oxidative stress, two risk factors for cognitive decline. Additionally, in a mouse model of AD, calorie restriction decreased the amount of amyloid plaque (Cox et al., 2019). Calorie restriction and intermittent fasting are effective strategies for improving cognition in humans as well as in mice models of AD and aging. Although the extent and practicability of calorie restriction in humans are still unclear, intermittent fasting is one tactic that may be used to enhance cognitive function as one age.

REFERENCES

1. Agahi, A., Hamidi, G. A., Daneshvar, R., Hamdieh, M., Soheili, M., Alinaghipour, A., Esmaeili Taba, S. M., & Salami, M. (2018). Does severity of Alzheimer's disease contribute to its responsiveness to modifying gut microbiota? A double-blind clinical trial. *Frontiers in Neurology, 9*, 662.
2. Ahluwalia, V., Wade, J. B., Heuman, D. M., Hammeke, T. A., Sanyal, A. J., Sterling, R. K., Stravitz, R. T., Luketic, V., Siddiqui, M. S., Puri, P., Fuchs, M., Lennon, M. J., Kraft, K. A., Gilles, H., White, M. B., Noble, N. A., & Bajaj, J. S. (2014). Enhancement of functional connectivity, working memory and inhibitory control on multi-modal brain MR imaging with rifaximin in cirrhosis: Implications for the gut-liver-brain axis. *Metabolic Brain Disease, 29*(4), 1017–1025.
3. Ahmadi, S., Wang, S., Nagpal, R., Wang, B., Jain, S., Razazan, A., Mishra, S. P., Zhu, X., Wang, Z., Kavanagh, K., & Yadav, H. (2020). A human-origin probiotic cocktail ameliorates aging-related leaky gut and inflammation via modulating the microbiota/taurine/tight junction axis. *JCI Insight, 5*(9), e132055.
4. Ambrosini, Y. M., Borcherding, D., Kanthasamy, A., Kim, H. J., Willette, A. A., Jergens, A., Allenspach, K., & Mochel, J. P. (2019). The gut-brain axis in neurodegenerative diseases and relevance of the canine model: A review. *Frontiers in Aging Neuroscience, 11*, 130.
5. Badal, V. D., Vaccariello, E. D., Murray, E. R., Yu, K. E., Knight, R., Jeste, D. V., & Nguyen, T. T. (2020). The gut microbiome, aging, and longevity: A systematic review. *Nutrients, 12*(12), 3759.
6. Baloni, P., Funk, C. C., Yan, J., Yurkovich, J. T., Kueider-Paisley, A., Nho, K., Heinken, A., Jia, W., Mahmoudiandehkordi, S., Louie, G., Saykin, A. J., Arnold, M., Kastenmuller, G., Griffiths, W. J., Thiele, I., Alzheimer's Disease Metabolomics, C., Kaddurah-Daouk, R., & Price, N. D. (2020). Metabolic network analysis reveals altered bile acid synthesis and metabolism in Alzheimer's disease. *Cell Reports Medicine, 1*(8), 100138.
7. Barbosa, T. M., & Levy, S. B. (2000). The impact of antibiotic use on resistance development and persistence. *Drug Resistance Updates, 3*(5), 303–311.
8. Barcena, C., Valdes-Mas, R., Mayoral, P., Garabaya, C., Durand, S., Rodriguez, F., Fernandez-Garcia, M. T., Salazar, N., Nogacka, A. M., Garatachea, N., Bossut, N., Aprahamian, F., Lucia, A., Kroemer, G., Freije, J. M. P., Quiros, P. M., & Lopez-Otin, C. (2019). Healthspan and lifespan extension by fecal microbiota transplantation into progeroid mice. *Nature Medicine, 25*(8), 1234–1242.

9. Berti, V., Walters, M., Sterling, J., Quinn, C. G., Logue, M., Andrews, R., Matthews, D. C., Osorio, R. S., Pupi, A., Vallabhajosula, S., Isaacson, R. S., de Leon, M. J., & Mosconi, L. (2018). Mediterranean diet and 3-year Alzheimer brain biomarker changes in middle-aged adults. *Neurology, 90*(20), e1789–e1798.
10. Biagi, E., Franceschi, C., Rampelli, S., et al. (2016). Gut microbiota and extreme longevity. *Current Biology, 26*(11), 1480–1485.
11. Boehme, M., Guzzetta, K. E., Wasén, C., & Cox, L. M. (2023). The gut microbiota is an emerging target for improving brain health during ageing. *Gut Microbiome, 4*, e2. https://doi.org/10.1017/gmb.2022.11.
12. Castro-Mejía, J., Khakimov, B., Krych, L., et al. (2020). Physical fitness in community-dwelling older adults is linked to dietary intake, gut microbiota, and metabolomic signatures. *Aging Cell, 19*(3), e13105.
13. Cekanaviciute, E., Yoo, B. B., Runia, T. F., Debelius, J. W., Singh, S., Nelson, C. A., Kanner, R., Bencosme, Y., Lee, Y. K., Hauser, S. L., Crabtree-Hartman, E., Sand, I. K., Gacias, M., Zhu, Y., Casaccia, P., Cree, B. A. C., Knight, R., Mazmanian, S. K., & Baranzini, S. E. (2017). Gut bacteria from multiple sclerosis patients modulate human T cells and exacerbate symptoms in mouse models. *Proceedings of the National Academy of Sciences of the United States of America, 114*(40), 10713–10718.
14. Chakraborty, A., Banerjee, S., Mukherjee, B., & Poddar, M. K. (2020). Calorie restriction improves aging-induced impairment of cognitive function in relation to deregulation of corticosterone status and brain regional GABA system. *Mechanisms of Ageing and Development, 189*, 111248.
15. Claesson, M., Jeffery, I. B., Conde, S., et al. (2012). Gut microbiota composition correlates with diet and health in the elderly. *Nature, 488*, 178–184.
16. D'Amato, A., Di Cesare Mannelli, L., Lucarini, E., Man, A. L., Le Gall, G., Branca, J. J. V., Ghelardini, C., Amedei, A., Bertelli, E., Regoli, M., Pacini, A., Luciani, G., Gallina, P., Altera, A., Narbad, A., Gulisano, M., Hoyles, L., Vauzour, D., & Nicoletti, C. (2020). Faecal microbiota transplant from aged donor mice affects spatial learning and memory via modulating hippocampal synaptic plasticity- and neurotransmission-related proteins in young recipients. *Microbiome, 8*(1), 140.
17. Dal-Pan, A., Pifferi, F., Marchal, J., Picq, J. L., Aujard, F., & Consortium, R. (2011). Cognitive performances are selectively enhanced during chronic caloric restriction or resveratrol supplementation in a primate. *PLoS One, 6*(1), e16581.
18. Eastwood, J., Walton, G., Van Hemert, S., Williams, C., & Lamport, D.

(2021). The effect of probiotics on cognitive function across the human lifespan: A systematic review. *Neuroscience and Biobehavioral Reviews, 128*, 311–327.

19. Erny, D., Hrabe de Angelis, A. L., Jaitin, D., Wieghofer, P., Staszewski, O., David, E., Keren-Shaul, H., Mahlakoiv, T., Jakobshagen, K., Buch, T., Schwierzeck, V., Utermohlen, O., Chun, E., Garrett, W. S., McCoy, K. D., Diefenbach, A., Staeheli, P., Stecher, B., Amit, I., & Prinz, M. (2015). Host microbiota constantly control maturation and function of microglia in the CNS. *Nature Neuroscience, 18*(7), 965–977.
20. Gill, S. R., et al. (2006). Metagenomic analysis of the human distal gut microbiome. *Science, 312*(5778), 1355–1359.
21. Magrone, T., & Jirillo, E. (2013). The interaction between gut microbiota and age-related changes in immune function and inflammation. *Immun Ageing, 10*(1), 31.
22. van Tongeren, S. P., et al. (2005). Fecal microbiota composition and frailty. *Applied and Environmental Microbiology, 71*(10), 6438–6442.
23. Wilmanski, T., Diener, C., Rappaport, N., et al. (2021). Gut microbiome pattern reflects healthy ageing and predicts survival in humans. *Nature Metabolism, 3*, 274–286.
24. Wu, L., Zeng, T., Zinellu, A., et al. (2019). A cross-sectional study of compositional and functional profiles of gut microbiota in Sardinian centenarians. *mSystems, 4*(4), e00325–19.
25. Yatsunenko, T., Rey, F. E., Manary, M. J., et al. (2012). Human gut microbiome viewed across age and geography. *Nature, 486*, 222–227.

INDEX

C

D

E

F

G

H

I

J

K

L

M

N

O

P

R

S

T